Introduction to Population Ecology

Larry L. Rockwood

Blackwell
Publishing

BLACKWELL PUBLISHING
350 Main Street, Malden, MA 02148-5020, USA
9600 Garsington Road, Oxford OX4 2DQ, UK
550 Swanston Street, Carlton, Victoria 3053, Australia

The right of Larry L. Rockwood to be identified as the Author of this Work has been
asserted in accordance with the UK Copyright, Designs, and Patents Act 1988.

First published 2006 by Blackwell Publishing Ltd

6   2012

Library of Congress Cataloging-in-Publication Data
Rockwood, Larry L., 1943–
    Introduction to population ecology / Larry L. Rockwood.
        p. cm.
    Includes bibliographical references (p.) and index.
    ISBN : 978-1-4051-3263-3 (pbk. : alk. paper)
1. Population ecology—Textbooks.
2. Ecology—Textbooks. I. Title.
    QH352.R63 2006
    577.8′8—dc22

                                                                        2005013101

ISBN : 978-1-4051-3263-3 (paperback)

A catalogue record for this title is available from the British Library.

Set in 10/12.5pt Minion
by Graphicraft Limited, Hong Kong
Printed and bound in Malaysia
by Vivar Printing Sdn Bhd

The publisher's policy is to use permanent paper from mills that operate a sustainable
forestry policy, and which has been manufactured from pulp processed using acid-free
and elementary chlorine-free practices. Furthermore, the publisher ensures that the text
paper and cover board used have met acceptable environmental accreditation standards.

For further information on
Blackwell Publishing, visit our website:
www.blackwellpublishing.com

Introduction to Population Ecology

Contents

Preface

This book is intended for undergraduate or graduate students who have had a basic course in ecology and who are ready for a more advanced examination of population ecology. That is, junior and senior undergraduates and graduate students. My motivation for writing this book originated from 30 years of teaching population ecology to an audience consisting mostly of MS and PhD students in our·Environmental Science and Policy program. Most of these students work in environmental consulting or engineering firms, county, national and international agencies (EPA, NASA, World Bank), and governmental units such as the US Fish and Wildlife Service or the National Park Service. Other students are technicians at the National Zoo or the Smithsonian Natural History Museum who want to advance their careers. This is a challenging audience, most of whom bring intense interest and real-life experience to the classroom. Yet few are headed for a research career at a major university and their patience with theory for the sake of theory is thin. Adapting to the situation, the goal of my course was not to train theoretical ecologists, but rather to develop in my students an appreciation for, and an understanding of, the basic principles of population ecology and the application of these principles to solving the problems faced by wildlife managers, environmental consultants, Fish and Wildlife bureaucrats and the like. Such individuals should be able to integrate the principles of population ecology with the challenges presented by their work in conservation or environmental biology.

I have never found a textbook that I felt hit the right level of topic coverage and mathematical sophistication. Most population texts have not yet done an adequate job of integrating metapopulation biology into the study of population ecology. Furthermore, most books still emphasize competition and predator–prey relationships as the only interactions worthy of detailed consideration. I have always been a firm believer in giving more coverage to herbivore–plant and mutualistic interactions. To that list I have now added parasite–host interactions. In terms of topic coverage, some unique features of this book are: (i) coverage of metapopulation ecology, including not only a separate chapter but also an integration of metapopulation concepts into chapters on competition, parasite–host, and predator–prey relationships; (ii) a discussion of stochastic, in addition to deterministic, models in several chapters; (iii) discussions of population viability analysis (PVA), especially in Chapter 1.

With regard to mathematical sophistication, population ecology can be taught at many levels, depending upon the mathematical background of the students (and of the professor!). I am not attempting to produce a book for future mathematical or theoretical ecologists. On the other hand, population ecology requires models and requires the use of mathematics. Accordingly, although I have used advanced algebra, I have tried to keep it as simple as possible. Although many ecological models are written as differential equations, this book does not require extensive knowledge of differential equations or calculus. Wherever possible I have provided graphs, simulations or other aids to help students understand how the equations work. Students will be made aware of the assumptions of the models and asked to evaluate how they could be applied to organisms or situations with which they are familiar. The book includes sample problems to illustrate the models as well as sample exercises. For example, Chapter 1 includes four sample exercises embedded in the chapter, several problem sets in Appendix 1, eight illustrative tables, and eight figures. Sample simulations illustrating discrete, exponential, and stochastic population growth can be found at www.blackwellpublishing.com/rockwood. A key to symbols used in each chapter has been provided for easy reference.

In Chapter 4 there are several sections on population projection of age- (or stage-) structured populations, in which the use of matrix algebra is necessary. Although it is possible to understand population projection without knowledge of matrix algebra, students will have a better understanding if they grasp the rudiments of matrices: how to multiply a matrix by a column vector, for example. Therefore, you will find a description of the basics of matrix algebra in Appendix 2, complete with examples.

This book was originally intended primarily for undergraduate and graduate students at universities and colleges where the faculty must teach a wide variety of courses, or in departments of environmental science where the students are oriented around solving practical environmental problems, as opposed to making their mark by developing new theory. However, I now believe this book can be used for any population-oriented course that goes beyond the general treatments found in ecology textbooks.

Finally, an ecology book devoid of field biology is an empty vessel. By that I mean that most ecologists want to see how theory translates into a better understanding of the natural world. Therefore throughout the book I have included both laboratory and field studies that illustrate population principles, from exponential growth to predator–prey population oscillations. I have attempted to draw these field examples from all types of ecosystems, from the tundra to the tropics, although I admit that the vast majority are from terrestrial rather than aquatic ecosystems.

This book is designed for a typical 14-week semester found in most universities in the United States. For shorter courses, the chapters on population regulation and life-history strategies (Chapters 3 and 6) could be omitted from Part I. From Part II a shorter course could omit the chapters on mutualism and parasite–host relationships without losing the major themes developed in the remaining chapters.

The first part of the book examines the fundamental properties of single-species populations, focusing on the processes of birth, death, immigration, emigration, and local extinction. We begin with populations having simple life histories and no age structure. Chapter 1 examines the properties of population growth with no restraints (density-independent growth), while Chapter 2 looks at the limitations on population growth resulting from intraspecific competition and density dependence. The third chapter explores the concept of population regulation. In Chapter 4 we examine populations with age (or stage) structures and the suite of properties associated with age-dependent growth. Chapter 5 is a relatively detailed examination of metapopulations and spatial ecology. The chapter concludes with a discussion of the major role metapopulation dynamics now play in the field of conservation biology. Chapter 6 is an examination of life-history strategies and introduces power laws and the controversial "metabolic theory of ecology." This chapter also reviews some of the classic life-history theories of Cole, MacArthur and Wilson, Lack, and Grime.

The theme of Part II is interspecific interactions. Chapter 7, on competition, emphasizes resource and spatial competition as well as the usual treatment of Lotka and Volterra. Chapter 8 emphasizes the "cost of mutualism" to each of the species involved. Chapter 9, on host–parasite interactions, includes sections on metapopulations and on social parasites, as well as a description of the classic host–microparasite, or SIR, model. Chapter 10 includes an extensive historical review of predator–prey theory and brings the reader up to date with some of the most well-known predator–prey interactions often described uncritically in ecology textbooks, such as the hare–lynx and moose–wolf relationships. In Chapter 11 herbivore–plant interactions are dealt with from the perspective of both theoretical models and chemical ecology. The relationships among plant, herbivores, and predators are explored in three-trophic-level models.

I would like to thank the many students and anonymous reviewers who have read and commented on this book. Specifically I wish to thank Thomas Wilson, Tom Akre, and Helene Jorgensen. Both the content and organization have greatly benefited from their suggestions. Portions of this book were developed while on study leave from George Mason University at Oxford University; I thank Yehuda Lukacs for that opportunity. I would like to thank Hannah Berry, Ward Cooper, and Rosie Hayden at Blackwell Publishing for their patience and encouragement. Most of all I am grateful for the support of my family. My wife Jane has provided me with emotional support, advice, and even logistical support during this very long process. Without her you would not be reading this book.

Part I
Single-species populations

What is population ecology? What distinguishes the study of populations from the study of landscapes and ecosystems? The answers lie in scale, focus and traditions. In population ecology the scale is a group or groups of taxonomically or functionally related organisms. The emphasis is on fundamental properties of these populations: growth, survivorship, and reproduction. The tradition is based on the interplay of theory, laboratory testing, and, ultimately, fieldwork. The competition and predator–prey equations of Lotka (1925) and Volterra (1926, 1931) stimulated the laboratory work of Gause (1932, 1934), Park (1948, 1954), Huffaker (1958), and others. Elton (1924), Errington (1946), Lack (1954), Connell (1961a, 1961b), Paine (1966), Krebs *et al.* (1995), and many others brought population ecology into the field, where its theoretical underpinnings are constantly tested. In the age of personal laptop computers and the internet, data can now be analyzed, sent around the world, and experiments redesigned, without ever leaving the field site. Increasingly sophisticated experimental design and statistical rigor constantly challenge new generations of scientists. Indeed, much of the training of modern ecologists is in methodology.

Yet why do we become ecologists in the first place? Is it because of our love of computer programs and statistics? For most of us, that would be, "No." More likely it is because of a love of the organisms that we find in natural ("wild") places. We love the sounds, the smells, the feel, the **being** in nature. Perhaps it is also because of our love of the **idea** of nature and of places not yet under the total domination of *Homo sapiens*. Nothing quite matches a day (or night) in the field for an ecologist, and we are usually eager to communicate these experiences to other people. Contrast an ecologist to a typical urban dweller like Woody Allen. In one of his movies Woody complains that he hates spending nights in the country because of the "constant noise of the crickets." Yet, he and his urban counterparts find the constant traffic noises of New York City soothing. Most population ecologists have a different view.

Population ecology is, in a primitive sense, an organized way of communicating our ideas about nature to others. Population ecology, with its emphasis on groups of individuals and their survival and reproduction, their relationships with their competitors and their predators, is rooted both in fieldwork and in natural history. As such it appeals to us at a very fundamental level. Instead of (or perhaps in addition to) swapping tales around the campfire at night, we communicate by publishing in journals or books.

Furthermore, without the basic data from population studies, most landscape and ecosystem studies would either be impossible to carry out, or would lack fundamental meaning. The advantage of ecosystem studies is the comprehensiveness of the approach. However, the disadvantage is the complexity of interactions among species and our lack of understanding of community organization. Everyone can agree that we need a better understanding of interspecific interactions, and this is the role of population ecology. To develop laws of ecosystem functioning, we first need to comprehend how individual populations behave. From there we can develop an understanding of interactions among populations. Therefore it seems to me that studies at the landscape and ecosystem level must be informed by data first gathered by population ecologists.

But this all sounds rather grand and theoretical. In the real world, knowledge of population ecology is absolutely necessary for conservation biologists, wildlife managers, and resource biologists. They are often faced with problems of preserving biodiversity or a wild living resource without adequate information. How can they best decide whether to limit or even shut down a fishery, and for how long? Is it necessary or wise to allow wolf (*Canis lupus*) hunting in Alaska in order to increase the caribou (*Rangifer tarandus*) herd? Has the introduction of wolves into Yellowstone actually decreased the elk (*Cervus elaphus*) herds? What are the causes of reptile and amphibian declines throughout much of the world? Although an ecosystem approach may be helpful and necessary to answer many of these questions, basic population data are also necessary. But more than data are necessary; we must understand how populations with different life histories grow and/or are limited. We need a fundamental understanding of the roles of competitors, parasites, and predators, and of their potential effects on a given population.

When John James Audubon was in the state of Kentucky in 1813, he witnessed the passing of a great flock of passenger pigeons (*Ectopistes migratoris*). This flock blackened the sky for more than three days as they passed overhead. Later Audubon estimated their numbers at between 1.1 and 2.5 billion birds (Souder 2004). Yet the last passenger pigeon in the wild was shot in 1900; the last individual in captivity died in 1914; and the species was extinct. How can a population decrease so swiftly, even if one acknowledges the role of hunting and habitat destruction?

Red grouse (*Lagopus lagopus*) go through population cycles every 4–5 years. The numbers oscillate over three orders of magnitude (Hudson *et al.* 1998), and these oscillations are synchronized over large geographical areas (Cattadori *et al.* 2005). Yet the population recovers regularly. On the other hand, when tawny owls (*Strix aluco*) were studied in Oxford, the number of mating pairs remained steady, at 17–30 pairs, even though their major rodent prey species oscillated from 10 to 150 per acre (Southern 1970). What are the differences between red grouse and tawny owls? Differences in reproductive parameters, developmental time, or survivorship? The fact that red grouse are primarily herbivores and owls primarily predators? Their competitors, parasites, predators? These are questions that only knowledge of population ecology allows us to answer.

When the moose (*Alces alces*) population recently crashed on Isle Royale in Lake Superior, Michigan, was the cause wolf predation? Parasites? Over-browsing of the vegetation? Wildlife scientists throughout much of the United States have complained for many years that white-tailed deer (*Odocoileus virginianus*) are over-browsing their habitats and causing changes in the vegetation. If so, why don't these deer populations crash? Is the recent movement of coyotes (*Canis latrans*) into the eastern United States and puma (*Felis concolor*) into the Midwestern United States the result of these large white-tailed deer populations? If not, what explains these dispersals from the "wild west" to the more urbanized areas of the USA east of the Mississippi River? One goal of this book is to give you the background and weapons that will allow you to address these questions.

In the twentieth century, the principles of population ecology, as we understood them, were applied to agriculture, forestry, wildlife management, fisheries, and conservation biology. Exploitation of populations in the name of "maximum sustainable yield" was based on the flawed logistic equation and/or inadequate data. Before the days of environmental impact statements, however, politicians and engineers largely ignored advice based on ecological science. While this situation has changed, ecologists, in order to remain credible, must work to develop better theoretical approaches and methodologies. And applied ecologists must be able to recognize which of several possible theoretical approaches applies to the population or community of concern. The purpose of this book is to help guide future wildlife refuge managers, EPA officials, or other applied ecologists through the workings of basic population principles and theory so that they make wise decisions in the future.

In Part I of this book our goal will be to establish the fundamentals of population growth for single-species populations. After determining these basic properties, we will examine how intraspecific competition affects population characteristics. We will also consider the evolution of different types of life histories and discuss whether a biological population is naturally "regulated."

Once we have an understanding of how single populations grow and sustain themselves in particular environments, we can begin to examine how interactions with populations of other species affect their life histories. In Part II we will progress to an examination of interspecific interactions such as competition, predation, parasitism, and mutualism. As we move through these interactions, we can evaluate their relative importance in population growth and regulation.

1

Density-independent growth

- The general laws and fundamentals of population growth
- Density-independent versus density-dependent growth
- Discrete or "geometric" growth in populations with non-overlapping generations
- Exponential growth in populations with overlapping generations
- Applications to invasive species and human populations
- Stochastic models of population growth and population viability analysis

1.1 Introduction

What is a population?

The basic definition of ecology, **the scientific study of the relationships between organisms and their environment**, is rather vague and the word **environment** requires an explicit definition. An alternative definition of ecology, **the scientific study of the distribution and abundance of organisms** (Krebs 1994, Andrewartha 1961), is more germane to population ecology. In population ecology we want to know what factors most likely control the growth rates, abundances, and distributions of biological populations.

As used here, a **population** (synonymous with **biological population**) consists of **a group of interbreeding organisms found in the same space or area** (i.e. they are **sympatric**) at **the same time.** It is presumed that these individuals form a functional unit in that they interact with one another and there is interbreeding among the individuals of the population. A **closed population** is one in which we expect no immigration or emigration of individuals from outside of the population. In reality, unless we are considering a population on a remote island, a mountaintop, or an isolated cave, populations are not closed to immigration or emigration. And unless we have successfully marked all individuals in a population, we are usually unaware of which individuals might be recent immigrants. Turchin (2003) integrates these ideas in his definition of a population: "a group of individuals of the same species that live together in an area of sufficient size to permit normal dispersal and migration behavior, and in which population changes are largely determined by birth and death processes."

A local population differs from a species or a **species population**, in that we are dealing with a group of individuals interacting in a particular time and space. White-tailed deer (*Odocoileus virginianus*) from northern Wisconsin and the Piedmont of Virginia, according to the biological species concept, are the same species as long as they produce viable offspring when they are interbred. But they would belong to different and distinct ecological populations. Actually, a population is often defined by the investigator(s) and may be somewhat arbitrary.

Fundamental principles and the use of mathematical models

What are the fundamental principles that dictate how populations grow? Population ecology is by necessity a quantitative discipline, and in order to answer questions about populations, mathematically oriented ecologists have derived a variety of predictive models. The first section of this book will examine growth models for populations of single species.

The diversity of life has led to a fantastic array of life histories. Just as the mass of a single bacterium is several orders of magnitude smaller than the mass of an elephant, population characteristics, such as generation time, also differ by several orders of magnitude.

Accordingly, no one model of population growth suits all organisms or all environments. This fact is both frustrating and stimulating. A search for a single set of models that applies to all life forms is pointless. On the other hand, the construction of quantitative models forces us to examine our assumptions about particular populations in an organized and explicit manner. Models, whether quantitative or qualitative, often produce unexpected results that may run counter to our intuitive sense of how things work. The work of Copernicus, Galileo, and others that culminated in the formal quantitative models of Newton showed that the solar system and the universe function in ways that were not at all intuitively obvious. A dissection of the life histories of both the emperor goose (*Chen canagica*) (Morris and Doak 2002) and the Amboseli baboon (*Papio cynocephalus*) (Alberts and Altmann 2003) populations, using a matrix population model, have shown us that adult survivorship has a greater impact on growth rates than either juvenile survivorship or fertility: a conclusion impossible to reach without the proper population model. As Atkins (1999) commented, "Quantitative reasoning (gives) spine to otherwise flabby concepts, enabling them to stand up to experimental verification." Models stimulate observations and experiments that allow us to learn more about our natural world.

A general rule of systems is that as one progresses from lower to higher levels of organization, properties are added that were not present at the lower levels. Thus an individual organism is not just a collection of physiological systems. Similarly, a population has properties not evident from the study of individuals. Populations have growth rates, age distributions, and spatial patterns. They also have allelic frequencies and other genetic properties. The first list of properties is within the province of **population ecology**; the latter is part of the discipline of **population genetics**. The two areas combined are known as **population biology**. Although this book deals only with population ecology, much of what I have written is based on the theory of evolution, which relies on principles of population genetics.

The models used here will be largely based on relatively straightforward algebra. However, matrix algebra and differential calculus will be introduced. For more sophisticated mathematical treatments the reader should consult Roughgarden (1998), Case (2000), Vandermeer and Goldberg (2003), or Turchin (2003). I will emphasize the assumptions of the models and discuss them in qualitative terms. Proofs or derivations, where needed, have been minimized, but sample problems and graphs are used to illustrate the workings of the models.

A perfect model would be general, realistic, precise, and simple (Levins 1968). As discussed above, the diversity of life has ruled out the perfect model. In order to attempt generality and simplicity, precision and reality are often sacrificed. If students are able to understand how population models are built, they will then be able to evaluate their reality. It should become evident that most models, while lacking precision, do illuminate basic population trends.

The general laws of population ecology

Sutherland (1996) wrote that "population ecology suffers from having no overall a priori theory from which explanations and predictions can be devised." He continued that "behavioral ecology has such a theory – evolution by means of natural selection – which yields the prediction that individuals will maximize fitness." I take this to mean that the discipline loosely known as evolutionary ecology has an a priori theory. Population ecology, however, should be treated as an extension of evolutionary ecology. Therefore, we should ask ourselves under what circumstances might a characteristic such as the low fecundity of the wandering albatross (*Diomedea exulans*), or a phenomenon such as the population cycles known for snowshoe hares (*Lepus americanus*), have evolved.

By contrast to Sutherland, Turchin (2001, 2003) asserts that population ecology is a vigorous and predictive science and does have a set of foundational principles that are almost equivalent to the laws of Newton. He has listed these three fundamental concepts: (i) populations tend to grow exponentially, (ii) populations show self-limitation (or bounded fluctuations), and (iii) consumer–resource interactions tend to be oscillatory. In the first case, without density-dependent feedback from the environment, all populations show a nonlinear, exponential growth pattern. Turchin (2001) calls this "the exponential law," and sees a direct analogue to the law of inertia proposed by Newton. The exponential law provides a starting point for more complex mathematical descriptions of population dynamics. The second theorem or principle, self-limitation, is based on the idea that per capita population growth decreases with resource depletion. The usual form of this idea, the logistic equation, fails as a law because of its simplistic assumptions (see Chapter 2). Nevertheless, it remains useful as a starting point. Finally, the tendency of consumer–resource interactions (such as predator–prey) to produce oscillations is explored at length in later chapters.

1.2 Fundamentals of population growth

If we were trying to understand the growth rate and thus the potential rate of spread of an invasive species, or if we wanted to calculate the potential for long-term survival of the Florida panther (*Felis concolor coryi*) (Seal and Lacy 1989), what sort of information do we need? How do we gather it? What do we do with the data? What models are appropriate? Here we begin to address these questions.

As a first approximation, population growth is determined by a combination of four processes: **reproduction** (sexual or asexual), **mortality, immigration,** and **emigration**. The addition of new individuals through reproduction, termed **fertility** or **fecundity**, may be via sexual reproduction (i.e. live births, hatching of eggs, seed production) or through asexual reproduction (i.e. binary fission, budding, asexual spores, clonal spreading of higher plants). The distinction between fecundity and fertility is traditionally as follows.

1 **Fecundity** is the potential reproductive output under ideal circumstances. This limit is set by the genotype. That is, reproduction is limited by genetic potential, not by the environment.
2 **Fertility**, by contrast, is the actual reproductive performance under prevailing environmental conditions. The fertility rate, by definition, is less than the fecundity rate.

The distinction between these two terms is often not rigidly adhered to, but it is useful to keep it in mind.

Both fecundity and fertility are expressed as rates. That is, the mean number of offspring produced per individual (or per thousand individuals in human demography) in the population, per unit time. Often these values are also expressed for a given unit of area. For example, according to the Population Reference Bureau (Washington, DC), the fertility rate of the human population of the world declined from 28 per thousand in 1981, to 22 births per thousand in 2001. Meanwhile, the birth rate in North America moved slightly downward from 16 per thousand in 1981 to 14 per thousand in 2001 (Anonymous 1981–2004). In populations such as humans, however, which breed over a period of 30 years without respect to seasons, we need to know the fertility rate for each age category in order to accurately predict population growth. All references to human birth and death rates in this chapter are per year.

The second fundamental factor that affects population growth is **mortality**. Mortality must also be expressed as a rate. That is, the mean number of deaths per individual (or per thousand), per unit time, per unit area. As above, unless the population has a **stable age distribution** (meaning that **the proportion of the population in each age class remains constant over time**), in order to predict future population changes we would need to know the death rate for each age category. Again, using data from the Population Reference Bureau, the human death rate for the world in 2001 was 9 per thousand, a decrease from 11 per thousand in 1981. In North America, the comparable figures are 9 per thousand in 1981 and 9 per thousand in 2001 (Anonymous 1981–2004).

In populations with age distributions (age structures), growth is also affected by the actual number of individuals in the different age categories. We will explore the effects of age distributions in detail in Chapter 4. At present it is sufficient to note that basic data on the overall birth and death rates may not produce an accurate picture of population growth in the short term. For example, examine the population figures for Europe and Asia in 2001 (Table 1.1), again data from the Population Reference Bureau. Not only are the birth and death rates different, but also their age distributions are different. In Asia, 30 percent of the population is under 15 years of age, while in Europe the comparable figure is a mere 18 percent.

A measure of population growth is **the intrinsic rate of increase**, r. We will discuss r in more detail later. For now, we define r as the growth rate per individual (or **per capita**)

Table 1.1 Statistics for human populations of Asia and Europe in 2001. All data are from the Population Reference Bureau (Anonymous 1981–2004). Birth and death rates are per thousand; r is per individual.

Region	Population size (millions)	Birth rate (per thousand)	Death rate (per thousand)	Rate of increase per individual (r)	Percent of population less than 15 years of age
Asia	3720	22	8	0.014	30%
Europe	727	10	11	−0.001	18%

per time unit (for example, per year) in a population, estimated as $b - d$, where b is the birth rate per individual per year, and d is the death rate per individual per year. The rate of growth per individual is:

$$r = b - d \tag{1.1a}$$

If the birth and death rates are expressed per thousand, as in human demography, the growth rate is:

$$r = \frac{b - d}{1000} \tag{1.1b}$$

From Table 1.1 we see that Asia had a positive growth rate, whereas Europe actually had a negative projected growth rate in 2001. If the intrinsic rate of increase of these two populations suddenly converged on the same value (a decrease in the Asian birth rate and an increase in Europe's fertility rate, combined with similar changes in the death rates), the population growth of Asia would still be greater than that of Europe for several decades, due to the higher abundance of reproductive individuals. Asia has a shorter generation time, which would affect population growth for a number of years. The estimated growth rate parameter, r (Eqn. 1.1), ignores the age distribution and generation time and actually assumes a stable age distribution (defined above). By **age distribution** we simply mean the **proportion** of the population in each age category, not the actual number per category.

Two other factors affect population growth: **immigration** and **emigration**.

1 The **immigration** rate is the number of individuals that join a population per time interval due to immigration. Ideally we should know the ages of individuals as they join the population.
2 The **emigration** rate is the number of individuals that leave the population per time interval. Again, it would be useful to know the age of the individuals that have left the population.

Unfortunately, gathering accurate information on immigration and emigration is extremely difficult in biological populations, and these factors are often ignored. When a population is termed **closed**, it is thought of as having negligible immigration and emigration. In the last two decades, however, there has been a shift in emphasis from the study of single populations to "metapopulation" ecology. Since the concept of a meta-population was developed by Levins (1969, 1970), major advances in both theory and field studies have taken place, particularly within the past 15 years (Hanski 1999). Levins originally defined a metapopulation as a "population of populations." In his view, local populations exist in a fragmented landscape of suitable and unsuitable habitats or "patches." Each local population is prone to extinction, but extinction may be balanced by immigration from other populations in the metapopulation landscape. The long-term survival of the metapopulation depends on the balance and interplay between extinction and immigration. Immigration and extinction are also key elements of the MacArthur and Wilson (1967) theory of island biogeography. However, MacArthur and Wilson were primarily concerned with the number of species in the community, while the meta-population concept focuses on populations of single species. Another difference is that MacArthur and Wilson were concerned with the relationship between islands, where extinction could occur because of small population size or stochastic events, and a source of species (the mainland) in which extinction would not normally occur. By contrast, in a metapopulation, extinction may occur in any patch and colonization can occur from any one patch to another. The applications of metapopulation studies to conservation biology are obvious, and have resulted in an explosion of publications. We will explore metapopulation dynamics in Chapter 5. Suffice it to say that, after decades of being ignored, immigration, emigration, and local extinction are now the subject of many theoretical and field studies (Hanski 1999).

As already noted, a population is rooted in a time and a place. This means that popu-lation sizes or population growth rates are **scaled** for a particular time unit and for a specific spatial unit. When life histories of different organisms are compared (Chapter 6) it becomes obvious that generation times vary across several orders of magnitude. The space needed to sustain one population of elephants may support a metapopulation of butterflies or several separate populations of lichens. Therefore, we are forced to ask, what is the appropriate scale of an ecological investigation (Peterson and Parker 1998)? That is, over what time spans and/or over what spatial scales should ecological investigations be conducted? As we explore simple models of population growth we should be aware of their limitations, and the extent to which they are applicable to long periods of time and/or to large landscapes.

In summary, a population is affected by its rates of fertility, mortality, immigration, and emigration, by its recent history (through its age structure), and by its generation time, which is determined by its life history. Growth rate is also determined by the envir-onment, and by how sensitive the population is to changes in the environment. By environ-ment, we mean not only the physical environment, but also interactions of the population with other species in its habitat.

1.3 Types of models

In developing a model of a population we usually begin with the present population; that is, the population at time = 0, and project it t time units into the future. The populations at these times are expressed as N_0 and N_t, respectively. There are two types of population equations. Each has advantages and disadvantages.

In **difference equations**, populations are modeled using specific, finite, time units. The time units are usually realistic, in that populations are measured in the field once (or perhaps several times) per year, but not continuously. Difference equations are most often used to model populations that have "discrete," rather than continuous, growth (see below). A basic equation summarizing the ideas presented in the previous section might look like this:

$$N_{t+1} = N_t + (B - D) + (I - E) \tag{1.2a}$$

where
N_t = the population size at time, t
N_{t+1} = the population size one time unit later
B = the number of births and D = the number of deaths in the population during the time interval between t and $t + 1$
I = the number of immigrants and E = the number of emigrants during this same time interval

This equation can be rewritten as:

$$N_{t+1} = N_t + (B + I) - (D + E) \tag{1.2b}$$

In most population studies it is assumed that immigration and emigration rates are insignificant compared with birth and death rates (Turchin 2003, but see Hanski 1999). Equation 1.2b can be simplified, and the numbers of births and deaths are converted to per capita (per individual) rates, b and d, respectively. The difference between b and d becomes the single growth parameter, R, known as the **net growth rate per generation** or **net reproductive rate**. Alternatively, the difference between b and d also equals λ (lambda), the growth rate per time period, usually per year. λ can be calculated for all types of population models and is known as the **finite rate of increase**. The usual form for the difference equation (using R) is shown as:

$$N_{t+1} = N_t(b - d) = N_t R \tag{1.2c}$$

In **differential equations**, it is assumed that population growth is "continuous" and populations are being continuously monitored. Models based on differential equations have a long history in the biological literature, including the earliest models of competitive, predator–prey, and host–parasite relationships (Lotka 1925). A simple differential equation for population growth is:

$$\frac{dN}{dt} = rN \tag{1.3}$$

Here dN/dt measures the instantaneous growth of the population, N. On the left side of the equation, the symbol d is used to indicate change in N per change in the time interval, t. The intrinsic rate of increase, r (Eqn. 1.1a), measures the per capita birth rate minus the per capita death rate during these same small time intervals. In a sense, r measures the probability of a birth minus the probability of a death occurring in the population during a particular time interval.

1.4 Density-independent versus density-dependent growth

If a population invades a new environment with "unlimited" resources, no competitors, and no predators, fertility rates will be high (approximating fecundity rates) and death rates will be relatively low. Under these conditions, the population grows either "geometrically" or "exponentially" depending upon its life history. This is known as **density-independent growth**. This simply means that the growth-rate parameter of the population is not affected by its present population size. In both geometric and exponential models, the growth rate is determined by a fixed parameter (R, λ, or r) that is not modified by competition for resources. Population growth is often curtailed by the environment even if the population is undergoing density-independent growth. Major disturbances or catastrophes such as fire, wind storms, landslides, and floods significantly reduce certain populations and may even cause local extinctions. By contrast, in Chapter 2 we will examine models of **density-dependent growth**. In these models, it is assumed that the population encounters a limiting resource (food, water, nest sites, available nitrogen, space, etc.), which limits its growth. In these models the growth parameter is modified and the net growth rate eventually approaches zero at a carrying capacity. The realized growth rate is said to depend on the density of the population, hence the term **density-dependent growth**.

1.5 Discrete or "geometric" growth in populations with non-overlapping generations

The use of an appropriate model depends first on the life history of the organism. So you first need basic information on the life cycle of the species. In this first model of density-independent growth, the population has a life history with discrete, non-overlapping generations. That is, there are no adult survivors from one generation to the next. Examples include annual plants, annual insects, salmon, periodical cicadas, century plants, and certain species of bamboo. In most of these cases the organism passes through a dormant period as a spore, a seed, or an egg, and/or a juvenile stage such as a larva or pupa. Once the adults reproduce, they perish, and the future of the population is based on the dormant or juvenile stage of the organism. As noted above, when modeling such populations we usually collapse fertility and mortality into one constant, R, the **net replacement rate** or **net growth rate per generation** – or λ, the **finite rate of increase**, when measuring growth per specific time period. When we are discussing annual plants or insects, λ, the growth rate per year, and R, the growth rate per generation, are identical, since generation time equals one year. However, in some populations, such as the periodical cicada (*Magicicada septendecim*), generation time equals 13 or 17 years, and in these cases it is useful to make a distinction between the growth rate per generation and a finite rate of increase. That is, $R \neq \lambda$, when T, the generation time, $\neq 1$ year.

To find R we often count one life stage of the population in successive years. For gypsy moths (*Lymantria dispar*) we estimate R by counting egg masses in successive years (see Example 1.1). R is estimated from the ratio of egg masses at time $t + 1$ versus time t. For the periodical cicada (Example 1.2), however, we would have to wait 17 years between generations before we could estimate R. The overall model is based on finding successive estimates of the growth rate based on:

Example 1.1

Gypsy moths (*Lymantria dispar*) are annual insects in which breeding takes place in early to mid summer. After the females lay their eggs, all adults die. The eggs hatch the following spring into larvae that feed on the leaves of tree species, especially species of oaks (*Quercus*). After a number of larval stages and a pupal stage, the adults emerge. After mating, females lay their eggs and die. Since generation time equals one year, Equations 1.4 or 1.5 may be used. In order to determine population growth in this species, we need to determine R. Assume that a local gypsy moth technician makes annual egg-mass counts in a local forest. She finds that in 2003 there are, on average, 4 gypsy moth egg masses per hectare and each mass contains an average of 40 eggs, for a total of 160 eggs per hectare. When she returns to the same forest in 2004, she finds 5 egg masses with an average of 40 eggs, or a total of 200 eggs per hectare. The local spraying program regulations state that spraying with Bt® (*Bacillus thuringiensis*) begins whenever egg masses reach 1000 per hectare. Assuming egg-mass density continues to increase at a constant rate, what is the predicted population for the year 2006? In what year would spraying be required?

Answers

In order to determine the net growth rate R, we find the ratio of $N_{t+1}/N_t = 200/160 = 1.25$. In the year 2006, three years have passed since the original survey in 2003. Using Equation 1.4:

$$N_{2006} = N_{2003}R^3 = (160)(1.25)^3 = 312.5$$

We therefore expect around 312 eggs per hectare in 2006.

We can now ask the question, if R continues at 1.25, in what year must spraying commence? Since we wish to solve for t, and time is an exponent in Equation 1.3, it is more convenient to use Equation 1.6:

$$\ln N_t = \ln(1000) = \ln(160) + \ln(1.25)(t)$$

$$6.91 = 5.08 + 0.223t$$

$$1.83/0.223 = t$$

$$t = 8.2 \text{ years}$$

Since the population only reproduces once a year, we cannot use a fraction of a year in the answer. Eight years after 2003, that is, in the year 2011, the number of egg masses is expected to be 954. By regulation, this does not trigger the spraying regime. One year later, however, the egg mass density would be 1192, and spraying would begin in 2012.

Example 1.2

The periodical cicada (*Magicicada septendecim*) has a most unusual life history (Borror *et al.* 1989). The juvenile stages spend 17 years underground feeding on plant roots. The population in a given area emerges synchronously from the ground as adults. After a great deal of racket, the males and females mate, and females lay their eggs in slits they have made in small branches of trees and shrubs in the forest. The adults then die, leaving the eggs as the next generation. The eggs hatch within a month. The nymphs drop to the forest floor and burrow underground, where they spend the next 17 years feeding and growing. The periodical cicada is obviously affected by disturbances within the forest habitat. Assume that in 1987 a survey found 500 adult female cicadas per hectare. The forest was selectively logged in the 1990s and a survey in 2004 found that the cicada population had dropped to 200 per hectare. More logging is planned during the next 20 years in this forest. Assume the population continues to decline at the same rate. If we define the minimum viable population for cicadas as 10 females per hectare, in what year is the population no longer viable? By minimum viable population we mean that the probability of extinction has become unacceptably high (Shaffer 1981, Miller and Lacy 2003). Random environmental perturbations or inability of males and females to find each other would likely cause this population to become extinct. See the section on population viability analysis in section 1.10 below.

Answer

First we must realize that only Equations 1.4 and 1.6, using net growth rate per generation, are applicable. But we also need to remember that generation time is 17 years. To find R, take the ratio of $200/500 = 0.40$. Since $R < 1$ we note that this population is decreasing. In order to find when the population is not viable, we solve Equation 1.6:

$$\ln 10 = \ln 500 + (\ln 0.40)t$$

$$2.3 = 6.2 + (-0.9)t$$

$$-3.9 = -0.9t$$

$$t = 4.3 \text{ generations}$$

Again, we cannot use fractions. After four generations, the population is projected to drop to between 12 and 13. After five generations, it declines to around 5 per hectare and is, by definition, no longer viable. Five generations, times 17 years per generation, equals 85 years. The population is not viable 85 years after the first survey in 1987. That is, in the year 2072. Evidently, however, action to conserve this forest cannot wait until 2072.

$$R_1 = N_1/N_0$$

$$R_2 = N_2/N_1$$

$$R_3 = N_3/N_2 \text{ etc.}$$

If we find that R remains more or less constant over time (that is, if these ratios of N_{t+1}/N_t remain constant), then we have:

$$N_1 = N_0R$$

$$N_2 = N_1R = (N_0R)R = N_0R^2$$

$$N_3 = N_2R = (N_0R^2)R = N_0R^3$$

and so on, leading to Equation 1.4:

$$N_t = N_0R^t \tag{1.4}$$

or

$$N_t = N_0\lambda^t \tag{1.5}$$

Note that the population grows whenever	R or $\lambda > 1$
the population is stationary (there is no growth) whenever	R or $\lambda = 1$
the population decreases whenever	R or $\lambda < 1$

The population grows according to the law of discrete or **geometric growth** (Fig. 1.1), when $R > 1$. Equations 1.4 and 1.5 can be rewritten using logarithms to make the growth curves linear. In Equations 1.6 and 1.7 we can use log to the base 10, or we can use natural logs (designated by ln) to the base e. Since other models use natural logs, we have used them in the equations below (and in the examples above).

$$\ln N_t = \ln N_0 + (\ln R)t \tag{1.6}$$

or

$$\ln N_t = \ln N_0 + (\ln \lambda)t \tag{1.7}$$

In each case $\ln N_0$ is the y-intercept and $\ln R$ or $\ln \lambda$ is the slope of a linear relationship between $\ln N$ and t (time), with time as the independent variable (x-axis). In Fig. 1.2, the value of $R = 1.2$ and the slope is therefore $\ln(1.2)$ or 0.18.

1.6 Exponential growth in populations with overlapping generations

In the previous section we dealt with a special kind of life history, one in which generations were distinct and non-overlapping. If the adults and juveniles are present

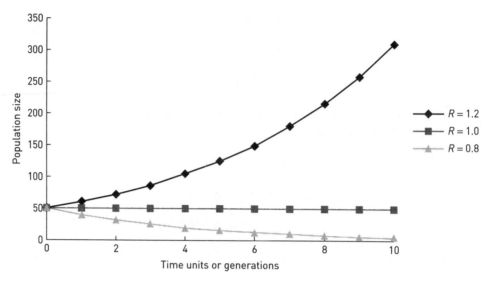

Figure 1.1 Discrete or "geometric" growth in a population with non-overlapping generations.

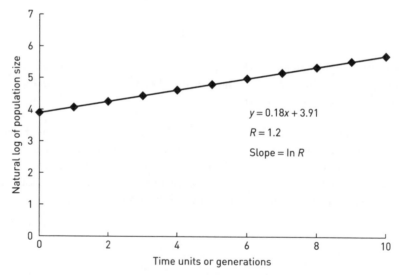

Figure 1.2 Natural log of growth in a population with discrete generations.

simultaneously and they interact with one another, our previous model is inappropriate. Instead we must use a model originally developed for a population capable of continuous growth, such as a *Paramecium* or a human population. That is, a population in which there is no distinct breeding season. Notwithstanding poetry about springtime and theories about phases of the moon, human babies are born throughout the year. In spite of the fact that this growth model is not strictly applicable for seasonal breeders such as deer, it is general enough that it is used whenever a population has a stable age distribution. (Recall that an age distribution refers to the proportions of the population belonging to different age classes, and that a stable age distribution is one in which these proportions remain constant from year to year.) In order to have a stable age distribution, fertility and mortality rates must remain constant for an extended period of time. We can approximate human population growth rates using the model, but we should recall that because birth rates around the world increased following World War II and then decreased after 1960, few human populations are in a stable age distribution.

The basic form of this model is the differential equation shown earlier as Equation 1.3: $dN/dt = rN$, where r is the **intrinsic rate of increase** or the **instantaneous growth rate**.

r is calculated by finding the difference between the instantaneous per capita birth rate and the instantaneous per capita death rate. The parameter r can be compared to the interest rate in a bank account which is continuously compounded. Such a rate is the continuous growth rate per dollar in an interest-bearing account, while r is the continuous growth rate per individual in a population.

The equation is easily solved by taking the integral from 0 to t of both sides of the equation, as follows:

$$\int_{N(0)}^{N(t)} \frac{dN(t)}{N} = r \int_0^t dt$$

which becomes: $\ln N(t) - \ln N(0) = rt - r0 = rt$

After exponentiation of both sides of the equation, we have: $N(t)/N(0) = e^{rt}$

Rearranging, we get Equation 1.8. This solved form is the one usually used in making population projections to some arbitrary time t in the future.

$$N_t = N_0 e^{rt} \tag{1.8}$$

where e is the base of natural logs.

In the above equations, the population grows if $r > 0$
the population is stationary if $r = 0$
the population is negative if $r < 0$

When r is positive, the growth is known as **exponential**; if r is negative the population is in exponential decline (Fig. 1.3).

We can make the equation linear by taking the natural logs of both sides of Equation 1.8, yielding:

$$\ln N_t = \ln N_0 + rt \tag{1.9}$$

When we graph $\ln N$ versus time, we again have a linear relationship, with $\ln N_0$ as the y-intercept and r as the slope of the line (Fig. 1.4).

Doubling time

A convenient statistic, often used by population ecologists and human demographers (demography is the study of population statistics), is doubling time. That is, how long will it

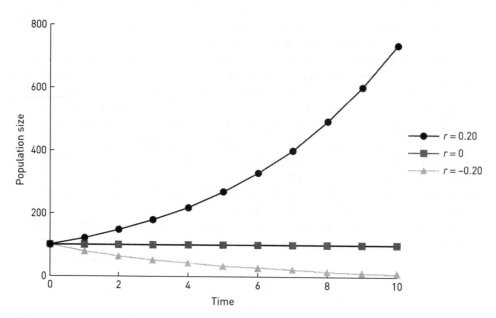

Figure 1.3 Exponential growth pattern in a population with overlapping generations and continuous breeding.

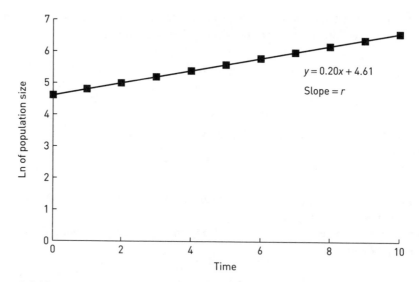

Figure 1.4 Natural log of growth in a population with overlapping generations and continuous breeding.

take a population to double from its present population size? Equation 1.8 can be rearranged to: $N_t/N_0 = e^{rt}$. We want to solve for the time at which the ratio $N_t/N_0 = 2$. So we have: $2 = e^{rt}$. Taking the natural log of both sides of the equation yields, $\ln 2 = rt$, where t is now doubling time. Since $\ln 2 = 0.693$, if we solve for t we end up with:

$$\text{Doubling time} = 0.693/r \tag{1.10}$$

Therefore if we know the intrinsic rate of increase we can easily find the projected doubling time of a population. Remember, however, that we are assuming that the population is not affected by its age distribution, and that r is a constant during this time period. That is, birth and death rates remain unchanged.

Doubling time probably has little meaning if r is very close to zero. Doubling time is undefined if $r = 0$. An r-value of 0.001, for example, would predict a doubling time of 693 time units; but it is extremely unlikely that r would remain a constant for such a long period of time. For a negative r-value ($d > b$), the result will be a negative number. The absolute value of this number is the time it will take the population to be reduced to half of its present size. Instead of "doubling time" the result is "halving time."

1.7 Exponential growth in an invasive species

During a hurricane in 1962, five captive mute swans (*Cygnus olor*) escaped into the Chesapeake Bay, in Maryland. Since they were pinioned and therefore flightless, their chance of survival during the winter was considered negligible and no attempt was made to capture them. One pair, however, successfully nested. By 1975 the descendents of this original pair numbered approximately 200, and by 1986 totaled 264. By 1999 the estimated population of mute swans in the Chesapeake Bay was 3955 (Anonymous 2003, Sladen 2003, Craig 2003). In 2001 the Maryland Department of Natural Resources, in an effort to control the swan population, began shaking (addling) mute swan eggs or covering them with corn oil to terminate embryo development. Mute swans were also removed from Federal National Wildlife Refuges. The result was a decline to 3624 in 2002 (Anonymous 2003). As shown in Fig. 1.5, prior to these control efforts, the population was growing exponentially with an intrinsic rate of increase of 0.17 and a doubling time of four years! (As an exercise, try using Equation 1.10 to verify the doubling time.)

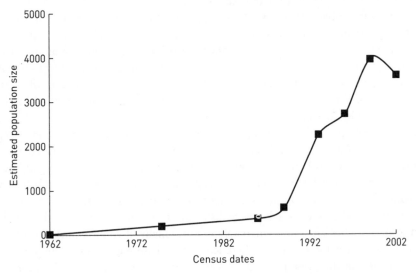

Figure 1.5 Mute swan (*Cygnus olor*) population in the Chesapeake Bay since 1962.

So what's the problem? Swans are considered graceful, even "majestic," and are thought of as harmless by their admirers. However, mute swans, in addition to being a non-native species, have become permanent residents. That is, they do not migrate as do other swan species. Recent data show that an average adult swan eats 3.6 kg of submerged aquatic vegetation (SAV) a day (Craig 2003). This is occurring at a time when biologists are struggling to re-establish SAV in the Bay. Is it necessary to control the mute swan population? If so, how?

The Fund for Animals took the US Fish and Wildlife Service to court to stop its plan to kill 525 swans in 2003 (Craig 2003). The debate evidently will continue for the indefinite future.

1.8 Applications to human populations

Few biological populations grow either geometrically or exponentially for long. As we will explore in the sections on intraspecific competition and logistic growth, as populations grow, resources become scarce. The resultant changes in birth and/or death rates slow growth. The human population of the world, however, has continued to grow since around 1650; it reached 6.0 billion by late 1999, and 6.3 billion by 2003 (Fig. 1.6a). Many scientists question how long this growth can be sustained. While most ecologists insist that human population growth must cease in the near future, some economists (Simon 1996) see no reason for limits to the human population. In the next section we will use data from the Population Reference Bureau (Anonymous 1981–2004) to illustrate how Equations 1.8 to 1.10 may be used in population projections.

Recall from Equation 1.9 that if we graph natural log of population growth versus time we can determine the intrinsic rate of increase by finding the slope of the graph. In Fig. 1.6b we have plotted the natural log of human population growth against time. The slope of this line, as determined by the statistical technique of linear regression and computed for us in an Excel™ spreadsheet, is 0.007. This is the best fit for the intrinsic rate of increase for the human population from 1650 to 2003.

If we examine Table 1.2, in which human populations in 2003 are broken down by continental regions, the strengths and weaknesses of this simple model become apparent. Most striking are the immense differences among populations. While the human population as a whole is growing twice as fast in 2003 as compared to the period of 1650 to the present (contemporary $r = 0.013$, historical $r = 0.007$), Europe has a negative r, while that of Africa is 0.024, almost twice the global growth rate. Secondly, over 60% of the human population resides in Asia.

Clearly, although human population growth is of global concern, it is a highly regional problem. From Table 1.2 you should be able to see that r is readily calculated as the difference between the birth and death rates. Secondly, you should try calculating projected doubling times based on Equation 1.10. You will find that the data published by the Population Reference Bureau differ slightly from your calculations. They are using more sophisticated models and are taking age distributions into account. Nevertheless, the differences in doubling times are remarkably minor. Finally, if you examine the last column you will also notice another great difference among these populations. The percentage of the population in the pre-reproductive years (15 years or younger) varies from 42% in Africa to a low of 17% in Europe.

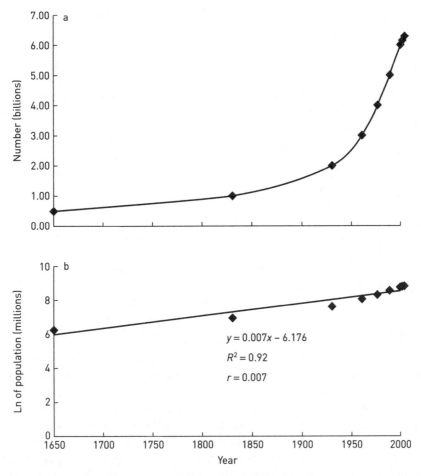

Figure 1.6 Human population growth since 1650: (a) world population, in billions; (b) natural log of population growth, in millions.

In his book *The Skeptical Environmentalist*, Bjorn Lomborg (2001) is rather sanguine about human population growth. He accepts the demographic transition model, which states that rapid growth has occurred because of a rapid drop in the death rate (due to modern methods of sanitation, improved food growth and distribution, better medical care, etc.) and that eventually, with improved standards of living and wealth, birth rates drop to match the low death rates. Indeed, in most European countries, human population growth has slowed, and even gone negative. In 2003, 20 countries out of 43 in Europe had a growth rate of zero or negative, including all 10 Eastern European countries. As noted above, the population growth rate (r-value) for Europe as a continent is negative. As for the future, Lomborg accepts a "medium variant forecast" from the UN. This prediction is zero population growth for the world by the year 2100. However, by then the world population is projected to be 11 billion. Consider that the world population was only one billion in 1850, two billion in 1950, and 6.3 billion in 2003. Lomborg is correct when he says that 60% of growth is from just 12 countries. Perhaps the world outside of Africa and Asia will not necessarily suffer a catastrophe from human population density,

Table 1.2 2003 human population data from the Population Reference Bureau
(Anonymous 1981–2004).

Region	Population size (millions)	Birth rate (per thousand)	Death rate (per thousand)	Rate of increase per individual (r)	Doubling time (years)	Percent under 15 years
World	6314	22	9	0.013	53	30%
Africa	861	38	14	0.024	29	42%
North America*	323	14	8	0.005	139	21%
Latin America†	540	23	6	0.017	41	32%
Asia	3830	20	7	0.013	53	30%
Europe	727	10	12	−0.002	NA	17%
Oceania‡	32	18	7	0.011	63	25%

* North America = the United States and Canada.
† Latin America includes Central and South America and the Caribbean Islands.
‡ Oceania includes Australia, New Zealand and the South Pacific Islands.
Countries of th former USSR have been distributed between Asia and Europe.

but what will happen in China, India, Pakistan, Bangladesh, and Nigeria, for example, in
the next 100 years? The 2003 data sheet from the Population Reference Bureau predicts
that China's population will stabilize at about 1.4 billion (compared to its present estimated population of 1.289 billion) by 2050. By 2050, however, the PRB predicts a population for India of 1.6 billion (compared to present population of 1.069 billion). The
question on the mind of the concerned biologist: Will there be any room for natural
habitats on a planet with 11 billion or, worse yet, 15 billion people?

Examine Table 1.3, which describes overall human demographic trends since 1981.
Lomborg (2001, p. 47) states that world population growth, in numbers per year, reached

Table 1.3 World human demographic trends since 1981. All data from
the Population Reference Bureau (Anonymous 1981–2004).

Year	World population estimate (billions)	Birth rate per thousand	Death rate per thousand	r per individual	Projected growth in numbers per year (millions)	Actual average growth per year during specified time period (millions)
1981	4.492	28	11	0.017	77.0	
1985	4.845	27	11	0.016	78.1	1981–85: 88.3
1987	5.026	28	10	0.018	91.3	1985–87: 90.5
1989	5.234	28	10	0.018	95.1	1987–89: 104.0
1991	5.384	27	9	0.018	97.8	1989–91: 75.0
1995	5.702	24	9	0.015	86.2	1991–95: 79.5
2000	6.067	22	9	0.014	85.5	1995–2000: 73.0
2003	6.314	22	9	0.013	82.6	2000–03: 82.3

Table 1.4 Human demographic trends in North America since 1981. Data from the Population Reference Bureau (Anonymous 1981–2004).

Year	Population estimate (billions)	Birth rate per thousand	Death rate per thousand	r per individual	Projected growth in numbers per year (millions)
1981	0.25	16	9	0.007	1.76
1985	0.26	15	8	0.007	1.83
1987	0.27	15	9	0.006	1.62
1989	0.27	16	9	0.007	1.90
1991	0.28	16	9	0.007	1.97
1995	0.29	15	9	0.006	1.75
2000	0.31	14	8	0.006	1.86
2003	0.323	14	8	0.005	1.62

a peak in 1990 at 87 million per year. Population Reference Bureau data agree on the time but not the number (over 100 million added in the period 1987–89). Absolute growth has averaged about 87 million per year in the latter part of the twentieth century, according to Population Reference Bureau data; Lomborg used the figure of 76 million, but this applies only to the 1990s. The 2003 Population Reference Bureau data sheet projects world population as 7.9 billion in 2025 and 9.2 billion in 2050. Lomborg's comparable numbers are "almost 8 billion" in 2025 and 9.3 billion in 2050.

Population growth in North America (Table 1.4) is rather variable, but reached a relative peak in 1991–92 when around two million people were added to the population per year. The data from 2003, however, reflect the fact that the 2000 census for the United States came in at almost seven million more than expected. Meanwhile, the US birth rate has fallen to 2.034 births per female (replacement rate is 2.10 births per female) (PRB, Anonymous 1981–2004).

Human population growth is greatest in Asia (Table 1.5). Peak absolute growth was in the period 1989–91, when around 58 million people were added per year. It declined unsteadily in the late twentieth century and is now about 50 million people per year. The r-value has declined steadily to 0.013 in 2003.

1.9 The finite rate of increase (λ) and the intrinsic rate of increase (r)

Both the intrinsic rate of increase (r) and the finite rate of increase (λ) are used commonly to track population growth and to compare growth rates among populations of the same species found in different environments, as well as among different species. Consequently it is important to understand the relationship between λ and r. As defined in Equation 1.5, λ is the growth rate per time period (usually per year) and is based on the ratio N_{t+1}/N_t. If the population lacks an age distribution or has a stable age distribution (SAD), the finite rate of increase, λ, is a constant. The population as a whole and each age class will grow as:

Table 1.5 Human demographic trends in Asia since 1981. Data from the Population Reference Bureau (Anonymous 1981–2004).

Year	Population estimate (billions)	Birth rate per thousand	Death rate per thousand	r per individual	Projected growth in numbers per year (millions)
1981	2.61	29	11	0.018	47.4
1985	2.83	28	10	0.018	51.4
1987	2.93	28	10	0.018	53.2
1989	3.06	28	9	0.019	58.7
1991	3.16	27	9	0.018	57.4
1995	3.38	24	8	0.016	54.5
2000	3.68	22	8	0.014	51.9
2003	3.83	20	7	0.013	50.1

$$\frac{N_{t+1}}{N_t} = \lambda \tag{1.11}$$

Rearranging Equation 1.8 and setting $t = 1$, we have: $N_{t+1}/N_t = e^{rt} = e^r$. Thus, when $t = 1$ and when there is a stable age distribution we have:

$$\lambda = e^r \tag{1.12}$$

and

$$r = \ln \lambda \tag{1.13}$$

1.10 Stochastic models of population growth and population viability analysis

All of the population models we have examined to this point are **deterministic models**. The models specify conditions leading to an exact outcome based on the parameters of the models. But natural systems are unlikely to be deterministic; rather they are more likely to be **stochastic**. In particular, small isolated populations are subject to stochastic processes because chance events can dominate their long-term dynamics. In stochastic models population parameters vary according to some kind of a frequency distribution. This distribution has a "central tendency" (a mean), but also has a range of variability around the mean. For example, in a deterministic model, if we know the present population size and the proper growth parameter, we forecast an exact expected population size for a specific time in the future. In a stochastic model, we would instead predict a range of possible population future sizes, with assigned probabilities.

Future population size in a small population is strongly influenced by **demographic stochasticity**, which is driven by variations in the fates of different individuals within a given year. For example, although the average female within a population may have 2.0

Example 1.3

A *Paramecium caudatum* population is cultured in the laboratory and sampled on a daily basis. Population sizes, based on 0.5 ml samples, are shown below. The population grows exponentially between days 0 and 3. Find the intrinsic rate of increase (r) for the population

Growth of a *Paramecium* population. Numbers are based on daily 0.5 ml samples.

Time in days	Number (N) per 0.5 ml	Natural log of N ln N	Per capita growth $\frac{N_{t+1} - N_t}{N_{t+1}}$
0	14	2.64	–
1	41	3.71	0.66
2	116	4.75	0.65
3	193	5.26	0.40
4	244	5.50	0.21
5	290	5.67	0.16
6	331	5.80	0.12
7	363	5.89	0.08

Answer

Since we want to know the value of the maximal rate of increase (the density-independent rate of increase), we examine growth only during the first three days (see Fig. 2.1 in the next chapter). From Equation 1.9 ($\ln N = rt + \ln N_0$) we know that to find r we need only convert column 2 to natural logs (column 3). Then find the slope between days 0 and 3. To find the slope we can use the formula:

$$r = (y_2 - y_1)/(x_2 - x_1). \quad \text{Thus,}$$

$$r = (5.26 - 2.64)/(3 - 0), \quad \text{and}$$

$$r = 0.87$$

Using Excel™, a linear regression on the same data yields the value of r as 0.89.

Example 1.4

The birth rate for Latin America in 1978 was 33 per thousand, while the death rate was 10 per thousand. (a) What was the intrinsic rate of increase, assuming a stable age distribution? (b) If the population size was 344 million, what was the projected population in 1982? (c) Between 1982 and 1990 the population increased from 377 million to 415 million. What was the r during that time? (d) Given this r-value, what was the doubling time? (e) What is λ?

Answers

a Given $r = b - d$, we have $r = 33/1000 - 10/1000 = \mathbf{0.023}$
b From 1978 to 1982 is four years. Therefore:

$$N_4 = N_0 \times (e^{0.023*4}) = 344 \text{ million} \times (e^{0.092})$$

$$= 344 \text{ million} \times 1.096 = \mathbf{377.15 \text{ million}}$$

c From 1982 to 1990 is 8 years. Therefore: 415 million = 377 million(e^{8r})

Simplifying: $415/377 = 1.10 = e^{8r}$

Taking natural logs: $\ln 1.10 = 8r$

Or, $0.096/8 = r = \mathbf{0.012}$

d Doubling time = $0.693/r = 0.693/0.012 = \mathbf{57.7 \text{ years}}$
e $\lambda = e^r = e^{0.012} = \mathbf{1.012}$

female offspring, some individuals may not reproduce at all, while others have a litter size of 4.0. Demographic stochasticity has effects not only on birth and death processes, but also on sex ratio. In the above example, some females may give birth only to males in a given year. Another important influence on population growth is **environmental stochasticity**, which is temporal variation in the population due to unexpected events, often tied to the physical environment, such as droughts, hail storms, fires, and landslides, but which may also include diseases. Environmental stochasticity can affect both large and small populations.

More realistic growth models, therefore, make forecasts based on probabilities, rather than predicting a single outcome. For example, weather forecasters no longer simply predict rain, but instead predict a certain probability of rain. Similarly, it would be prudent for population models to predict an expected population size, but allow for other population sizes to occur with particular probabilities. Again, this approach is especially important in small populations, and over short time intervals. If the population is large and the time frame is very long, the expected population sizes dictated by deterministic models become highly probable.

Stochastic models are the basis for the quantitative approach to conservation biology known as **population viability analysis** (PVA). Although it is beyond the scope of this book to explore stochastic models and PVA in detail, there are excellent discussions of these models in Morris and Doak (2002) and Beissinger and McCullough (2002). For more information on stochastic models, see also Pielou (1977) and Nisbet and Gurney (1982).

PVA is so important because many wildlife populations that were once numerous, widespread, and occupied contiguous habitats are now small, restricted in distribution, and isolated from each other. The problem with small, isolated populations is that they are increasingly subject to stochastic processes and increasingly likely to go locally, if not globally, extinct. The purpose of population viability analysis is to predict the likely future status of a population or collection of populations (Morris and Doak 2002). PVA is a set of analytical and modeling approaches for assessing the future course and risk of extinction of a population (Beissinger and McCullough 2002). PVA examines how (i) genetic, demographic, and environmental stochasticity, (ii) catastrophes and "bonanzas," and (iii) spatial variation affect the future of the population.

Demographic and environmental stochasticity were defined above. Small populations are also affected by genetic processes such as (i) genetic drift resulting in the loss of genetic diversity in the population, (ii) inbreeding depression, and (iii) monopolization by a small number of males in a polygynous mating system. The biggest concern is the rate of loss of heterozygosity and its effects on the future fertility and mortality rates of the population.

PVA also attempts to anticipate how rare events which result in extremely low survival and/or reproduction (catastrophes) or their opposite (bonanzas) might affect the future course of a population. Catastrophes can be local or regional events of low probability with significant density-independent effects. For example, one of two remaining whooping crane (*Grus americana*) populations in the United States was decimated by a hurricane in 1940 and this population went extinct soon thereafter. The only remaining population of the black-footed ferret (*Mustela nigripes*), at Shirley Basin in Wyoming, was being decimated by an outbreak of distemper, while the prairie dogs (*Cynomys ludovicianus*), its prey species, were suffering from the plague. In 1986, conservation biologists, fearing extinction unless action was taken, captured the last remaining 18 ferrets to start a captive breeding program. The captive population grew rapidly, and by 1992 biologists determined the captive population was large enough to sustain a reintroduction program. Currently black-footed ferrets have been reintroduced into six areas in their historic range.

Finally, variation in fertility and mortality can also be spatial. That is, if a population is subdivided into different locations, vital statistics can vary depending on the location of the subpopulation. Again, we cannot explore these topics in detail here. But the following paragraphs explore the consequences of demographic stochasticity for density-independent growth.

In a simple stochastic approach we specify probabilities for births and/or deaths rather than using an exact population average. For example, suppose the arithmetic average litter size of a small mammal population is 1.167 females per female per year, but the actual number of females produced per year varies from zero to two (for simplicity, we follow the traditional practice of only counting females). We then must determine the probability that a given female produces zero, one, or two female offspring. For a given number of females at time = zero, we can then make predictions as to the likelihood of various numbers of offspring in the next year.

In the following simple example, assume that adults die after reproduction, but all individuals in a given litter survive. However, litter sizes (B_i) vary from 0 to 2 with the probabilities shown in Table 1.6. The value of λ is based on the arithmetic average of the litter

Table 1.6 Probability that an individual female will have 0, 1, or 2 female offspring, and the expected net reproduction.

Probability, p_i, of having a given litter size, B_i	Litter size (B_i) = the number of female offspring per year	Expected net reproduction = p_iB_i
0.167	0	0
0.500	1	0.500
0.333	2	0.667
		$\lambda = 1.167$

sizes $= \sum p_iB_i$. The expected finite rate of increase for the population as a whole is therefore the sum of the last column ($\lambda = 1.167$).

For N females, there are, therefore, finite probabilities that the next generation will produce anywhere between 0 and $2N$ female offspring in the next generation. The probability that a population of N females goes extinct in the next year, for example, is $(0.167)^N$. For a population of six females the probability that the population will go extinct in the next year is $(0.167)^6 = 2.17 \times 10^{-5}$. For a population of one female, the probability equals 0.167. Similarly, the probability that the population will double in one year is $(0.333)^N$. A radical population shift such as extinction or doubling in one year is likely only in very small populations.

In Fig. 1.7 the probabilities from Table 1.6 are applied to a population of three females at time $= 0$. One time unit later ($t = 1$), the population size has a possible range of values from 0 to 6. The most likely outcome is $\lambda N = 1.17 \times 3$, or 3.51. In reality there cannot exist fractions of individuals, so the population, one time unit later, is equally likely to remain at three or grow to four females.

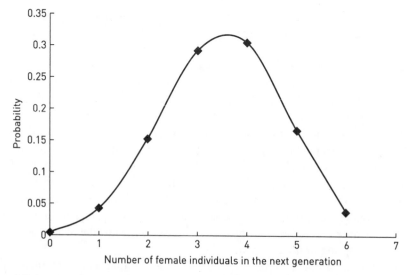

Figure 1.7 Stochastic growth in a population of three females, based on the parameters of Table 1.6.

As noted by Pielou (1977) and others, the probability that a population will go extinct can be estimated by Equation 1.14:

$$P_{0,t} = \left(\frac{d}{b}\right)^{N_0}$$ (1.14)

where
$P_{0,t}$ = the probability of extinction at time t
d = per capita death rate and b = per capita birth rate

For any finite population there is a probability of one that the population will go extinct, given enough time, unless the birth rate is higher than the death rate ($b > d$, $\lambda > 0$). Even then, there is a finite non-zero probability of extinction in any generation. Again, this chance of extinction is heavily influenced by the size of the population, with the smallest populations the most likely to go extinct.

As pointed out by Morris and Doak (2002), adding variability to population statistics does not simply mean that population growth is more variable; it means that populations do worse than they would without variation. The use of an arithmetic mean, as in the example above, overestimates growth most of the tim . As Morris and Doak (2003, p. 25) state, "using simple arithmetic averages to characterize the population growth rate in a variable environment is not just a simplification, it is actually wrong."

When variation is added the most likely result is that the population will grow according to the **geometric mean**, rather than the **arithmetic mean**. The geometric mean of a set of numbers is always less than or equal to the arithmetic mean, and the difference between the two increases as the variability in the data increases.

For example, assume that a population with an initial population size of 50 grows for 100 time periods ($t = 100$), with an arithmetic mean value for λ of 1.05. With no variation, using Equation 1.5, we get the predicted population size of:

$$N_{100} = N_0\lambda^{100} = 50(1.05^{100}) = 6575.$$

Now assume that we allow λ to vary between 0.90 and 1.20, with equal probabilities ($p_i = 0.50$ for each). We have:

$$N_{100} = 50(0.90^{50})(1.20^{50}) = 50(0.005)(9100) = 2345$$

This is the **most likely outcome** and is based on the geometric, rather than the arithmetic, mean. As shown in Table 1.7, the arithmetic mean $= \sum_{i=1}^{n} p_i\lambda_i$ where p_i = probability of a given λ_i. In the above case, $p_1 = 0.50$ for λ_1 ($= 0.90$), and $p_2 = 0.50$ for λ_2 ($= 1.20$).

Therefore the **arithmetic mean** $= (0.50 \times 0.90) + (0.50 \times 1.20) = 1.050$

However, the **geometric mean** $= \prod_{i=1}^{n}\lambda_i^{p_i} = 0.90^{0.5} \times 1.20^{0.5} = 0.949 \times 1.095 = 1.039$

As stated above, the geometric mean is always less than or equal to the arithmetic mean, and in this case the geometric mean of 1.039 is less than the arithmetic mean of 1.050. If

Table 1.7 Calculating the arithmetic versus the geometric mean for population projections.

Probability, p_i	λ_i	$p_i\lambda_i$	$\lambda_i^{p_i}$
0.25	0.60	0.15	0.880
0.25	0.80	0.20	0.946
0.50	1.40	0.70	1.183
		Arithmetic mean	Geometric mean
		$= \displaystyle\sum_{i=1}^{n} p_i\lambda_i$	$= \displaystyle\prod_{i=1}^{n} \lambda_i^{p_i}$
		$= 1.050$	$= 0.985$

we use the geometric mean instead of the arithmetic mean in Equation 1.8, we have the most likely outcome when λ varies between 0.90 and 1.20 with equal probabilities:

$$N_{100} = (50)(1.039^{100}) = 2345$$

which is the same result we found above, but is much less than the projected population of 6575 using the arithmetic mean.

Let us try another example. Assume that $\lambda = 0.60$ 25% of the time, $\lambda = 0.80$ 25% of the time, and $\lambda = 1.40$ 50% of the time (Table 1.7). The arithmetic mean is, again, 1.05. Based on the arithmetic mean, we expect the population to grow since $\lambda > 1.00$. However, the geometric mean is less than one, and the most likely result is that this population will decline.

The geometric mean, however, provides us only with the "most likely" outcome when population parameters vary. In fact, if the population parameters are allowed to vary randomly, many different outcomes are possible. For example, in Fig. 1.8 we see the results

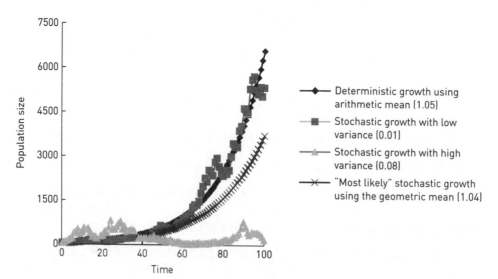

Figure 1.8 Deterministic versus stochastic growth with high and low variance. Initial population size = 50; $\lambda = 1.05$, except where noted.

Table 1.8 Results of 20 simulations of population growth for a deterministic model versus two stochastic models, one with low and one with high variability. In all cases the initial population size was 50 individuals, the arithmetic mean finite rate of increase (λ) was 1.05, and the simulation was run for 100 time units. In the low-variability simulations, λ was allowed to vary between 0.90 and 1.20 $(\bar{x} = 1.05 \pm 0.1)$; in the high-variability simulation, λ was allowed to vary between 0.55 and 1.55 $(\bar{x} = 1.05 \pm 0.3)$. In the stochastic simulations, growth rates were randomly generated using the Excel™ RAND functions. Note that the deterministic result is greater than the stochastic result/low variability in 15 of 20 simulations and greater than the stochastic result/high variability in 19 of 20 simulations.

Simulation number	Deterministic result	Stochastic result (low variability)	Stochastic result (high variability)
1	6575	3729	5772
2	6575	4156	28
3	6575	5972	1004
4	6575	3631	13
5	6575	5516	291
6	6575	5700	13
7	6575	2363	201
8	6575	3796	1
9	6575	5821	246
10	6575	2845	44
11	6575	7107	5244
12	6575	2113	3169
13	6575	19,561	106
14	6575	3910	640
15	6575	2509	122
16	6575	13,731	8
17	6575	3706	1917
18	6575	6304	53
19	6575	15,570	4
20	6575	12,972	8450
Average for the 20 simulations	6575.0	6550.6	1366.3

of one simulation. A comparison of growth using the arithmetic and geometric means yields the expected results. Stochastic growth with low variability (variance around the mean is 0.01), shows growth, but with obvious variation. The end result of growth with high variation (variance of 0.08 around the mean) is a population of only 178 individuals (N_0 was 50) after 100 time units.

Although this result is "typical" there are many other possible outcomes. Table 1.8 presents the results of 20 different simulations of population growth for a deterministic and two stochastic models (low versus high variability). The basic result is that the deterministic model, using the arithmetic mean for λ of 1.05, produced a larger final population

size than did the stochastic/low-variability model in 15 of the 20 simulations. The final
population size for the deterministic model was greater than that of the stochastic/
high-variability model in 19 of 20 simulations. The low-variability result is larger than
the high-variability result in 18 of 20 simulations.

In summary, a stochastic model generates a frequency distribution of probabilities that
particular population numbers will appear in the next generation. There will always be a
finite probability that the population will go extinct, but the most likely outcome (i.e., the
highest probability) will be that $N_{t+1} = N_t\lambda$, using the geometric mean for λ.

1.11 Conclusions

In this chapter we have explored models illustrating the Turchin (2001)
first law of population ecology. That is, biological populations tend to grow
exponentially. Populations with discrete or continuous generations, as well
as populations with age structures, all obey the exponential law. As will be
detailed in Chapter 4, populations with age structures must first achieve
a stable age distribution before growing according to the exponential law.
The exponential law even applies to populations undergoing demographic
stochasticity as described in section 1.10 above (Turchin 2001). And we do
not have to assume a constant environment. If the environment varies such
that per capita birth and death have a stationary probability distribution,
we still obtain exponential growth or decline in the population (Maynard
Smith 1974).

Accordingly, if the environment does not affect the population in a systematic manner, all types of biological populations show exponential
growth. Traditionally, ecologists have treated populations with discrete
generations differently from those with overlapping generations. Difference
equations such as 1.4 and 1.5 have been used in the first case. By contrast
differential equations (1.3) and their solved forms (1.8) have been employed
to describe populations with overlapping generations. In both cases we
use the finite rate of increase, λ, or the intrinsic rate of increase, r, as a
common currency for comparing population growth potentials.

However, populations do not grow forever. Eventually individuals begin to
run out of space, food, water, or other resources and/or become increasingly subject to predation or disease. This is where the second principle,
that of self-limitation, comes into play. In the next chapter we will examine
this principle, and the models, traditionally known as density-dependent
models, that attempt to implement it.

2

Density-dependent growth and intraspecific competition

- Density dependence in populations with discrete generations
- Density dependence in populations with overlapping generations
- Nonlinear density dependence of birth and death rates and the Allee effect
- Time lags and limit cycles
- Chaos and behavior of the discrete logistic model
- Adding stochasticity to density-dependent models
- Laboratory and field data
- Behavioral aspects of intraspecific competition

2.1 Introduction

One of the great philosophical divides between ecologists and many economists is the application of the ecological principle of self-limitation to human populations. The late University of Maryland economist Julian Simon, long the bête noir of the environmental movement, was no believer in the ecological notion of a **carrying capacity** for humans. In his book *The Ultimate Resource*, Simon (1996) proposed that human ingenuity and technology would always triumph over any limiting resource. He had public disputes with ecologists such as Norman Myers and Paul Ehrlich. Simon famously won a series of ongoing bets with Ehrlich on whether certain raw materials would run out by specific dates. Now, as human population growth has ceased or gone negative in many European countries (Anonymous 1981–2004), publications decrying the coming "population crash" and its ramifications have materialized in the popular media. Basically, ecologists see self-limitation of all biological populations as inevitable, while most economists, especially those in the United States, see economic growth as both certain and beneficial.

The concept of a carrying capacity for biological populations is connected with the logistic equation, found in all ecology text books, and also introduced formally as Equation 2.8 later:

$$dN/dt = rN\left(\frac{K-N}{K}\right).$$

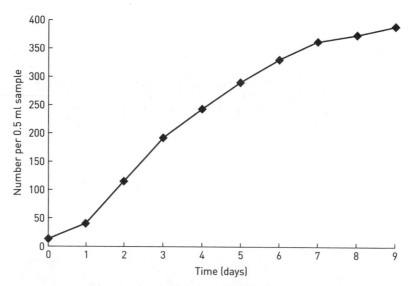

Figure 2.1 Population growth in a *Paramecium* population.

The logistic was originally formulated by the French mathematician Verhulst in pre-Darwinian times (1838), but was not applied routinely to biological populations until Pearl and Reed (1920) rediscovered it. Pearl (1927) then promoted the application of the logistic to a variety of biological populations. See Kingsland (1995) for an interesting review of this history.

Yet Turchin (2003) and many other population biologists now assert that, though the logistic is useful as a general framework, this equation is fundamentally flawed when applied to biological populations. The logistic model is not a general law of population growth, but is rather a special case. If an ecologist wants to win the argument with an economist about human (or any) population limitation he/she needs to understand the assumptions and flaws of logistic or logistic-like models. The goals of this chapter are: first, to describe density-dependent growth models for both discrete and continuously breeding populations; second, to examine the assumptions of these models; third, to investigate how violations of these assumptions shape the behavior of populations.

In the first chapter we assumed **density-independent growth**: that is, population growth unlimited by competition for resources. Most biological populations, however, do not long sustain such growth. Even in an isolated laboratory population, growing without competing species or predators, realized growth slows and ceases. Examine the *Paramecium* population history presented as Table 1.6 and plotted in Figure 2.1. Population growth slows after day 2 and almost ceases by day 7. Our experience in the laboratory is that a *Paramecium caudatum* population will stop growing at about 400 per 0.5 ml sample. Based on data such as this, one of the basic assumptions of most ecological models is that populations do not have unlimited resources, and that eventually the population encounters a limiting resource (or perhaps a parasite or predator) which restricts population growth. This is by no means a new idea. In 1840 Liebig, in his law of the minimum, asserted that under steady-state conditions the population size of a species is constrained by whatever resource is in shortest supply. According to the **logistic model**, population growth ceases when the population reaches the **carrying capacity** of the

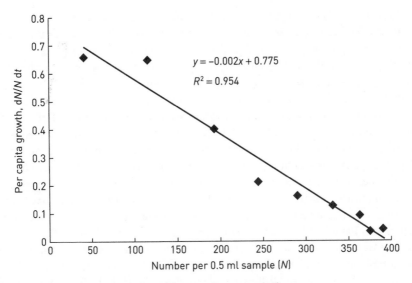

Figure 2.2 Per capita growth in a *Paramecium* population.

environment for that population. This is a **density-dependent growth** model in which the carrying capacity is identified by the symbol K. For a given species, in a specific environment, carrying capacity is defined as the number of individuals that can be maintained indefinitely.

One way to visualize density-dependent growth is to graph the per capita growth rate versus population size. Figure 2.2 plots the growth rate per individual in the *Paramecium* population versus population size. From Table 1.6, per capita growth is found by dividing the growth between time intervals t and $t + 1$ by the population size at time $t + 1$. Notice that even though the population is growing through day 6, the trend per individual is steadily downward in a more or less linear fashion. Where this line intersects the x-axis, per capita growth has fallen to zero. The value of this point $(N, 0)$ is an estimate of the carrying capacity, K. Note that the linear regression on these data indicates that K should equal around 390, which is in agreement with Fig. 2.1. The R^2 value means that the model has explained 95% of the variance in the data.

Since the logistic model is really based on competitive interactions, we should define **competition** before proceeding further. A formal definition of competition is: **a biological interaction between two or more individuals for a resource in short supply.** When the interaction is between individuals of the same species it is termed **intraspecific** competition; when between individuals of different species it is known as **interspecific** competition. A **resource** is any substance or factor in the environment that determines growth, survivorship, or reproduction of individuals in the population. Therefore, depletion of this resource decreases growth, survivorship, or reproduction. For competition to be meaningful, the resource must be in short supply now, or in the immediate future. Plants may compete for space, light, water, or nutrients, while animals often compete for food, nesting sites, hiding places, or mates. Certain aspects of the environment, such as temperature, are not resources per se, and cannot be competed for. On the other hand, if a lizard needs to raise its body temperature it will seek out a rock on which to bask in the sun. If there are limited numbers of basking sites, they become resources in short supply, and may be competed for.

Paradoxically, perhaps, the ultimate effect of competition is a decrease in fitness. Thus competition is said to be a **reciprocally negative interaction**. All individuals that engage in competition may lose energy and/or time that they could have invested in their own growth, survivorship, or reproduction. When sports teams or animals engage in competition, we identify a winner and a loser (throwing out the occasional tie). Male elk (*Cervus elaphus*) and bighorn sheep (*Ovis canadensis*) engage in some amazing combats. The winner mates with the female(s) and his fitness is increased, **relative to the losing male**. But, if that male had been able to mate with the females without combat, his long-term fitness would be greater still, since he would have conserved the energetic costs (and risks of injury) associated with combat. This theoretical point, however, ignores situations where competitive interactions between males are necessary for stimulation of reproductive activities.

Competition also differs in its manifestations. We recognize here two basic forms of competition: interference and depletion. The term **interference competition** seems to have originated with Park (1962). A similar concept is **encounter interference** (Schoener 1983). In interference competition access to the resource is blocked by behavioral or chemical means. Interference competition applies to territoriality, guarding behaviors, and, by this definition, allelopathy. In allelopathy, plants secrete chemicals that accumulate in the environment and prevent other plants from germinating or growing within this area. A similar phenomenon is the secretion of antibiotics by fungi that prevent growth of bacteria within a certain radius of the colony. In ants, when a high-quality bait such as a chunk of tuna is placed on the forest floor, one species often recruits soldiers to form a ring around the tuna. If they deny access to all workers and soldiers except those from their own colony, they are engaging in interference competition.

Depletion competition involves the simple removal of the resource without active interference. This is the same idea as **exploitation** competition (Park 1962) and **consumption** competition (Schoener 1983). All of these terms refer to situations in which plants or animals consume resources to the detriment of competitors, but without directly interfering with access to the resources. This is a sort of "first come, first served" type of competition. Sutherland (1996) compares depletion competition to "drinking the pub dry." We will avoid here the terms "scramble" and "contest" competition (Nicholson 1954), neither of which is biologically realistic.

Intraspecific competition manifests itself through density-dependent modifications in (i) birth and death rates, (ii) growth rates, and (iii) adult size, especially in organisms with determinant life cycles. That is, the eventual size of an adult beetle, for example, is largely determined by the feeding rates and sizes of the larval stages (within genetic constraints). Finally, intraspecific competition is reflected in complex behavior patterns such as male–female interactions. These latter topics will be explored near the end of this chapter.

In the following sections (2.2 and 2.3) we will do a simple derivation of equations that describe density-dependent growth for populations with discrete and with continuous growth. We will analyze how these equations work and what they might tell us about how populations behave in nature.

2.2 Density dependence in populations with discrete generations

As we saw in Figs 2.1 and 2.2, in a density-dependent population we expect growth to slow and eventually stop as a population increases, and reaches the carrying capacity. For populations with discrete generations, we can begin with Equation 1.4: $N_t = N_0 R^t$.

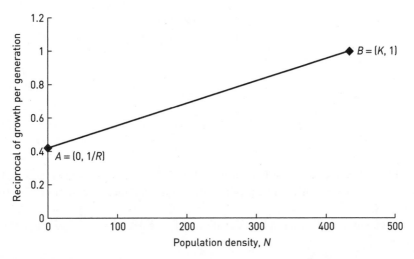

Figure 2.3 Reciprocal of growth per generation versus population density.

To incorporate intraspecific competition into a model, we simply modify the growth-rate factor, R. One approach is to graph the reciprocal of increase per generation, N_t/N_{t+1} versus N_t (Fig. 2.3). If a population is very small (virtually zero), the population is assumed to grow at the maximal rate, R. Rearranging Equation 1.3, we have $N_t/N_{t+1} = 1/R$. Point A is therefore $(0, 1/R)$. The carrying capacity, K, occurs when $N_t/N_{t+1} = 1.0$: that is, when there is no population change from one generation to the next. Point B, then, is $(K, 1)$. If we assume that population growth per generation follows a straight line between points A and B, we have the y-intercept at $1/R$ and the slope is therefore:

$$\frac{1 - (1/R)}{K - 0} = \frac{1 - (1/R)}{K}$$

The general linear equation $y = a + mx$ becomes:

$$N_t/N_{t+1} = 1/R + \left(\frac{1 - (1/R)}{K}\right)N_t$$

Rearranging and providing the common denominator RK,

$$N_t/N_{t+1} = \frac{(K) + \left(1 - \dfrac{1}{R}\right)(N_t R)}{RK} = \frac{(R - 1)(N_t) + K}{RK}$$

Therefore: $N_t = N_{t+1}\dfrac{(R - 1)(N_t) + K}{RK}$

And, $N_{t+1} = (N_t)\dfrac{RK}{(N_t)(R - 1) + K}$

If we divide the numerator and the denominator of the right side of the equation by K, we get:

$$N_{t+1} = (N_t)\frac{(RK/K)}{[(N_t)(R-1)/K] + (K/K)}$$

Finally:

$$N_{t+1} = \frac{N_t R}{1 + \frac{(N_t)(R-1)}{K}} \tag{2.1}$$

Equation 2.1 is known as the **Beverton–Holt** (1957) model, well known among fishery scientists, and is very similar in behavior to the traditional logistic equation (Gurney and Nisbet 1998).

By convention, and to simplify Equation 2.1, we let $a' = (R-1)/K$. Equation 2.1 becomes:

$$N_{t+1} = \frac{N_t R}{1 + a'N_t} \tag{2.2a}$$

A good way to see how this equation encompasses density dependence is to distinguish R_I, the density-independent growth parameter, from R_A, the density-dependent or "actual" growth parameter. In this case, Equation 2.2a becomes:

$$N_{t+1} = N_t R_A \tag{2.2b}$$

$$R_A = R_I(1 + a'N_t)^{-1} = \frac{R_I}{1 + a'N_t} = \frac{R_I}{1 + N_t\left(\frac{R_I - 1}{K}\right)} = \frac{R_I}{1 + \left(\frac{N_t R_I - N_t}{K}\right)} \tag{2.3a}$$

$$R_A = R_I\left[1 + \left(\frac{N_t R_I - N_t}{K}\right)\right]^{-1} \tag{2.3b}$$

Equation 2.3b tells us that the maximal or density-independent growth rate, R_I, is modified by the population size at time t relative to the carrying capacity, K. For example, if N is very small, the actual growth rate, R_A, is virtually equal to R_I.

If $N = K$, however, and if we replace N_t by K, the expression inside the bracket collapses to R_I. Therefore, $R_A = R_I \times R_I^{-1} = 1.0$. This means that $N_{t+1} = N_t(1.0) = N_t$. So if $N = K$ there is no growth in the population and $N_{t+1} = N_t$.

In Table 2.1 notice how Equation 2.3b modifies the R_A and the actual population size with time. The population size after 13 generations for the density-dependent population is about half that of the density-independent population, and the actual R steadily drops toward the no-growth value of 1.00.

We must remember, however, that equation 2.3b is based on the two points, $(0, 1/R)$ and $(K, 1)$, from Fig. 2.3. Furthermore, we assumed a straight line would connect these two points. This, in turn, is based on the assumption of **exact density dependence** or "exactly compensating" density dependence (Silvertown and Doust 1993). This assumption appears unrealistic. Hassell (1975) therefore proposed that we could relax this assumption of exact or linear density dependence by simply modifying equation 2.3b

Table 2.1 Density-independent growth compared to density-dependent growth using Equation 2.3b. $N_0 = 100$, $R_1 = 1.2$, $K = 1000$ in all cases.

Time, t	Density-independent net reproductive rate, R_1	Density-dependent net reproductive rate, R_A	N_t in the case of density-independent growth	N_t in the case of density-dependent growth
0	1.20	1.20	100	100
1	1.20	1.17	120	118
2	1.20	1.17	144	138
3	1.20	1.16	173	161
4	1.20	1.16	207	187
5	1.20	1.15	249	217
6	1.20	1.14	299	249
7	1.20	1.14	358	285
8	1.20	1.13	430	323
9	1.20	1.12	516	364
10	1.20	1.11	619	408
11	1.20	1.10	743	452
12	1.20	1.09	892	498
13	1.20	1.08	1070	543
14	1.20	1.07	1284	588

and replacing −1 with the exponent: $-b^*$ (Eqn. 2.4). Exact compensation (linear density dependence) occurs when $b^* = 1$, producing a slope of −1, but **overcompensation** ($b^* > 1$, implying a slope < -1) is the result when plant yield, for example, drops more rapidly than expected with increases in density.

$$R_A = R_I \left[1 + \left(\frac{N_t R_I - N_t}{K} \right) \right]^{-b^*} \tag{2.4}$$

Undercompensation occurs when $b^* < 1$, and means that population size drops more slowly than expected (as compared to exact compensation) as density rises. From Fig. 2.4 you can see that while the actual value of R declines in the density-dependent model, it declines fastest when $b^* > 1$ (that is, **overcompensation**), and declines more slowly when $b^* < 1$ (**undercompensation**). The thick line in Fig. 2.4 is a linear regression showing how R is reduced along a linear path when $b^* = 1$.

Figure 2.5 illustrates the time path of population growth under the conditions specified in Fig. 2.4. Population growth is obviously most rapid with density-independent growth and is slowest with density-dependent growth and overcompensation ($b^* = 1.6$).

The preceding model can be applied directly to the **law of the constant final yield**, well known from botanical and agricultural research. The law essentially states that agricultural yield per area will increase with plant density up to the maximum or "final" yield. Thereafter, increasing the number of plants per area simply reduces the average size per plant (or animal) without increasing total yield (Fig. 2.6). In simple terms we can write:

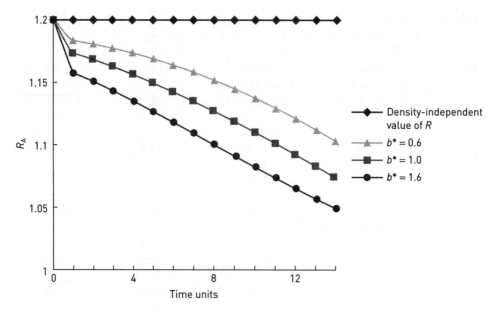

Figure 2.4 Effect of density on the actual rate of increase, R_A.

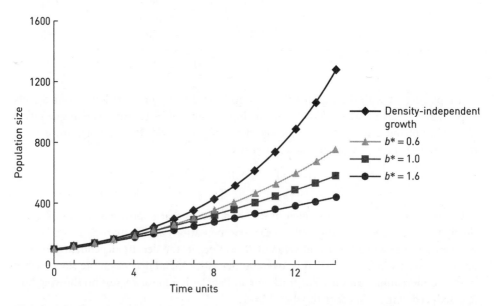

Figure 2.5 Density-independent and density-dependent growth with various values of b^*.

$$C = N\bar{w} \tag{2.5}$$

where
C = the final constant yield in kilograms per area,
N = the density of plants, that is, the number per unit area, and
$\bar{w}$ = mean mass per plant in kilograms.

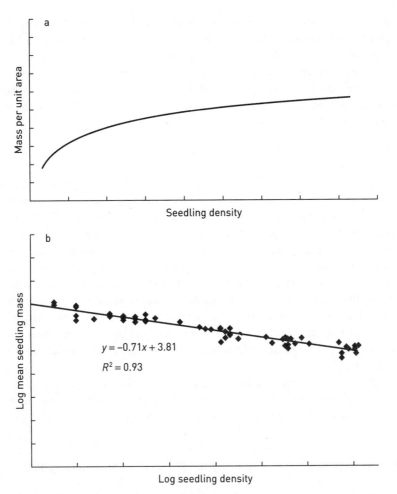

Figure 2.6 The law of the constant final yield for plants and sessile animals. (a) Yield (mass) per unit area increases with density until it reaches a threshold, after which there is no increase in total yield. (b) As density increases, mass per individual decreases linearly on a log–log scale.

This equation, however, only provides information about the end point of a dynamic process. Just as $R_A = R_I(1 + a'N_t)^{-1}$ in Equation 2.3a, the actual mean mass per plant, $\bar{w}$, can be expressed as a function of its maximum potential mass when grown under density-independent conditions.

$$\bar{w} = w_m(1 + a'N)^{-b*} \tag{2.6}$$

where
w_m = maximum potential mass per plant
a' = a carrying capacity parameter.

In this case, a' is often interpreted as the amount of area needed for each individual plant to achieve its maximum growth potential, and $b*$ provides a mechanism for different reaction rates to density. As before, when $b^* = 1$, there is exact density-dependent compensation and the law of the final constant yield is obeyed in a linear fashion. Combining Equations 2.5 and 2.6, we have an equation that describes the effects of density-dependent growth on yield under a variety of conditions. Plant populations may vary between plots, and different plant parts (grain yield or above ground biomass, for example) may respond differently to changes in density. Equation 2.7 produces different-shaped curves as the value of b^* is varied.

$$C = N w_m (1 + a'N)^{-b*} \tag{2.7}$$

The shapes of the curves would be similar to those of Figure 2.5, except that the y-axis would be output in mass rather than population density. Part of the dynamic process leading to the law of the constant yield involves an increase in mortality (self-thinning) over time as populations increase in density. As a new population is established in a suitable habitat, self-thinning follows several steps. (i) As individuals grow, they increase in size (mass). (ii) When a critical density is reached, known as the thinning limit, density-dependent mortality begins; this step occurs earlier in populations with higher initial density. (iii) Eventually the population reaches a stage where any increase in the mass of some individuals is offset by mortality of other members of the population. Total mass no longer increases and the final constant yield in mass per unit area has been reached (Figure 2.6). The point of final yield is reached more quickly in populations with higher initial densities.

Although the increase in mortality with population density is usually assumed to be linear in models such as the logistic (see below), we can introduce nonlinear responses, as shown in Equations 2.4, 2.6, and 2.7 above, and in Equations 2.11 and 2.13 below.

2.3 Density dependence in populations with overlapping generations

The logistic equation

The more familiar treatment of density dependence is to examine growth curves such as Figure 2.1, and apply a modification of the differential equation: $dN/dt = rN$. The resultant equation, known as the logistic, can be derived as follows. Examine Fig. 2.2 once more. The y-axis is per capita growth rate $(dN/dt)(1/N)$. Since the per capita growth rate $= r$ when N is very small, we can identify a point A as the y-intercept $(0, r)$. When $N = K$, according to theory, per capita population growth stops. Therefore point B, along the x-axis, would be the point $(K, 0)$. As in the previous example, if we assume that populations respond in a linear manner to population density, we use these two points to describe a straight line. The slope of this line is:

$$(0 - r)/(K - 0) = -r/K$$

The y-intercept is r and we have:

$$(dN/dt)(1/N) = (-r/K)N + r$$

Rearranging, we now have:

$$dN/dt(1/N) = r\left(1 - \frac{N}{K}\right) = r\left(\frac{K - N}{K}\right)$$

Multiplying both sides of the equation by N reveals the usual form of the logistic.

$$dN/dt = rN\left(\frac{K - N}{K}\right) \tag{2.8}$$

Although the logistic is in the form of a differential equation, it is fairly easy to understand how it affects population growth. Again, it is useful to examine what the equation does to the growth rate, r. As above, we will distinguish between r_a, the actual growth rate as modified by carrying capacity, and r_m, the density-independent growth rate. r_m has also been called r-max or the Malthusian parameter. r-max represents the maximal growth rate of a genotype as it interacts with the environment without competition.

$$r_a = r_m\left(\frac{K - N}{K}\right) \tag{2.9}$$

When the population is very small, $N \approx 0$, and $\left(\frac{\overset{\bullet}{K} - N}{K}\right) \approx 1.$ Therefore, $r_a \approx r_m$.

When $N = 0.5K$, then the expression $\left(\frac{K - N}{K}\right) = 0.5$,   and $r_a = (0.5)r_m$.

When $N = K$, the expression $\left(\frac{K - N}{K}\right) = 0$ and $r_a = 0$.

Finally, when $N > K$, the expression $\left(\frac{K - N}{K}\right) < 0.$ Therefore r_a is negative and the population drops back toward K.

The differential form of the logistic equation can be integrated and solved, resulting in the following:

$$N_t = \frac{K}{1 + e^{a - rt}} \tag{2.10a}$$

where a is a constant of integration.
   Dividing both sides of the equation by K yields:

$$\frac{N_t}{K} = \frac{1}{1 + e^{a - rt}}$$

Taking the inverse:

$$\frac{K}{N_t} = 1 + e^{a - rt}$$

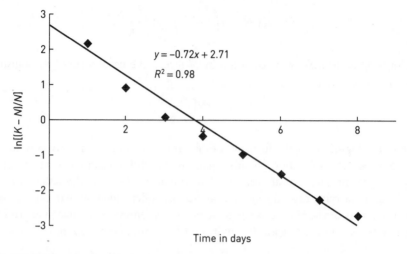

Figure 2.7 Estimate of actual r for a *Paramecium* population using the logistic equation.

Manipulating, we get:

$$\frac{K - N_t}{N_t} = e^{a - rt}$$

Finally, taking the natural log of both sides gives us:

$$\ln\left(\frac{K - N_t}{N_t}\right) = a - rt \tag{2.10b}$$

This expression is useful because it becomes the equation for a straight line with $a =$ y-intercept and the slope equal to $-r$ (see Fig. 2.7). When $t = 0$ the y-intercept, a, becomes:

$$a = \ln\left(\frac{K - N_0}{N_0}\right)$$

This gives us another form of Equation 2.10a, in which a, the constant of integration, is replaced by the y-intercept:

$$N_t = \frac{K}{1 + \left[\left(\dfrac{K - N_0}{N_0}\right)(e^{-rt})\right]} \tag{2.10c}$$

Furthermore, if we graph $\ln\left(\dfrac{K - N_t}{N_t}\right)$ versus t (Fig. 2.7), the absolute value of the slope of the line approximates r. This allows us to estimate the actual value of r over a specified time period. For example, if we use the example, once again, of the *Paramecium*

Table 2.2 Estimating r_a, (r actual) from the solved form of the logistic equation. Data are for the *Paramecium* population in Table 1.6. K is approximated at 400. See Figure 2.7.

Time in days	Number per 0.5 ml sample, N	$\ln\left[\dfrac{K - N_t}{N_t}\right]$
0	14	3.32
1	41	2.17
2	116	0.90
3	193	0.07
4	244	−0.45
5	290	−0.97
6	331	−1.57
7	363	−2.28
8	375	−2.71

population in Table 1.3 and Figure 2.1, we can determine its actual growth rate, r_a, over the eight days of the experiment. We simply add a column for the expression $\ln\left(\dfrac{K - N_t}{N_t}\right)$ (Table 2.2). To do this, however, we must have an approximation for the carrying capacity, K. In Table 2.2, K is estimated as 400.

In Fig. 2.7 a linear regression identifies the slope as −0.73 (R^2 is the proportion of the variance explained by the linear model). The actual r is therefore 0.73, as compared to the r_{max} of 0.89 calculated in Chapter 1.

In understanding how the logistic affects population growth, it is instructive to examine population growth, dN/dt, as a function of population size. Since $dN/dt = rN\left(\dfrac{K - N}{K}\right)$, it also equals $rN\left(1 - \dfrac{N}{K}\right)$, which equals $rN - (rN^2)\left(\dfrac{1}{K}\right)$. If we set dN/dt equal to zero, there are three solutions to this equation: $r = 0$, $N = 0$, or $N = K$. If we assume that $r > 0$ we are left with two solutions ($N = 0$ and $N = K$). That is, $dN/dt = 0$ when $N = 0$ and when $N = K$. The result of plotting dN/dt versus N results in a parabola (Fig. 2.8). Maximum growth (dN/dt) occurs where $N = K/2$, which is 500 in this case, since we have set $K = 1000$. The problem with this solution is that, since maximum growth theoretically occurs at half carrying capacity, harvesting of wild living resources was managed with that number as a goal. This has led to the decimation of many populations since stochastic and density-independent mortality were not accounted for.

Assumptions of the logistic equation

How much trust can we put in either the traditional logistic equation or the Beverton–Holt equations? Is the typical logistic growth curve actually found in biological populations? Laboratory studies on growth of protozoan populations such as *Paramecium caudatum*, yeast, *Drosophila*, grain beetles and diatoms (Gause 1932, 1934, Vandermeer 1969, Pearl 1927, Crombie 1945, Park *et al.* 1964, Tilman 1977), do consistently show a logistic growth

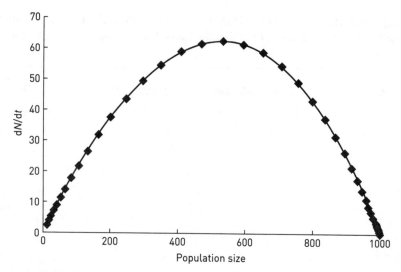

Figure 2.8 Population growth as a function of *N* based on the logistic equation.

curve. A number of field populations have also followed logistic growth fairly closely. Examples include Tasmanian sheep (*Ovis aries*) (Davidson 1938), wildebeest (*Connochaetes taurinus*) (Deshmukh 1986), willows (*Salix cinerea*) (Alliende and Harper 1989) and barnacles (*Balanus balanoides* and *Chthamalus stellatus*) (Connell 1961a, 1961b). However, there are many more cases where populations grow cyclically or unpredictably and generally do not display logistic growth. An examination of the assumptions of the logistic equation explains why many populations display non-logistic growth patterns.

Assumptions of the logistic equation:

1 The carrying capacity is a constant;
2 population growth is not affected by the age distribution;
3 birth and death rates change linearly with population size (it is assumed that birth rates and survivorship rates both decrease with density, and that these changes follow a linear trajectory);
4 the interaction between the population and the carrying capacity of the environment is instantaneous: that is, the population is "sensitive" to the carrying capacity with no time lags;
5 abiotic, density-independent factors do not affect birth and death rates (no environmental stochasticity);
6 crowding affects all members of the population equally.

Considering all of the above, it is not surprising that populations in the field do not often stay at a given density for long periods of time. In the laboratory, when we grow a *Paramecium* population, its growth curve often fits the logistic since: (1) it is maintained in a constant environment, which should have a constant carrying capacity; (2) it reproduces via binary fission and has no age structure; (3)–(6) are seemingly irrelevant or satisfied. Once we step into the field and work with insects, vertebrates, or plants, several of these assumptions are violated.

Because these assumptions cannot be met, a natural population is unlikely to long remain "at equilibrium" with the environment. In the next two sections we will examine assumptions 3 and 4 in some detail, since their effects on population growth are less obvious and more interesting than might be expected.

2.4 Nonlinear density dependence of birth and death rates and the Allee effect

As mentioned above, density-dependent birth and death rates are assumed to vary linearly with density (Fig. 2.9). Although it is known from both laboratory (Smith and Cooper 1982) and field studies (Arcese and Smith 1988) that birth and death rates are often nonlinear (Fig. 2.10), such differences seem to have a minimal impact on natural populations. One major exception, however, is known as the Allee effect (Allee 1931). Allee proposed that many species have a **minimum viable population** (MVP) size. As described in Chapter 1, although Allee may have had a specific number in mind, below which death rates rise and/or birth rates collapse, a more modern view is that the probability of extinction has become unacceptably high when a population becomes small (Shaffer 1981, Miller and Lacy 2003), but there is no one specific number described as a MVP.

Why should there be higher death rates in very small populations? Proposals include: (i) group cooperation reduces losses from predators; (ii) group foraging for food is more efficient (foraging facilitation); and (iii) small populations are more subject to density-independent or stochastic extinctions as well as genetic effects such as inbreeding depression. Low birth rates in small populations could result from pollination failure in plants, male and female animals unable to locate each other, or the chance of a very unequal sex ratio (large number of males, few females).

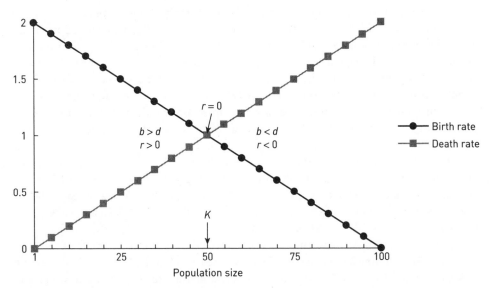

Figure 2.9 Linear response of birth and death rates to population density. b, birth rate; d, death rate; r, intrinsic rate of increase; K, carrying capacity.

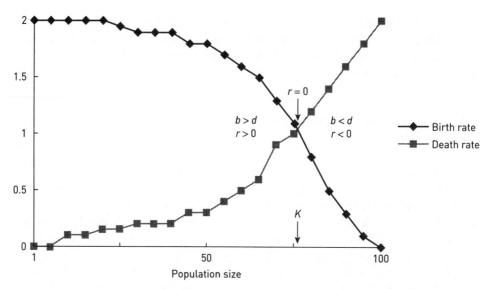

Figure 2.10 Nonlinear response of birth and death rates to population density.

For example, the common guillemot or murre (*Uria aalge*) nests in colonies. Breeding success in south Wales was found to be only 25% in the least dense populations as compared to an average of 75% in the densest populations (Birkhead 1977). The reason for this appears to be that predation on eggs and chicks by gulls is reduced in dense guillemot populations. A similar result was obtained in a study on lapwings (*Vanellus vanellus*) in which egg clutches lost to avian predators declined with an increase in the number of close neighbors (Berg *et al.* 1992). Nest parasitism also appears to increase in low populations. For example, small dickcissel (*Spiza americana*) populations are particularly hard hit by brown-headed cowbirds (*Molothrus ater*) (Fretwell 1986).

Other studies have found that cooperative hunters such as lions, hyenas, wolves and various fish species have much higher success rates when hunting in large groups and do poorly when population sizes fall (Caraco and Wolf 1975, Major 1978). This translates into a higher mortality rate in the smaller groups. From the perspective of the prey, a dense population is harder to surprise, and mortality from predation is lower in larger prey populations (Kenward 1978, Jarman and Wright 1993). Colonial nesting sunfishes have even been found to suffer lower rates of fungal infections on their eggs as compared to solitary sunfish (Cote and Gross 1993). Therefore higher density leads to higher, not lower, survivorship.

Although this is an oversimplification, we can illustrate the Allee effects graphically by identifying a minimum viable population size, **MVP** (Fig. 2.11). Below point MVP, the population declines to extinction. Above point MVP the population increases rapidly before slowing down as it approaches *K*. The value of *r* is positive above MVP and below *K*, but is otherwise negative (Figure 2.11).

The extinction of the heath hen (*Tympanuchus cupido*) is a likely example of the Allee effect. By 1870, hunting and habitat loss had restricted it to Martha's Vineyard off the coast of Massachusetts. In 1908 a 650 ha refuge was established and the population grew to about 2000 birds. In 1916, however, a fire swept across the island, destroying nests, eggs, and females on the nests. The following winter was severe and an unusually heavy concentration

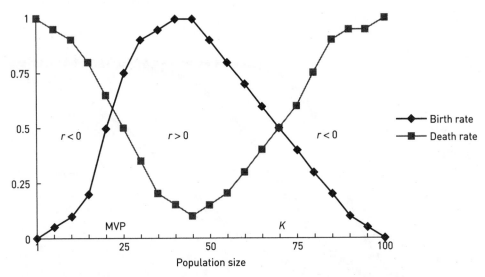

Figure 2.11 Birth and death rate versus population density, showing the Allee effect. MVP, minimum viable population.

of goshawks (*Accipiter gentilis*) arrived. The population was reduced to less than 150, of which most were probably males. By 1928 the population had declined to a single male that died in 1932. In this case, abiotic effects contributed significantly to the extinction of the heath hen. Although we can never be sure, it appears likely that something of a similar nature happened to the passenger pigeon (*Ectopistes migratoris*) described in the introduction to Part I. They were mercilessly hunted and no one imagined the possibility of extinction. But once their flocks were drastically reduced, they appeared unable to recover (Souder 2004).

Nonlinear modifications to the logistic

In order to evaluate the potential for a nonlinear feedback on the logistic population response, we can modify Equation 2.10c by adding the term b^*, as we did in Equation 2.4 and Figs 2.4 and 2.5. Recall that a value for b^* of 1.0 describes "exact compensation" and depends upon the linear response by the population to a carrying capacity. A $b^* > 1$ illustrates overcompensation and a $b^* < 1$ describes undercompensation. We can modify Equation 2.10c by adding a b^*-value:

$$N_t = \frac{K}{1 + \left[\left(\dfrac{K - N_0}{N_0}\right)(e^{-rt/b^*})\right]} \tag{2.11}$$

Figure 2.12 illustrates the effect of a nonlinear feedback on logistic growth. The population with the b^*-value of 0.6 grows the most rapidly, whereas the population with the b^*-value of 1.6 grows the most slowly. As in Fig. 2.4, when $b^* > 1$, small increases in density result in a rapid drop in the actual value of the growth rate, r, as a result of

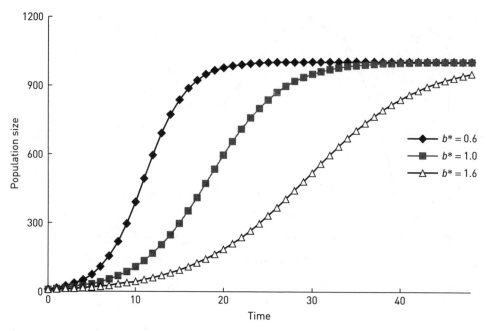

Figure 2.12 Effect of nonlinear feedback on logistic growth.

overcompensation. Similarly a $b*$-value < 1 results in **undercompensation** and a higher actual r-value.

The theta logistic model

As illustrated by Figs 2.22–2.26 in section 2.8, the life history of a population frequently does not respond to increases in density in a linear fashion, as assumed by the logistic equation. A well-known variation of the logistic model, known as the theta logistic, more elegantly introduces nonlinear density dependence than we did in the previous section. First, we must introduce another model from fishery science, the Ricker (1952) model, which is a useful discrete form of the logistic.

To find the Ricker, we begin by making a distinction between the actual rate of increase, r_a and the exponential rate of increase, r_m or r_{max}, as we did above in Equation 2.9.

$$r_a = r_m \left(\frac{K - N}{K} \right) = r_m \left(1 - \frac{N}{K} \right)$$

Now let us substitute r_a for r in the equation for exponential growth (Eqn. 1.8), which gives us:

$$N_t = N_0 \, e^{r_a t}$$

Next we transform this to a simple difference equation for adjacent time intervals N_{t+1} and N_t. Since this is one time step and $t = 1$, we can remove t from the exponent in the above equation.

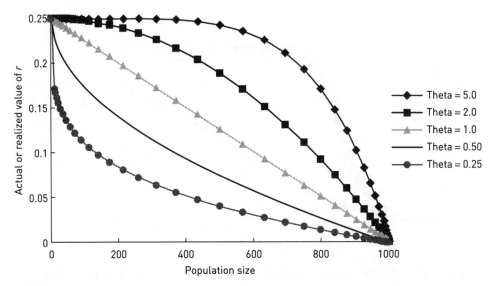

Figure 2.13 Behavior of the theta logistic. Effect of population size on the realized value of r. In all cases $r_{max} = 0.25$ and $K = 1000$.

Simultaneously we will substitute $r_m[1 - (N/K)]$ for r_a. The result is Equation 2.12, the Ricker equation:

$$N_{t+1} = N_t\,e^{r\left(\frac{K-N_t}{K}\right)} = N_t\,e^{r\left(1-\frac{N_t}{K}\right)}$$

(2.12)

This equation can be modified with the parameter θ (theta) as a superscript of the ratio N/K (Eqn. 2.13). The theta logistic was originally proposed by Gilpin and Ayala (1973). When $\theta = 1.0$, we have the traditional logistic growth response to density. When $\theta < 1.0$ density dependence is strong even when the population is far below the carrying capacity. By contrast, when $\theta > 1.0$ density dependence is weak until the population is close to the carrying capacity.

$$N_{t+1} = N_t\,e^{r\left(1-\left(\frac{N_t}{K}\right)^\theta\right)}$$

(2.13)

For example, in Fig. 2.13, we can examine the effect of θ on an actual or realized r-value. In each case the r_{max} is 0.25, the carrying capacity is 1000 and the initial population size equals 10. As predicted from the logistic, when $\theta = 1$ the decline in the actual r-value is linear as the population increases. When θ is less than 1.0 the actual r-value decreases rapidly with population size. By comparison, if θ is greater than 1.0 we can see that r remains close to r_m until the population gets much closer to the carrying capacity.

If we examine population growth versus time for the same theta values, the expectation is that growth will be suppressed at low population levels when theta is less than 1.0, but that the population will approach the carrying capacity quickly when theta is greater than 1.0. These predictions are borne out by Fig. 2.14.

As shown by Saether *et al.* (2002) the theta logistic is a powerful model for analyzing variation in density dependence among bird populations, and is the basis for other

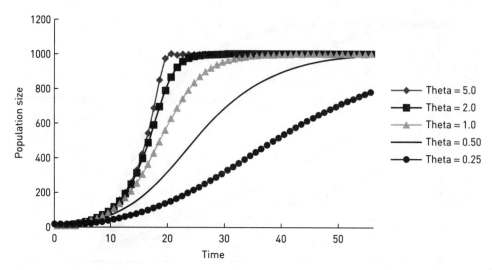

Figure 2.14 Behavior of the theta logistic. In all cases $r_{max} = 0.25$ and $K = 1000$.

population models (for example, predator–prey interactions) in which we do not want to assume a linear relationship between population density and survivorship, fertility, or r.

2.5 Time lags and limit cycles

Assumption 4 of the logistic equation, in which populations are assumed to respond immediately to carrying capacity, is highly unlikely for populations with great reproductive potential. In order to explore this possibility, we can introduce a "lag time" effect into the logistic equation. Using the discrete time form of the logistic (Eqn. 2.1), substituting λ for R, and remembering that $\lambda = e^r$, Equation 2.12 is an equivalent to Equations 2.1 and 2.10c. To introduce time lags, Equation 2.14 is modified as shown in Equation 2.15 (Pielou 1977).

$$N_{t+1} = \frac{\lambda N_t}{1 + \dfrac{N_t(\lambda - 1)}{K}} \qquad (2.14)$$

$$N_{t+1} = \frac{\lambda N_t}{1 + \dfrac{N_{t-T}(\lambda - 1)}{K}} \qquad (2.15)$$

A more familiar form of this same equation is simply:

$$dN/dt = rN_t\left(\frac{K - N_{t-T}}{K}\right) \qquad (2.16)$$

in its continuous form, and

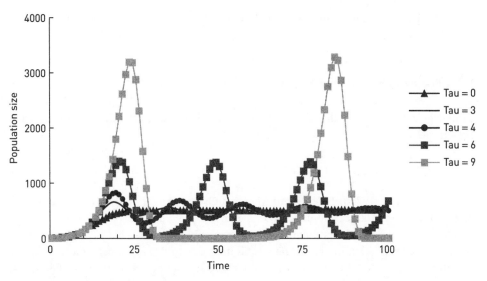

Figure 2.15 Logistic growth with time lags. In all cases $r_{max} = 0.30$ and $K = 500$.

$$N_{t+1} = N_t \, e^{r\left(\frac{K-N_{t-T}}{K}\right)}$$

(2.17)

in its discrete form. This is a simple modification of equation 2.12, the Ricker model.

Where N_t appears, it is modified by the Greek letter tau (T). The population responds to the carrying capacity based on what the population size was *tau* time-units in the past. Robert May and others (May and Oster 1976, May 1981a) have shown that lag time, combined with the intrinsic rate of increase (r), produces a predictable and interesting series of modifications to logistic growth. The product of r and T determines the behavior of the population. As summarized below, long time delays before the population reacts to carrying capacity, combined with a high growth potential, lead to population behaviors that wander further and further from the stable point at K predicted by the logistic equation.

- If $0.37 > rT > 0$, the population follows the logistic equation, and the population achieves a stable number (or **stable point**) at the carrying capacity with no oscillations.
- If $1.57 > rT > 0.37$, the population is temporarily oscillatory, but the oscillations dampen to a stable point at the carrying capacity.
- If $2.0 > rT > 1.57$, the population undergoes permanent oscillations around the carrying capacity. This is called a **limit cycle**.
- If $rT > 2.0$, the oscillations are so violent that the population goes extinct.

Figure 2.15 displays five simulations based on Equation 2.17. In series 1, *tau* $= 0$ and we have the usual logistic growth curve. In series 2 the product of $rT = 0.90$, and in series 3 the product is 1.20. We expect temporary oscillations converging on a stable point at the carrying capacity in both of these cases. In series 4 the product of $rT = 1.80$ and we have a stable limit cycle. Finally, in series 5, the product of $rT = 2.70$. We expect extinction, and although the simulation shows two population cycles, in reality the population is extinct after 30 time units.

2.6 Chaos and behavior of the discrete logistic model

Time lags are implicit in the discrete logistic model. We can actually remove *tau* from Equation 2.17 and return to Equation 2.12:

$$N_{t+1} = N_t\, e^{r\left(\frac{K-N_t}{K}\right)} = N_t\, e^{r\left(1-\frac{N_t}{K}\right)}$$

As May (1975a, 1975b, 1981a), and May and Oster (1976) have shown, if the growth rate (r or λ) is very large, populations behave in unusual and unexpected ways.

Recall that if $b^* > 1$, the population shows overcompensation. That is, there is a larger than expected reduction in growth rate or biomass due to density dependence. As shown in Table 2.3, if the combined values of r and b^* produce a net rate of increase with a large reproductive potential, the population moves from a stable equilibrium at the carrying capacity to fluctuations which ultimately reach chaos when the net r is large enough. For example, when r is less than 2.0 (R or $\lambda < 7.39$), the population moves to a stable point (Fig. 2.16), although note that when $r = 1.5$ there is a small oscillation before the population settles in at the carrying capacity. At r-values between 2.0 and 2.53 (Fig. 2.17, Table 2.3) the population regularly cycles between two points. For r-values between 2.53 and 2.66, the population cycles among four points (Fig. 2.18). An eight-point cycle is produced by r-values between 2.66 and 2.69. Finally, at r-values > 2.69 (R or $\lambda > 14.761$) the behavior of the population is known as **chaos** (Fig. 2.19). That is, the population never enters into a predictable pattern. Over short periods, these chaotic fluctuations would be indistinguishable from seemingly random responses to the environment. A deterministic model, then, can produce results that appear to be stochastic and, if one were looking for biological causation for the behavior of such populations, one would be confused indeed.

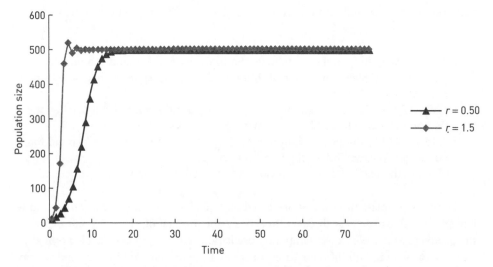

Figure 2.16 Behavior of the discrete logistic model: stable equilibrium point when $r < 2.0$.

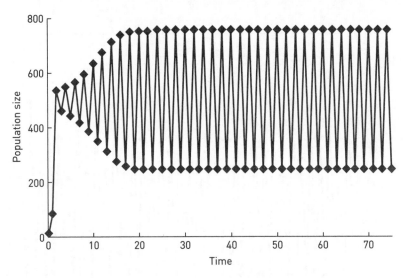

Figure 2.17 Behavior of the discrete logistic model: two-point cycle when $r = 2.20$.

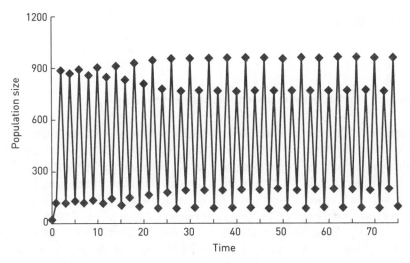

Figure 2.18 Behavior of the discrete logistic model: four-point cycle when $r = 2.60$.

Table 2.3 Behavior of the discrete logistic model based on the net growth parameter, r, with equivalent values of R or λ. Net growth is influenced by the nonlinear feedback parameter b^*, when $b^* \neq 1$. Adapted from May (1975b), May and Oster (1976), and Alstad (2001).

Net growth rate, r	Equivalent value of R or λ	Behavior of the discrete logistic model
$2.000 > r > 0$	$7.389 > \lambda > 1.000$	Stable equilibrium point
$2.526 > r > 2.000$	$12.503 > \lambda > 7.389$	Two-point cycle
$2.656 > r > 2.526$	$14.239 > \lambda > 12.239$	Four-point cycle
$2.685 > r > 2.656$	$14.658 > 14.239$	Eight-point cycle
$r > 2.692$	14.761	Chaos

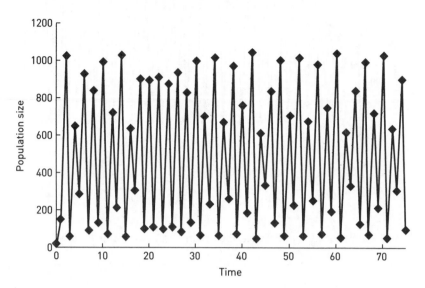

Figure 2.19 Behavior of the discrete logistic model: chaos when $r = 2.75$.

2.7 Adding stochasticity to density-dependent models

Just as we did in Chapter 1, we can perform stochastic simulations with density-dependent models. Using the Beverton–Holt model (Eqn. 2.1), Fig. 2.20 shows a deterministic growth curve for an initial population size of 100, a carrying capacity of 1000, and a deterministic value for lambda of 1.1. In order to simulate the effects of demographic stochasticity, we can add a random function in Excel™ that allows lambda to vary with a mean of 1.1 but with a variance of 0.03. One such result is shown in Fig. 2.20. In this particular case, when the population is small it does not grow very quickly, but it eventually reaches the carrying capacity. Notice that it takes over 100 time units to reach carrying capacity even though the deterministic population reaches K at around 50 time units. Running 25 stochastic simulations in this manner produces a range of population sizes after 100 time units of 128–1000 with a mean of 904 individuals. The lessons are basically the same as in the previous chapter: adding variability leads, in most cases, to a smaller population than that expected from a deterministic model.

Using the Ricker model (Eqn. 2.12), we next simulate both demographic stochasticity (adding variability to r) and environmental stochasticity (by allowing K to vary with time) in Fig. 2.21. In all cases the initial population size is 50, the deterministic r is 0.1, and the carrying capacity is 1000. We have also allowed both r and K to vary simultaneously in one series of simulations. In Fig. 2.21 we see that the population in which both r and K are allowed to vary goes extinct. The population with the stochastic r eventually reaches carrying capacity and the population with the stochastic K goes through several crashes. The results of these simulations depend on how much variability we allow for demographic versus environmental stochasticity. Suffice it to say to the combined effects of demographic plus environmental stochasticity raise the probability of extinction and, as above, variability normally produces smaller populations.

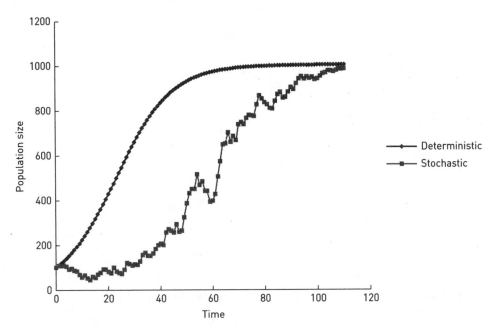

Figure 2.20 Deterministic versus stochastic growth in a population with an initial population size of 100, a carrying capacity of 1000, and a deterministic λ of 1.1. In the stochastic model, the deterministic average λ is 1.1 with a variance of 0.03.

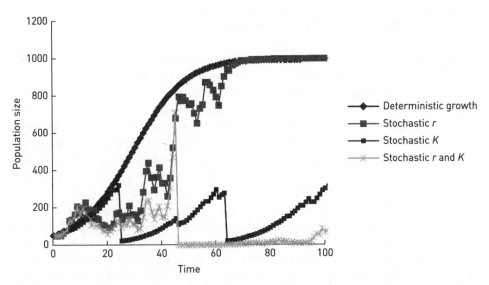

Figure 2.21 Effect of demographic stochasticity (r) and environmental stochasticity (K) on behavior of the Ricker model. Initial population size = 50, deterministic $r = 0.1$, carrying capacity $K = 1000$.

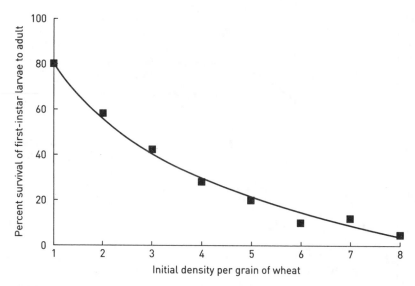

Figure 2.22 Survivorship of the grain beetle *Rhizopertha dominica* versus initial density of first-instar larvae.

2.8 Laboratory and field data

According to the assumptions discussed above, an increase in population density should lead to one or more of the following:

1 a linear increase in mortality;
2 a linear decrease in fertility;
3 a reduction in average growth rate; and
4 a reduction in the average size of adults.

Crombie (1942, 1944) showed that flour beetles (*Rhizopertha dominica*) raised in the laboratory were negatively affected by density. Both survivorship (Fig. 2.22) and fertility (Fig. 2.23) decreased with density, although in both cases the effects were nonlinear. Moreover, many large mammal populations have fertility and mortality patterns that show density dependence but are also nonlinear (Figs 2.24 and 2.25; Fowler 1981). Nonlinearity extends to bobwhite quail (*Colinus virginianus*) (Roseberry and Klimstra 1984) and clado-cerans (Smith and Cooper 1982). On the other hand, both elk (*Cervus elaphus*, called red deer in Europe) (Fig. 2.26, Houston 1982) and grizzly bears (*Ursus arctos*) (McCullough 1981) show a linear decrease in fertility with population density.

Two examples of reduction in growth rate with density will suffice here. The first is that of the growth of tadpoles of the frog *Rana tigrina*. Whereas it takes only two to three weeks for tadpoles to develop into mature frogs at densities of 5 to 10 (in a 2-liter aquarium), it takes almost ten weeks when there are 160 frogs in the same space (Dash and Hota 1980). In harp seals (*Phoca groenlandica*) sexual maturity is achieved when an individual reaches 87% of mean adult body weight. In low populations this occurs at between 4 and 5 years, whereas in dense populations sexual maturity is reached at between 6 and 7 years (Lett *et al.* 1981).

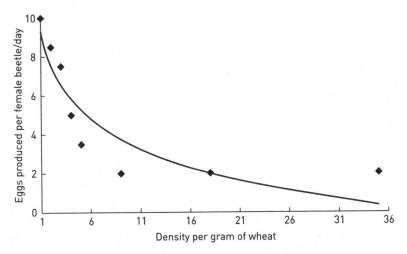

Figure 2.23 Fertility versus density in *Rhizopertha dominica*.

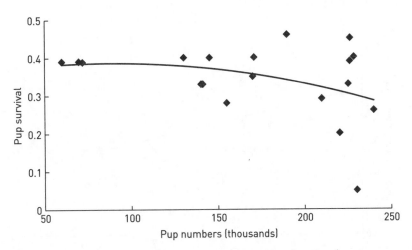

Figure 2.24 Survivorship of northern fur seal (*Callorhinus ursinus*) pups as a function of pups born. After Fowler (1981).

The body size of adults is also affected by density in many populations. For example, in highly dense reindeer (*Rangifer tarandus*) populations, mean jaw size of adults is 23 cm, whereas at low density the mean size is between 24 and 25 cm (Skogland 1983). As we saw in Section 2.2, the response to an increase in density among plant populations is a reduction in mean weight per individual. The same principle applies to sessile animal populations. When Branch (1975) examined populations of limpets (*Patella cochlear*), the most common diameter was 60 mm when there were 125 individuals per square meter. When the density was increased to 1225 per square meter, the most common size class was 20 mm. The total biomass obeyed the law on the constant final yield (Fig. 2.6a). Biomass increased with density up to 400 individuals per square meter, but then leveled off at 125 g per square meter for all densities from 400 to 1225.

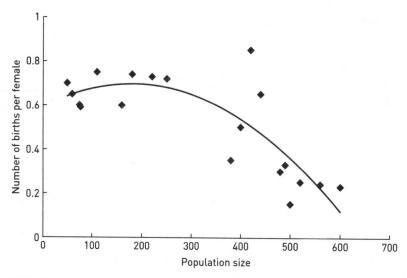

Figure 2.25 Birth rate of the American bison (*Bison bison*) as a function of population size. After Fowler (1981).

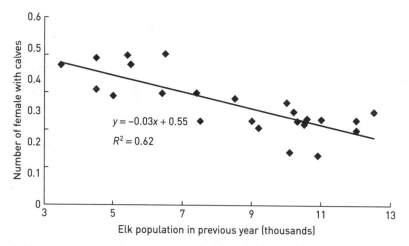

Figure 2.26 Recruitment of elk (*Cervus elaphus*) calves in Yellowstone National Park as a function of adult population size in the previous year. Based on Houston (1982).

2.9 Behavioral aspects of intraspecific competition

Castes in social insects

Passera *et al.* (1996) have shown that intraspecific competition may affect the ratio of castes among social insects. Colonies of the ant *Pheidole pallidula* increase their relative investment in soldiers when exposed to other colonies of the same species. The worker force is divided into two basic castes: small-headed minors (workers) who do most of the labor, and large-headed majors (soldiers) specialized for defense. In a laboratory experiment, when colonies were exposed to odors from a conspecific colony, they increased the number of soldier pupae. After seven weeks, those colonies exposed to conspecific odors averaged 40.1 soldiers as compared with an average of 22.6 soldiers in the control colonies.

Male–male competition in horned beetles

The complexities of intraspecific competition are illustrated through elegant work on horned beetles (Emlen 2000). Beetle horns are rigid extensions of the exoskeleton and have evolved repeatedly within this order of insects. These horns are generally only expressed in males and are used in combat with other males for access to resources and/or females. These resources are in discrete, readily defensible patches, and the horns allow males to defend these sites and to mate with the females found there. The jousting contests between males can be dramatic and the winner is rewarded with mating privileges.

These horns, however, do not come without cost. They constitute a large investment in energy, and as much as 10% of body mass may be devoted to them. Horn growth prolongs development time and risk of larval mortality, and there is a trade-off between horn development and the ability to fly. Finally, the development of horns leads to lowered visual acuity and to smaller eyes. Nevertheless, it appears that since larger horns are useful in gaining access to females, the reproductive benefits gained from large horns offset the costs of production and maintenance.

The ability of males to grow horns is based on larval nutrition. Both final adult body size and the size of horns in males depend on the amount of food they consume as larvae. Males exposed to poor conditions as larvae are small and do not produce horns, whereas well-fed males become large adults able to produce horns. Horn production shows an insignificant level of heritable genetic variation (Moczek and Emlen 1999).

These complexities and trade-offs are illustrated through studies of dung beetles of the genus *Onthophagus*. The basic life history of these beetles is as follows. After finding a patch of dung, a female digs a tunnel in the soil beneath the dung. She then buries dung below ground to provide nutrition for the larvae. Females spend days inside a single tunnel, pulling down pieces of dung to various blind ends of tunnel branches, each with its "brood ball" of dung. A single egg is laid at the top of a brood ball, and a single larva develops in isolation within the brood ball.

Females mate repeatedly with males during the time of egg laying. Male reproductive behavior consists of securing their unique access to females in the tunnels. The large, horned males guard the tunnels and fight other males trying to approach the females. Larger males with larger horns win these fights. For two males of the same size, the one with the larger horns usually wins. Thus long horns provide males with significant advantages.

However, all is not lost for a small male. Although they are not adept at guarding entrances to tunnels or winning jousts, small males have other tactics. They attempt to slip undetected past the large males, or they dig side tunnels that intercept a guarded tunnel well below ground. Thus hornless males may manage to sneak undetected into guarded tunnels and mate with the female. If such a "sneaker" male is caught, he is chased out by the resident male, who then returns to the female and mates with her. This dilutes or displaces sperm from the sneaker male. Sneaker males actually do best when they have no horns at all, since horns get in the way of sneaking! Success for a small male depends on rapid and undetected entry into tunnels, and horns cause vibrations as they scrape against tunnel walls, which would alert the resident male. Finally, as mentioned earlier, beetles with very large horns tend to have undeveloped eyes and/or a diminished flying ability due to smaller wing size and greater body weight. Since dung is a temporary resource, another component of fitness is the ability to disperse and locate new dung heaps. As the dung resource diminishes, some of the males, as larvae, will receive a less-than-optimum amount of food. They will be small and lack horns. Thus, although they would be poorly equipped to defend a female in a burrow, they are well adapted to disperse and find a new dung pile.

Both intraspecific competition and sexual selection are played out on the fields of manure.

Male–female competition in dunnocks

Another aspect of intraspecific competition is competition between the sexes as expressed through mating systems. Obviously ecological conditions can influence the mating behavior of individuals. As conditions change, a cooperative pair or an animal society may quickly dissolve into a set of competing individuals. The following study (Davies 1992, 1995) illustrates some of the complexities in what appears to be a simple pair-bonded mating system.

The dunnock (*Prunella modularis*), formerly known as the hedge sparrow, is not a true sparrow but an accentor. In Old English "dun" means brown and "ock" means little. Thus *P. modularis* is the archetypal little brown bird. It seems unremarkable as it shuffles about under the bushes collecting tiny insects for its young. The impression gained is of harmonious cooperation. The Reverend Morris (1856, in *A History of British Birds*) admired this species so much that he urged his parishioners to emulate its behavior. The Reverend, as quoted by Davies (1992), found this bird to be "unobtrusive, quiet and retiring, without being shy, humble and homely . . . sober and unpretending in its dress . . . while still neat and graceful . . ." Due to its extraordinary mating system, however, the Reverend Morris would hardly have been pleased had his congregation followed the example of these birds. Although a dunnock territory can contain one seemingly harmonious pair, a female in the territory next door may be mating with two males, or a male may be mating with two females.

The conventional view of pair formation in songbirds is that males first set up territories and advertise for mates by singing. Females then choose among male territories. However, in dunnocks females defend their own territories against other females, occupying exclusive areas with little overlap among neighbors. Females settle independently of the males, based on the quality of the territory. Males then compete to defend the females from other males.

The fact that males and females set up territories independently results in a wide variety of mating behaviors. In some cases, a single male defends one female territory (producing monogamy); at other times a male defends two adjacent females (producing polygyny). In other cases two males share the defense of one female (polyandry), or several adjacent females (polygynandry).

If two males share a territory with a single female, the dominant or alpha male (usually the older one) tries to evict the beta male. If he is unsuccessful, the two share defense of the joint territory. Some males wander for weeks before finding a permanent home. Sometimes a beta will overlap with two alphas. The larger the territory of the female, the harder it is for a male to monopolize her. The larger the male territory, the more likely he is to have two females.

Once a female has built her nest, she solicits copulations from males. Mating begins 3–7 days before the first egg is laid, and lasts up to completion of the clutch of 3–5 eggs. One egg is laid per day during this period. Monogamous males chase off neighboring males who are interested in mating. When two males share a territory, the alpha male follows the female everywhere to prevent her from copulating with the beta male. A female often maneuvers to throw off the alpha male. She then solicits the beta for mating. Females have many tricks, and seem intent on preventing exclusive mating by the alpha.

The act of copulation was described by Selous (1933) as "bizarre." "The hen elevated her rump and stood still, when the male, hopping up, made little excited and very wanton-looking pecks in this region, that is to say the actual orifice. There was actually no mistaking the nature and significance of the actions, rather lecherous, as it seemed to me. This is a very remarkable thing . . . but I do not understand it" (Selous 1933, pp. 107–9). During the pecking described by Selous, the female cloaca makes pumping movements, resulting in the ejection of a small droplet of fluid. As soon as this occurs the male copulates with her.

Davies (1992) analyzed the droplets and discovered that they contain masses of sperm from the previous matings. Like other birds, female dunnocks store sperm. Females with two males may copulate up to six times per hour. As a result, there will be a pool of sperm in the female's cloaca and vagina for much of the mating period. The male's pecking stimulates the female to eject this pool of sperm to make way for his insemination and give his own sperm a better chance of being stored.

After the excitement of the mating period, life on a dunnock territory becomes peaceful during incubation. Females incubate the clutch of eggs alone for 11–12 days. Males help with chick feeding for 11–12 more days in the nest and for two weeks after fledging until the young become independent.

The results of DNA tests on the blood of the chicks and the parents showed that the female is the mother of all chicks in the nest. In monogamous and polygynous territories, the male was the father of all chicks. When two males guarded a territory with one female, however, if the female mated with both males, paternity was mixed. The alpha male fathered 55% of the brood and the beta male 45%.

From the female perspective, when both males mated with her they both helped feed the brood. If she mated exclusively with the alpha male, only he helped feed the brood. Males, however, cannot recognize their own young. If they have mated with the female, they will help raise the young, even in cases where they did not happen to father any of the chicks. If a beta male gained a large share of the copulations he more readily helped with feeding of the chicks. Similarly, if an alpha male was removed experimentally, allowing the beta a larger share of the matings, the beta male worked harder to feed the chicks than did the alpha. Thus, males varied their parental effort in relation to their chance of paternity, not simply according to dominance rank.

For a female, polygyny is the least desirable situation. She must share the help of one male with another female, and some of her chicks often starve to death. A polyandrous female was the most successful because she had the help of two males in raising her young. This explains why the females sneaked around and tried to get the beta males to mate with them. They hoped for future help in rearing of the young.

For a male, the situation is reversed. In polygyny, although each female is less productive, the combined output of two females often exceeds that of one female in monogamy.

Thus the mating system reflects an intraspecific competitive battle between the sexes in dunnocks. Behavioral and genetic studies of other bird species show that the dunnock mating system is not unusual. However, extra-pair copulations and fertilizations vary across species for reasons only poorly understood (Petrie and Kempenaers 1998, Blomquist et al. 2002). Extra-pair paternity is much less frequent in non-passerine than in passerine birds (Birkhead et al. 2001). For example Blomquist et al. (2002) found that in western sandpipers (*Calidris mauri*) only 5% of all chicks were the product of extra-pair matings. Nevertheless the application of molecular techniques to behavioral studies has allowed us to ask new questions about the mechanisms of intraspecific competition.

Competition versus cooperative behavior within a group

Research on lions by Heinsohn and Packer (1995) illustrates the behavioral complexities displayed by animals that are simultaneously territorial and cooperative group-foragers. African lions (*Panthera leo*) engage in a wide variety of group-level activities from hunting to communal cub rearing. At the same time, the group defends a territory from other groups of lions. When prey is difficult to capture they hunt cooperatively, but cooperation breaks down when prey is easy to catch. Female lions nurse each other's young, but, more importantly, they jointly protect their young from males that are intent upon infanticide. The threat of attack by conspecifics is a driving force in lion sociality. Large prides dominate smaller ones, and solitary lions are often killed or injured during attacks by lions of the same sex.

Using playbacks of recorded roars, Heinsohn and Packer found that lions are able to distinguish pride members from strangers. They also found that certain females show a consistent behavior of lagging behind their companions during group activities, including hunting.

Female lions live in social groups (prides), which contain 3–6 related adults, their dependent offspring, and a group of immigrant males. The males defend the pride against incursions by other males; females defend their young against infanticidal males, and the territory from other females. At least two females are needed for a territory, and they advertise ownership by roaring. Using broadcasted roars, the investigators showed that some females become "laggards" early in life, and this behavior persists into adulthood. Laggards were those individuals that hung back, and approached the audio speaker only after the leaders had already responded. The order in which the individuals approached the speaker was the same throughout the playbacks. Because territorial fights often lead to injury or death, laggards were ensuring their safety, at least from initial attacks. They typically followed the leaders by 30 to 120 seconds.

In the theoretical game or model known as "prisoner's dilemma," in any single task two individuals benefit when they work together (mutual cooperation) but both lose when neither contributes (mutual defection). However, in this game, the greatest payoff for one individual comes from providing no help (cheating) to a partner who cooperates, while the lowest payoff results from helping out (cooperating) while the partner cheats. In a repeated series of encounters, however, cheaters are eventually punished by withdrawal of further cooperation by other individuals. In large groups the game gets more complicated, but cheaters can eventually be detected and punished.

In the case of the lion pride, female leaders approached the speaker slowly and stopped to look behind at the laggards. The leader females mistrust the laggards, but they do not directly punish them. Although lions might be tempted to reduce their risk of injury from territorial defense by withdrawing or not cooperating, they still need companions for territorial defense, to share the protection of the young from males, and for the capture of large prey. In addition, females are often closely related, usually sisters. Given their shared genes, laggards appear to be tolerated, and leaders continue to arrive alone at the speaker 30 seconds to two minutes before the laggards arrive.

Heinsohn and Packer classified the behavior of individual female lions as: (i) unconditional cooperators, who always led the response; (ii) unconditional laggards, who always lagged behind; (iii) conditional cooperators, who lagged least when needed most; (iv) conditional laggards, who lagged most when needed most.

The parallels with human societies are evident. How these behaviors developed and why they are tolerated in animal (as well as human!) societies will remain a fascinating topic for future investigations.

2.10 Conclusions

Models can be derived for density-dependent populations using both difference and differential equations for populations with discrete and overlapping generations. Although the equations differ in detail, all presume that the growth-rate parameter is dampened as the population approaches a carrying capacity. Modifications of the logistic include the introduction of the Allee effect and the inclusion of time lags. In populations with discrete generations there is an inherent time lag which produces an overshoot of the carrying capacity when the net growth rate is large enough. In continuously breeding populations, if we introduce a time-lag variable (*tau*), a large value of r combined with even a modest time lag can cause populations to exhibit a variety of behaviors commonly found in nature. These include limit cycles and dramatic growth phases followed by spectacular population crashes ("boom and bust" cycles). Therefore, when we study populations in nature we should never be surprised when many of them, particularly those populations with high growth potentials, do not remain constant from one year to the next. Finally, the logistic equation is merely a starting point for encompassing the idea of limitation to population growth. It is, as suggested by Turchin (2001), a special case assuming a linear relationship between population density and vital rates. Most importantly it does not allow for time lags, which we have seen have powerful effects on the behavior of populations.

Intraspecific competition has a major influence on the life history of a population. Fertility, mortality, growth and developmental rates, as well as behavior characteristics are all shaped by intraspecific competition, often in subtle and surprising ways.

3

Population regulation

3.1 Introduction

Are populations regulated? If so, how? What does population regulation really mean? These questions have been debated for many years and were at the core of a hotly contested difference of opinion between ecologists who emphasized the importance of density-dependent factors versus those who emphasized density-independent factors in population regulation (Turchin 1995). The logistic equation is, of course, based on density dependence. But how do we differentiate density-independent from density-dependent causation in populations with time lags, or in those dominated by stochastic processes?

The ecologists who emphasized the primary role of density-independent factors in population regulation were often those who worked with small animals, especially insects and rodents, and/or in habitats characterized by drought or short growing seasons. Among them were early influential ecologists such as Andrewartha and Birch, who worked on Australian grasshoppers, the distribution of which was determined by the length and intensity of the wet season. The northern boundary was determined by conditions that were too dry for their food plants, and the southern boundary was said to have too much moisture for the grasshoppers.

In the first major ecology textbook, Andrewartha and Birch (1954) concluded that abundance of a population was limited by the same conditions that limited its distribution. A major example used by Davidson and Andrewartha (1948) was the distribution and abundance of thrips (*Thrips imaginis*), tiny insects that feed on pollen and soft tissues of flowers in southern Australia. Populations were said to increase unchecked in the spring, with a sharp decline during the summer drought when flowers were scarce. Andrewartha and Birch (1954) asserted that thrip populations were checked by rainfall, not by their food supply, a potentially density-dependent factor. Advocates of density dependence, such as

Nicholson (1957), Solomon (1957), and Lack (1954, 1966), often cited the importance of competition in population regulation, especially in vertebrate populations.

Turchin (1995, 1999, 2003) and others (Ricklefs 1990) have asserted that this debate is largely artificial and that density dependence has been repeatedly and convincingly demonstrated (see examples from Chapter 2). At the same time, we know that the physical environment and other stochastic factors often reduce population size before density-dependent factors become operational.

When population regulation takes place, what is the dominant mechanism? Competition? Predation? Parasites? We will explore this question in the second half of the book. Note, however, that the answer may differ by trophic level. For example, Hairston *et al.* (1960) proposed that while herbivore populations were mainly limited by predation, producers (green plants), decomposers, and predators were usually limited by competition. This is such a broad generalization that is difficult to imagine how to test it. Nevertheless, this theory resurfaces regularly in modified forms.

3.2 What is population regulation?

One of the problems with this debate is a lack of agreement as to what a "regulated population" is. Given what we learned in the previous two chapters about the behavior of populations with time lags, with high reproductive potentials, or under the influence of demographic and environmental stochasticity, it is not realistic to expect a population to show a simple attraction to a specific number called the carrying capacity. In Chapter 2 we defined a **stable point** as a stable number at the carrying capacity with no oscillations. But we also recognized that the population could be temporarily oscillatory with oscillations dampening and moving toward a stable number or point. We also discussed populations that show regular oscillations around the carrying capacity (**stable cycles**), oscillations between two, four, and eight points, and chaotic behavior. All of these population behaviors are based on variations of the density-dependent logistic model.

Population regulation does not depend on a specific stable equilibrium point, but rather on a "long-term stationary probability distribution of population densities" (Turchin 1995) or a "stochastic equilibrium probability distribution" (May 1973). The key concept is that there is some mean population level around which a regulated population fluctuates. In addition, over time the population does not wander increasingly far away from this level. The variance of the population density is bounded (Royama 1977). All of these definitions relieve us of the expectation of a fixed carrying capacity or of a fixed stable point. The population is not expected to seek a stable point from which it does not wander. Instead, we allow for stochastic variation around the carrying capacity, **which can itself vary over time**. Hence, an ecological equilibrium is not a fixed or stable point, but a cloud of points. However, since it is often difficult to distinguish density dependence from stochastic noise (see below), many populations have been described as "density vague" (Strong 1986).

Fluctuations in the Dow Jones Industrial Average have been used as an example of an unregulated system, with rainfall patterns on Barro Colorado Island in Panama as an example of a regulated system (Turchin 1995). Some economists and meteorologists might argue with these examples, but the general idea is that the Dow Jones average, in spite of short-term ups and down, shows a trend upward over long time spans. A better example might be the carbon dioxide content of the atmosphere from 1850 to the present.

Although there are annual fluctuations, CO_2 concentration has steadily increased from around 280 parts per million (ppm) to over 360 ppm. By contrast, ice cores have shown that carbon dioxide concentration in the atmosphere from the year 1000 to 1750 was a good example of a regulated system. With regard to rainfall, records from 1900 to 2000 showed that rainfall fluctuated from year to year in Panama, but showed no overall pattern of increase or decrease.

Population regulation does not occur in the absence of density dependence. Thus a population showing pure exponential growth or decline would be unregulated. If population density had no effect on the per capita growth rate, there could be no range of population densities to which the population would return. In this context, Turchin (1995) uses the Murdoch and Walde (1989) definition of density dependence as "a dependence of per capita population growth on present and/or past population densities." While density dependence is a property of the overall population dynamics, which may involve time lags, no specific mechanism is necessarily responsible (Turchin 2003). That is, one aspect of the life history, such as birth rates, may not be density-dependent.

Although density dependence is necessary for population regulation, it is not a sufficient condition. For regulation to occur the following must also be true: (i) density dependence must be of the right sign (there must be a tendency to return to a carrying capacity); (ii) the return tendency must be strong enough to counteract potential disruptive effects of density-independent or stochastic factors; and (iii) the lag time over which the return tendency operates must not be too long (Turchin 1995).

Given more precise definitions of population regulation, and density dependence, Turchin (1995) found that the other remaining pieces needed are: (i) an acceptable statistical test of density dependence, and (ii) time-series data long enough to detect density dependence. From a review of the literature, Turchin (1995) concluded that whenever adequate data have been gathered for an appropriate period of time and tested with an appropriate method, density dependence has been found. This does not mean that all populations are regulated at all times. Lack of regulation doubtless occurs temporarily, but the probability of detectable density dependence will increase with the length of time that data are collected on a population.

3.3 Combining density-dependent and density-independent factors

In Section 2.7 we added stochasticity to density-dependent models. To simulate density-independent effects we simply added environmental stochasticity by allowing the carrying capacity to randomly vary. We can more explicitly add a stochastic density-independent factor by modifying the Ricker equation (2.12). Let us assume that a certain level of density-independent mortality happens every year, but it varies randomly. We can modify the Ricker equation as shown below (Eqn. 3.1) by simply subtracting the number of deaths from density-independent factors:

$$N_{t+1} = N_t \, e^{r\left(\frac{K - N_t - D}{K}\right)}$$

$$(3.1)$$

We can also select different levels of density-independent death rates. As shown in Fig. 3.1, the population growth is increasingly affected as we move from zero to increasingly higher amounts of density-independent growth.

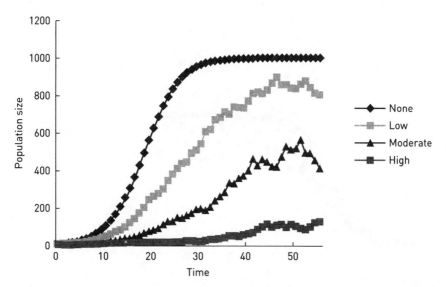

Figure 3.1 Effect of different levels of stochastic density-independent mortality on population growth, based on the Ricker equation (Eqn. 2.12). In all cases $r = 0.25$, $N_0 = 10$, and $K = 1000$. Population growth is increasingly affected as the strength of stochastic density independent mortality increases.

Instead of assuming that density-independent effects are uniformly negative, let us now assume that some years are good years (bonanza years) and others are bad years. That is, environmental stochasticity does not always have a negative effect on our population. We will simply introduce a certain amount of "environmental noise." That is, density-independent effects that can be either beneficial or detrimental. To simulate this we can still use Equation 3.1, but we now introduce stochastic variations that can be positive or negative, but with an average value of $D = 0$. In Fig. 3.2 compare the four stochastic simulations with the deterministic growth curve with no density-independent mortality. Obviously, if we increase the variations around D, the "cloud" of points around K gets larger. The density-independent factors would get increasingly important and obscure the density dependence of the population.

3.4 Tests of density dependence

How can we detect density dependence in the field? For a density-independent population, Tanner (1966) proposed that we can simply use the equation for discrete growth, $N_{t+1} = \lambda N_t$. After taking natural logs of both sides of the equation we can write:

$$\ln N_{t+1} = \ln \lambda + \ln N_t = (1.0)\ln N_t + \ln \lambda \qquad (3.2)$$

When we plot $\ln N_{t+1}$ versus $\ln N_t$, if λ is a constant, we should have a straight line with the slope of 1.0 and a y-intercept equal to $\ln \lambda = r$. But if there is density dependence and the growth rate slows with population size, when $\ln N_{t+1}$ is graphed against $\ln N_t$, a linear regression through the data should yield a slope less than 1.0.

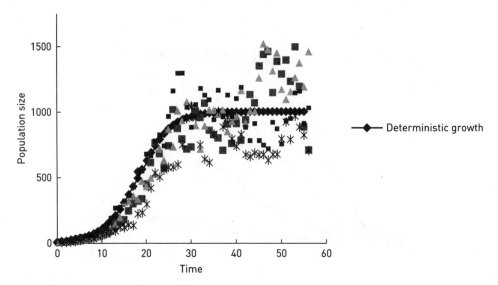

Figure 3.2 Effect of stochastic "environmental noise" on population growth. The deterministic growth curve has no density-independent effects. In all cases $r = 0.25$, $N_0 = 10$, and $K = 1000$.

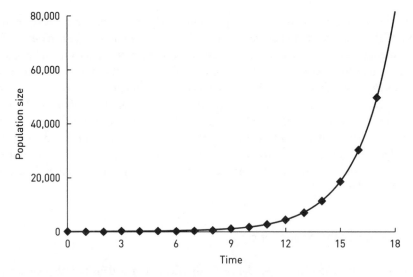

Figure 3.3 Exponential growth. Initial population size $= 10$ and $r = 0.50$.

For example, consider Figs. 3.3 and 3.4. In Fig. 3.3 we have an exponentially growing population with $\ln \lambda = r = 0.50$ and an initial population size of 10. In Fig. 3.4 we have graphed $\ln N_{t+1}$ versus $\ln N_t$. Since this population is not showing density dependence we get what we expect, a positive slope equal to 1.0. Now contrast this with a hypothetical yeast population showing density-dependent growth (Fig. 3.5, based on Pearl 1927). If we take natural logs and graph the data as we did in Fig. 3.4, we find that the slope of the

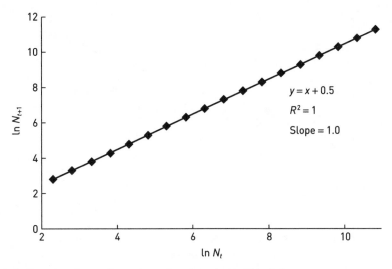

Figure 3.4 Density-dependence test for data from Fig. 3.3.

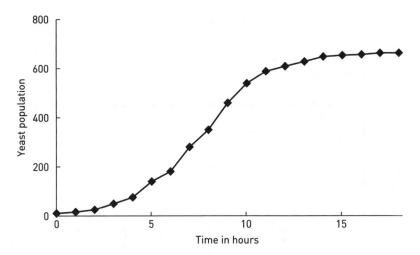

Figure 3.5 Hypothetical yeast population growing in the laboratory. Adapted from Pearl (1927).

line is indeed less than one (Fig. 3.6). The question remains, however, how far less than one should a regression slope be before we consider it significantly different?

Tanner (1966) examined 70 data sets for animal populations and found slopes significantly different from one in 63 of them. However, this method for detecting density dependence is fundamentally flawed. First, a linear regression assumes data points are independent. In this analysis the x-value in one time series becomes the y-value in the next time series. Second, measurement error in the population estimates inevitably leads to a slope of less than one. Therefore a slope of less than one is often just an artifact of measurement error and not evidence for density dependence. Finally, the expected relationship between N_{t+1} and N_t is not necessarily linear if there is environmental variability

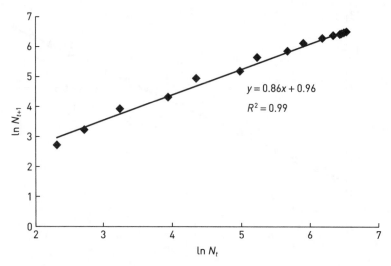

Figure 3.6 Density-dependence test for a yeast population. Based on data from Fig. 3.5.

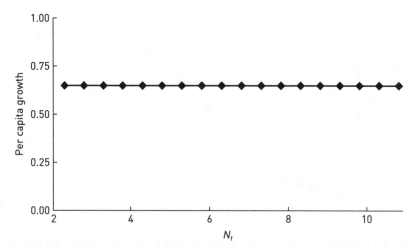

Figure 3.7 Per capita growth test for density dependence in a population with exponential growth and no carrying capacity. Based on data from Fig. 3.3.

or if the population has such a high growth rate that it approximates chaos (as described in Chapter 2).

A better test for density dependence is to examine the per capita growth rate versus population size (Turchin 1995, Case 2000). In the logistic equation we expect the per capita growth rate to have a negative slope when graphed against population number (Fig. 2.2 from Chapter 2). By contrast, in exponential growth, the per capita rate remains steady. For example, examine Figs 3.7 and 3.8. In Fig. 3.7 (based on exponential-growth data from Fig. 3.3), the slope equals zero, indicating no change in the per capita growth rate with population density. By contrast, for our yeast population graphed in Fig. 3.5 we find

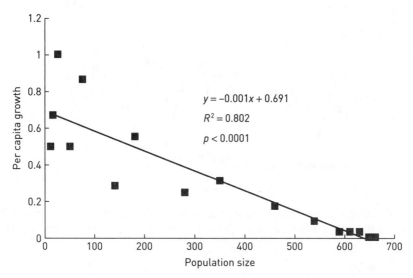

Figure 3.8 Per capita growth test for density dependence in a yeast population. Based on data from Fig. 3.5.

a significant negative slope for the same analysis (Fig. 3.8). Examining plots of per capita growth versus N have many advantages, including the detection of density dependence in populations with environmental noise (Case 2000).

Let us now look at some field data. The following is based on long-term Christmas Bird Counts of waterfowl populations in the Chesapeake Bay area of Maryland and on the Piedmont of Virginia. These data were obtained by Heath (2002) from the United States Fish and Wildlife Service and from the Virginia Society for Ornithology. Christmas Bird Counts (CBC) were initiated in the late 1800s as a replacement for the traditional Christmas hunt, and may be the oldest wildlife census in the world. The CBC, however, depends on volunteers, many of whom are not professional biologists, and the use of CBC data in scientific studies is problematic. Nevertheless, the CBC often represents the only long-term data on waterfowl in regions such as the Virginia Piedmont. In addition, Maryland and Virginia waterfowl populations have been affected by habitat loss, hunting pressure, and environmental degradation in the Chesapeake Bay. On the other hand, due to land-use changes including the creation of new reservoirs and wetlands, waterfowl populations may be increasing on the Virginia Piedmont (Heath 2002).

Indeed, if we examine CBC estimates of Canada geese (*Branta canadensis*) populations from 1958 to 2001, there is a distinct increase (estimated $r = 0.17$) in the Piedmont population, while the Coastal Plain population has no distinct trend other than that of increasing oscillations (Fig. 3.9). In Fig. 3.10 we have tried to test for density dependence in the Piedmont population by the first method, graphing $\ln N_{t+1}$ versus $\ln N_t$. The resultant regression is so close to one that the conclusion is inescapable that the Piedmont goose population is **not** density-dependent at this time. By contrast, in Fig. 3.11 we see that the regression slope departs radically from one in the Coastal Plain population. In fact the regression is so weak that the slope is not significantly different from zero. How do we interpret this? On the one hand we might conclude that the Coastal Plain population is very density-dependent. Or we might conclude that this is just an unreliable set of data.

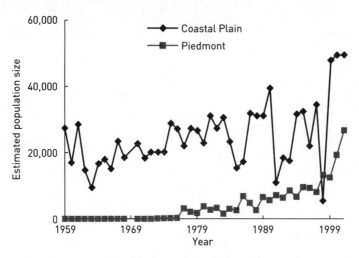

Figure 3.9 Canada goose (*Branta canadensis*) population on the Coastal Plain of the Chesapeake Bay and on the Virginia Piedmont. Based on Christmas bird counts.

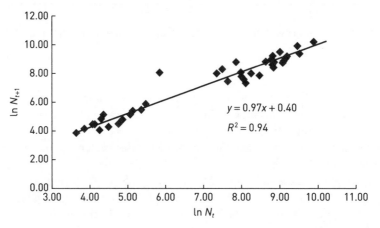

Figure 3.10 Test of density dependence of the Piedmont Canada goose population. Based on Christmas bird counts.

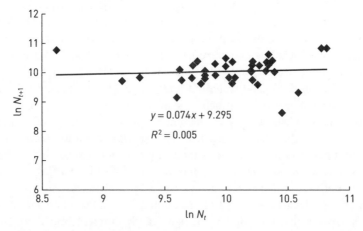

Figure 3.11 Test for density dependence of the Coastal Plain Canada goose population. Based on Christmas bird counts.

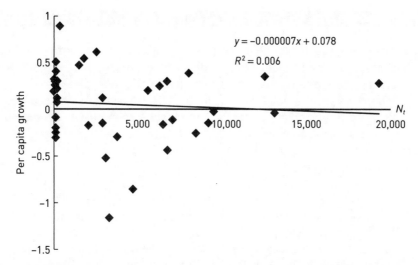

Figure 3.12 Per capita growth versus population size in the Piedmont Canada goose population.

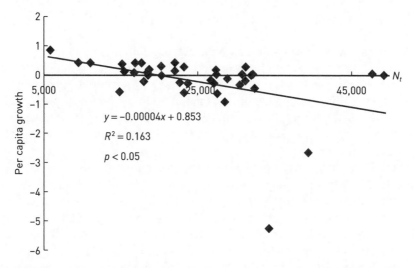

Figure 3.13 Per capita growth versus population size in the Coastal Plain Canada goose population.

Now, let's try the second method for determining density dependence, using per capita growth rates. From Fig. 3.12 we learn that the regression line for the Piedmont population is not significantly different from zero. This means that the Piedmont goose population has shown no decrease in per capita growth through 2001. Therefore, both tests tell us that the Piedmont population is not density-dependent at this time.

On the Coastal Plain (Fig. 3.13), although the slope is very small, it is negative and the regression is significant. Therefore, although the data are rather weak, it does appear that this population shows density dependence.

3.5 Conclusions

Populations are affected by their own life histories and vital rates, and by feedback from the environment. Such feedback components, which may involve time lags, are termed **endogenous** by Turchin (2003). As we have reviewed before, demographic stochasticity can also seriously affect population behavior in spite of density-dependent feedback. By contrast, **exogenous** factors refer to environmentally related density-independent factors that affect population density, but are not, in turn, affected by it (Turchin 2003).

Our main conclusions are:

1 The per capita rate of change, r, is affected by both endogenous and exogenous factors, and both should be examined when attempting to understand population behavior (Turchin 2003).
2 As stated above, negative feedback between the realized growth rate of the population and its density is a necessary but not sufficient condition for population regulation.
3 Population dynamics are nonlinear, and exogenous (density-independent) factors may dominate population behavior.
4 A more reliable method for detecting density dependence is to plot per capita growth against N, as opposed to $\ln N_{t+1}$ versus $\ln N_t$.

Assuming we can reliably demonstrate population regulation, is it due primarily to competition or to predation/parasitism? Is regulation in many populations simply a reflection of habitat loss? How often is regulation due to local processes as opposed to metapopulation dynamics (see Chapter 5) (Murdoch 1994)? Are populations under the influence of stochastic environmental factors to the extent that density-dependent population regulation is largely irrelevant?

Leirs *et al.* (1997) have shown that the population of the African rodent *Mastomys natalensis* is regulated by an interaction between stochastic and deterministic seasonality and nonlinear density-dependent factors. Using a variety of statistical techniques, they analyzed population data based on a multiple mark–recapture method. The best fit to the data came from a model encompassing both density-independent factors (previous three months' rainfall data), and density-dependent factors (a nonlinear demographic model). This example affirms that population regulation is often a combination of stochastic density-independent as well as density-dependent factors operating simultaneously.

4

Populations with age structures

- Survivorship
- Fertility
- Mortality curves
- Expectation of life
- Net reproductive rate, generation time and the intrinsic rate of increase
- Age structure and the stable age distribution
- Projecting population growth in age-structured populations
- The Leslie or population projection matrix
- Reproductive value
- Sensitivity analysis

4.1 Introduction

In the previous two chapters we either examined populations without distinct age classes or we specified that these populations had stable age distributions. We also assumed that models for continuous breeders would apply to seasonal breeders such as white-tailed deer (*Odocoileus virginianus*). We danced around the problem of forecasting population growth for populations with complex age distributions and the fact that the age distribution itself can govern the behavior of the population, at least in the short term. Knowledge of the age-specific survivorship and fertility patterns of a population allows us to understand what age categories are most important to the future survival of the population. If you wanted to help conservation biologists ensure the long-term survival of a sea turtle population, for example, what recommendations would you make? Would it be most effective to protect the beaches where the females lay their eggs? Limit predation on the eggs? Gather up the hatchlings, sequester and feed them for a year before releasing them? Or take steps to reduce the mortality of adults through monitoring and regulating fishing fleets? To answer these questions we will need to learn a number of skills that will allow us to extract growth rates from basic life-history data and to project population growth using different assumptions.

When age-specific fertility and survivorship data are gathered, we put them together in the form of a **life table**. From the information in the life table we are then able to calculate a variety of interesting statistics that acquaint us with the characteristics of the population. In addition to age-specific fertility and survivorship, we need information on the current number of individuals in each age class, since population growth in the short term is also strongly influenced by the latter.

The following example might fix in your mind the potential importance of age distribution on population growth. Imagine a group of several thousand young people attending a concert by the latest pop icon on an island off the coast of California. Suddenly a catastrophic series of earthquakes eliminates the entire local population while simultaneously California splits off from the rest of North America. Assume further that a few of the concertgoers are able to colonize the original island off California, and future population growth is now based on this group. Most of the concertgoers would obviously have been teenage girls (assume a few teenage boys also were dragged along). Further, assuming an abundance of food, this California island would rapidly be repopulated. Growth would, however, be irregular, since all of the girls would be the same age and reach menopause more or less simultaneously in the future. Growth would slow until their daughters began to reproduce. Now imagine the same scenario, except that survivors of the disaster were individuals attending an AARP (American Association of Retired Persons) convention. Presumably all, or almost all, of the females at the convention are over 55. What would the future of the California island population be in this case? Evidently age distribution can contribute to extinction of a population!

Many of the techniques we will examine in this chapter were developed for the life insurance industry and applied to human populations. For example, actuaries need to calculate the risk of insuring the life of their clients. Life tables were developed so that the probability that a 50-year-old pharmacist would live another 10 years could be determined; policy rates were then set accordingly. Such techniques were easily translated to animal populations, and the comparative study of survivorship among different groups of animals was initiated (Deevy 1947).

More problematic has been the application of life tables to plant populations. Rates of growth, reproduction, size, and mortality are not distinctly related to age in plants, as is the case for animals, but are highly variable and highly dependent on the local environment. This "phenotypic plasticity" can be demonstrated by growing genetically identical clones in different environments. Growth, size, and fertility, when measured against age, will be dramatically different (Silvertown and Doust 1993). The recommended solution is to develop a life-history table in which **stages** are used instead of age classes (Werner 1975, Werner and Caswell 1977, Hubbell and Werner 1979). For example, in her study of teasel (*Dipsacus sylvestris*), Werner (1975) used the following stages instead of age classes: (1) first-year dormant seeds; (2) second-year dormant seeds; (3) small rosettes; (4) medium rosettes; (5) large rosettes; (6) flowering plants. In teasel, plants die after flowering. Instead of calculating the probability that an individual would survive from one age class to the next, she calculated the probability that an individual of one life stage would survive to the next stage. We will also take a look at stage-based methods for animal populations later in this chapter.

Another problem in applying life tables to plants is that plant populations often spread through vegetative propagation. Grasses, for example, spread horizontally via rhizomes. New shoots arise from rhizomes and often separate from the original plant. A complex terminology has been developed to explain this phenomenon. For example, a **genet** is an individual that has arisen from a seed. A **ramet** is a new plant that is a clone but which has arisen through vegetative propagation and is now a completely independent plant with its own roots and shoots. Thus a population of grasses may consist of several genets, each of which has several ramets. Clonal populations may proliferate indefinitely without flowering. This has led to some fascinating life cycles such as that of the giant bamboo (*Phyllostachys bambusoides*), in which clones of perennial ramets proliferate, forming large populations that flower only once every 120 years (Janzen 1976). All of the clonal ramets flower simultaneously and then the entire population dies, leaving behind only seeds with which to found the next population. Studies conducted since the great 1988 fire in Yellowstone National Park have forced biologists to revise the conventional wisdom that aspen (*Populus tremuloides*) does not reproduce by seed, but spreads by cloning. Instead Turner *et al.* (2003) suggest that new genets of aspen as well as some of the perennial herbs of the forest floor are produced after fires, but recruitment of new individuals (ramets) during fire-free intervals is primarily through asexual reproduction.

In the sections below, as we refer to age-specific traits of survivorship or fertility, keep in mind that many plant and animal populations would often require a rather different approach in which we examine survivorship and fertility by size class or by stage in the life cycle.

4.2 Survivorship

The construction of a life table begins by gathering information on survivorship by age class. This sounds simple, but is easier described than actually done. For example, one method is to study a **cohort** of individuals all born at the same time, and follow the survivorship of these individuals until the last member of the cohort dies. At the beginning of such a study, it would be necessary to locate and mark all newborn individuals. Subsequently, one would need to verify when each individual died. Individuals that simply disappeared could not be assumed to have died; they might have emigrated. Studies of cohorts are obviously best done on small populations and on populations that move about in a predictable way. The advantage of studying a cohort is that one knows the exact age of each individual. The disadvantage is that such a study lacks generality, in that cohorts born in different years may have different survivorship schedules. In addition to the difficulties one might encounter in actually marking and gathering information on all members of a cohort, there are also practical problems. A cohort study on most species of turtles, for example, would require the entire professional life span of the investigator (picture a student waiting 50 years to finish her PhD dissertation). A life table developed in this manner is known as a **fixed-cohort, dynamic**, or **horizontal life table**.

A second approach is to locate and examine all of the dead individuals in a population during some defined period of time. We would need a method for estimating the age of the animals or plants at death. This approach to the construction of a life table assumes that the rates of survival in the population are fairly constant. If this is not the case, age-specific mortality rates will be confused with year-to-year variation in mortality of the overall population. Data gathered in this manner produce a **static, vertical**, or **time-specific life table**.

A third approach is to collect life-history data for several cohorts over as long a period as possible. In most populations there is a large difference between juvenile and adult

Table 4.1 Survivorship data for males born between 1800 and 1890, taken from the Fairfax City Cemetery, Fairfax, Virginia.

Age category	Number who died in the age category	Number alive at the beginning of the age class	S_x based on a cohort of 1000	Survivorship, l_x. (Proportion of original cohort alive at the beginning of the age category)
0–1	0	207	1000	1.000
1–4	1	207	1000	1.000
5–9	0	206	995	0.995
10–14	2	206	995	0.995
15–19	2	204	986	0.986
20–24	4	202	976	0.976
25–29	0	198	957	0.957
30–34	1	198	957	0.957
35–39	0	197	952	0.952
40–44	5	197	952	0.952
45–49	2	192	928	0.928
50–54	12	190	918	0.918
55–59	9	178	860	0.860
60–64	18	169	816	0.816
65–69	24	151	729	0.729
70–74	33	127	614	0.614
75–79	33	94	454	0.454
80–84	35	61	295	0.295
85–89	20	26	126	0.126
90–94	3	6	29	0.029
95–99	2	3	14	0.014
100–104	1	1	5	0.005
105–109	0	0	0	0.000
Total population	207			

survivorship. Therefore, even though survivorship data on adults are often relatively easy to gather, such data do not apply to the juvenile age classes. Depending on the age when reproduction begins, it is possible to find the growth rate of the population without specific data on juvenile survivorship, as described later in the chapter.

No matter how the data are gathered, the objective is to produce an estimate of age-specific survivorship and fertility. In human demography, age-specific survivorship is based on a theoretical cohort of 1000 individuals. If we let S_x equal the number of individuals surviving to age x, we set S_0 equal to 1000. Then, S_1 = the number of individual surviving to age 1, S_2 = the number surviving to age 2, etc. Table 4.1 is based on data gathered by an ecology laboratory from the Fairfax City (Virginia) cemetery. In this case 207 male grave-stones were examined and the ages at death calculated. (Data were gathered only from

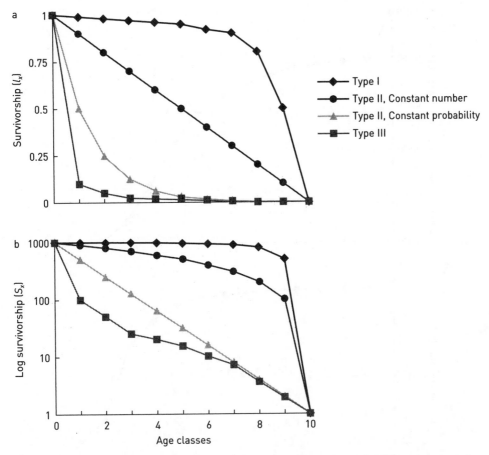

Figure 4.1 Survivorship "type" curves: (a) on an arithmetic scale; (b) on a log scale.

graves in which the birth date was between 1800 and 1890. Why was this done?) Because humans are so long-lived, the data were placed into five-year age intervals. The first step is to use the number who died in each age interval (column 2) to produce column 3, the number of survivors by age class. Next, we normalize the population to a theoretical cohort of 1000 by dividing each number by 207 and multiplying by 1000. This produces the S_x column. However, in most ecological studies we do not, in fact, use the S_x data. Instead, each number in the S_x column is divided by 1000 to produce the survivorship function, l_x. Each value in the l_x column stands for the proportion of the population that survives to a given age, x. It is measured from birth until the last or oldest member of the population dies. l_x is known as **age-specific survivorship** and can be thought of as the probability, at birth, of living to a specific age class. By definition, $l_0 = 1.00$.

The survivorship table is used to construct the survivorship curves found in all ecology textbooks. In a survivorship curve, age (x), the independent variable, is graphed against survivorship. The y-axis may be on a straight arithmetic scale; however, many authors prefer a log (base 10) scale for survivorship. Pearl (1927) introduced the idea that biological populations routinely fit one of three "types" of survivorship curves (Fig. 4.1a). The type I curve, known as the "death at senescence" curve, is characterized by excellent survivorship at all ages from birth until "old age," at which time the death rate rapidly accelerates

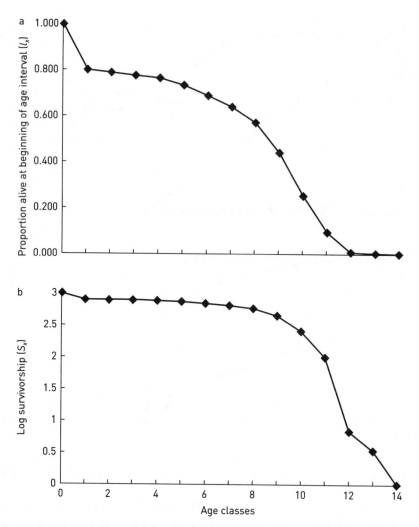

Figure 4.2 Dall sheep (*Ovis dalli*) in Denali National Park, Alaska. (a) Survivorship (l_x); (b) log of survivorship (S_x). After Deevy (1947).

and survivorship plummets. The type II curve is linear and assumes that either a constant number or a constant proportion of the population dies in each age interval. Examine Fig. 4.1. When survivorship is expressed on an arithmetic scale, a constant **number** of deaths per age interval produces a linear curve. When log to the base ten of survivorship is used (Fig. 4.1b), the constant **probability** of death per age interval produces a straight line. Finally, the type III curve applies to the vast majority of biological populations. In this curve there is very high mortality among the juvenile age classes while adult survivorship is relatively high. This is illustrated most dramatically in Fig. 4.1a, using the arithmetic scale for survivorship.

How realistic are these three survivorship "types?" Probably few populations exactly match any particular one. Furthermore, as found by Petranka and Sih (1986) for the salamander species *Ambystoma texanum*, survivorship curves may vary from year to year and place to place for the same species. (Recall our discussion in Chapter 1 of population viability analysis, in which we emphasized that demographic traits are subject to both temporal

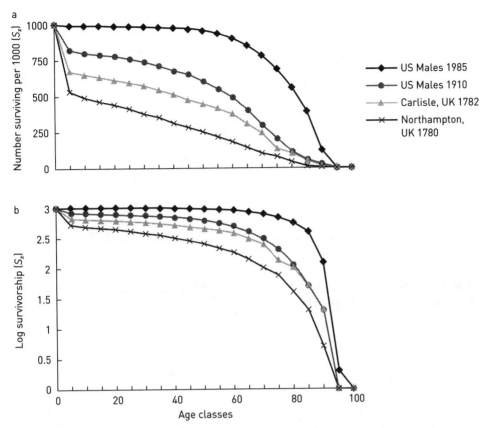

Figure 4.3 Human survivorship curves from two eighteenth-century English and two modern United States populations: (a) on an arithmetic scale; (b) on a log scale. After Lotka (1925) and Peters and Larkin (1989).

and spatial variability.) However, we can make some general comments. The least realistic of the three types is type I. A type I curve applies to laboratory populations of animals such as *Drosophila*. If provided with ample food, the population has a high rate of survivorship until the end of its maximum life span, when individuals die more or less simultaneously (Hutchinson 1978).

Natural populations of mammals such as Dall mountain sheep (*Ovis dalli*) (Deevy 1947), and many African ungulates (Caughley 1966), have a type I survivorship curve, although notice that 20% of the Dall sheep die in the first year of life (Fig. 4.2).

While modern human populations have a type I survivorship curve, in the not-so-distant past, living to a ripe old age was not assured (Fig. 4.3). By looking for the age where 500 of the original 1000 in a population are still alive, we have an idea of the average life expectancy at birth (Fig. 4.3a). For the modern (1985) US population, this figure is after the age of 80 (Peters and Larkin 1989). By contrast, for US males living early in the twentieth century, this figure was less than 60 (Lotka 1925). For eighteenth-century English populations, average life expectancy was less than 10 years in Northampton and around 40 years in Carlisle (Lotka 1925)! When these same values are plotted on a log scale, however, they all approximate a type I survivorship curve (Fig. 4.3b).

In order for an organism to have a type II survivorship curve, all stages of the life history must be more or less equally vulnerable to predation or other causes of death. Birds,

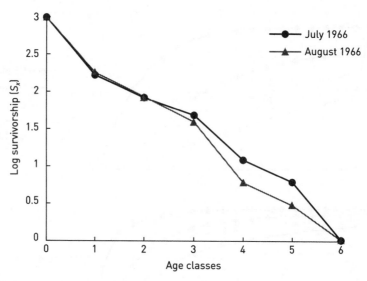

Figure 4.4 Log survivorship for two cohorts of white-crowned sparrows (*Zonotrichia leucophrys*). Based on Baker *et al.* (1981).

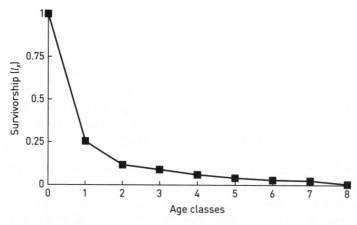

Figure 4.5 Survivorship curve for a gray squirrel (*Sciurus carolinensis*) population in North Carolina. Based on Barkalow *et al.* (1970).

especially the adult stages, are most commonly cited as having a type II survivorship curve. For example, when Gibbons (1987) examined longevity records of vertebrates in captivity, only birds displayed a type II survivorship curve on an arithmetic scale. In a study of white-crowned sparrows (*Zonotrichia leucophrys*) (Fig. 4.4) Baker *et al.* (1981) found a type II survivorship curve on a log scale, which indicates a more or less constant probability of death, irrespective of age. The maximum life span was 49 months in this species. Botkin and Miller (1974), however, argued that birds do not, in fact, have an age-independent mortality rate. The survivorship curve for the sooty shearwater (*Puffinus griseus*), based on an arithmetic scale, appears to be type II. However, Botkin and Miller showed that, on closer examination, while the mortality rate in the early age classes was 0.07 per year, there was an increase in the mortality rate of 0.01 per year. They concluded that mortality was not in fact age-independent in the sooty shearwater, nor indeed in most species of birds.

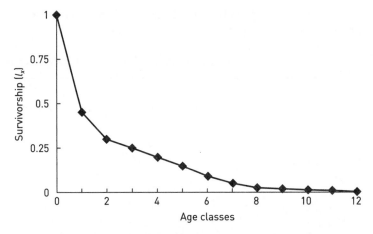

Figure 4.6 Survivorship in captive female golden lion tamarins (*Leontopithecus rosalia*). From J. Ballou (personal communication).

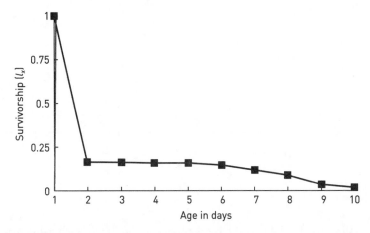

Figure 4.7 Survivorship schedule for *Phlox drummondii* on an arithmetic scale. Based on Leverlich and Levin (1979).

The type III survivorship curve, which features heavy mortality among young age classes followed by good to excellent adult survivorship, applies to most biological populations from barnacles to sea turtles to plants (Hutchinson 1976). Even medium-sized mammals, such as gray squirrels (*Sciurus carolinensis*) (Fig. 4.5; Barkalow *et al.* 1970) and golden lion tamarins (*Leontopithecus rosalia*) (Fig. 4.6; Jonathan Ballou, personal communication) have type III survivorship curves, as do most amphibians such as *Ambystoma tigrinum* (e.g. Anderson *et al.* 1971).

Actually, most species do not follow any one of the type curves precisely, especially when an arithmetic scale is used. For example, in *Phlox drummondii* (Leverlich and Levin 1979) survivorship drops from 1.00 to 0.67 in the first 63 days after germination. After 124 days survivorship is down to less than 0.30. Mortality is minimal thereafter until the plants are almost a year old (Fig. 4.7). Yet this plant, when its survivorship is plotted on a logarithmic scale, has been used as an example of a type I curve (Smith 1996). Therefore, although survivorship curves are extremely useful in order to visualize the large amount of data in a life table, there is little agreement as to what constitutes a survivorship "type," and nothing is gained by attempting to fit a life table to any of the three "type" curves.

Table 4.2 Fertility data from 1985 US Vital Statistics. Fertility is based on the average number of daughters born in five-year age intervals.

Age class	Mean number of female offspring per female, m_x
0–1	0
1–5	0
5–10	0
10–15	0
15–20	0.025
20–25	0.250
25–30	0.500
30–35	0.150
35–40	0.100
40–45	0.010
45–50	0
50–55	0
55–60	0
60–65	0
65–70	0
70–75	0
75–80	0
80–85	0
85–90	0

4.3 Fertility

The other half of the life table is the **fertility column**, m_x. Here each value represents the average number of female offspring produced per female of a given age. Again, gathering accurate data on fertility in the field is problematic for many populations. In order to simplify calculations, we count only the number of females. That is, the values are mean numbers of females by age class. Fertility, like survivorship, can be graphed as a function of age, and the resultant fertility curve is usually triangular or rectangular in shape. For example, Table 4.2 illustrates human fertility based on 1985 United States Vital Statistics (Peters and Larkin 1989). Age classes until age 15 are usually termed, "pre-reproductive." Ages 15 to 45 are considered the reproductive age classes. Figure 4.8 illustrates the usual triangular shape, with maximum reproduction occurring in the 25- to 30-year age classes. After age 45, fertility falls back to zero. These are the post-reproductive age classes (modern medical science, however, is pushing the normal boundary of reproduction past 45 years). On the other hand, the North Carolina gray squirrel population (Table 4.3) would have a rectangular shape if fertility were graphed against age.

As usual, when populations are sampled, data do not necessarily follow generalized trends. For example, in a study by Grant and Grant (1992) on the cactus ground finch (*Geospiza scandens*), the fertility schedule had a very irregular shape, dropping radically in the seventh year and showing an unexpected spike in the twelfth year (Fig. 4.9). In this

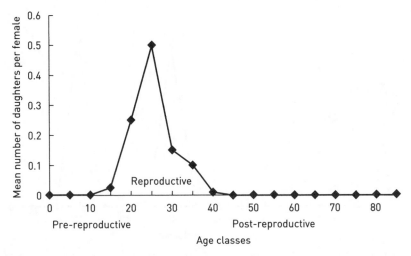

Figure 4.8 Human fertility for the population of the United States, 1985. From Peters and Larkin (1989).

Table 4.3 Life table for a gray squirrel population from North Carolina (Barkalow *et al.* 1970). (For explanation of symbols see text.)

Age	l_x	m_x	p_x	q_x	$l_x m_x$
0	1.000	0	0.253	0.747	0
1	0.253	1.28	0.458	0.542	0.324
2	0.116	2.28	0.767	0.233	0.264
3	0.089	2.28	0.652	0.348	0.203
4	0.058	2.28	0.672	0.328	0.132
5	0.039	2.28	0.641	0.359	0.089
6	0.025	2.28	0.880	0.120	0.057
7	0.022	2.28	0	1.00	0.050
8	0	0	–	–	0
					$R_0 = 1.119$

case, environmental variation dominated the fertility schedule. Amboseli baboons (*Papio cynocephalus*) also have an irregular fertility schedule (Alberts and Altmann 2003).

The sum of the m_x column defines the **gross reproductive rate** (GRR). This number is the average number of female offspring produced by a female that survives at least through the last reproductive age class:

$$\text{GRR} = \sum m_x \tag{4.1}$$

The two pillars of a life table are the survivorship (l_x) and fertility (m_x) columns. Long hours of fieldwork are necessary to gather the data in order to produce such a life table. Once the table is produced, many other calculations and projections are possible, although all of them assume that the survivorship and fertility columns remain constant.

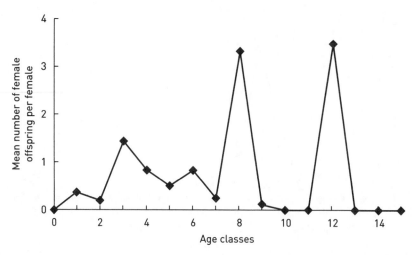

Figure 4.9 Fertility schedule of the cactus ground finch (*Geospiza scandens*). From Grant and Grant (1992).

From the l_x column we can develop two parallel columns, which provide information on how survivorship and mortality rates change with age. Consider Table 4.3. The l_x column is based on the probability, at birth, of surviving to a given age class. The p_x column, by contrast, is the **age-specific probability of surviving** to the next age class. That is, p_2 tells us the probability that an individual who has survived to the age of two will survive to be three years old. Similarly, p_4 would tell us the probability that a four-year-old lives to age five. These p_x values are critically important when we want to project future population growth, as will become clear later in this chapter. p_x is calculated according to the formula:

$$p_x = \frac{l_{x+1}}{l_x} \qquad\qquad (4.2)$$

For example, in Table 4.3, we see that $p_0 = l_1$, since $p_0 = 0.253/1.000$ $(l_0) = 0.253$. $p_1 = 0.116/0.253 = 0.458$, and so on. Notice that $p_7 = 0$, since no seven-year-old squirrel lives to be eight years old. p_8 is undefined and is the equivalent of dividing zero by zero.

The companion value to p_x is q_x, which is the proportion of the population that has survived to a given age, x, but which will **die** in the next time or age interval. q_x is simple to calculate since it is equal to $1 - p_x$ (Eqn. 4.3). This is based on the idea that $p_x + q_x = 1.0$. We recognize only two states of being, alive or dead.

$$q_x = 1 - p_x \qquad\qquad (4.3)$$

4.4 Mortality curves

Caughley (1966, 1977) and others found that mortality curves (q_x) for female mammals, such as Orkney voles (*Microtus arvalis*) (Leslie *et al.* 1955), toque monkeys (*Macaca sinica*) (Dittus 1977), buffalo (*Syncerus caffer*) (Sinclair 1977), Himalayan thar (*Hemitragus*

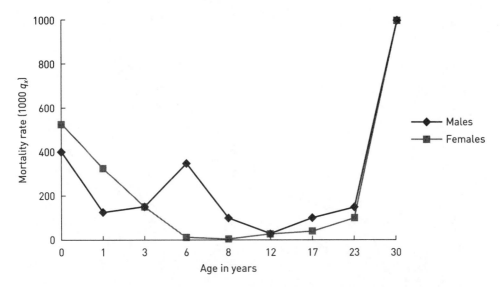

Figure 4.10 Mortality curves for male and female impala (*Aepyceros melampus*).
Data from Jarman and Jarman (1973); analysis from Ralls *et al.* (1980).

jemlahicus), domestic sheep (*Ovis aries*), Dall mountain sheep, and elk (*Cervus elaphus*) all
follow a U-shaped pattern (Fig. 4.10). The U-shape is the result of the fact that juvenile
phases usually have high mortality, but they are coupled with adult phases that have low
mortality. In the final phases of life mortality increases (the senescent phase). Ralls *et al.*
(1980) suggested that in polygynous species, although females show the U-shaped pattern,
males have a spike of mortality in sub-adult to young adult age classes. Impala (*Aepyceros
melampus*) and toque monkeys, for example, illustrate this pattern. These higher male
mortality rates are due to male–male competition for mates, and the tendency for males
to leave the natal group in many species (Ralls *et al.* 1980).

Interestingly, a similar pattern can be found for the United States human population
(Fig. 4.11). United States Vital Statistics for 1986 (Anonymous 1988) have a U shape for
females, with high mortality in the first year of life, followed by a low rate of mortality
until age 15. Thereafter the mortality increases throughout life. Males show higher mortality rates from the age of one onwards. More striking, however, is that the largest separa-
tion in mortality rates between the sexes occurs from the ages of 15 to 30. These age
classes are exactly those discussed by Ralls *et al.* (1980), in which male mammals suffer
greater mortality in the sub-adult to early-adult age classes. Do human males, like other
mammals, suffer these higher mortality rates due to male–male competition for females
and/or due to dispersal away from the parental home, as Ralls *et al.* (1980) have suggested?
Or are there other reasons, such as the suggestion that human male brains mature at a
slower rate than those of females (Thompson *et al.* 2000)?

4.5 Expectation of life

Another statistic of interest to demographers, ecologists, and even to non-scientists is the
age-specific expectation of life. The question to be addressed is, what is the average life

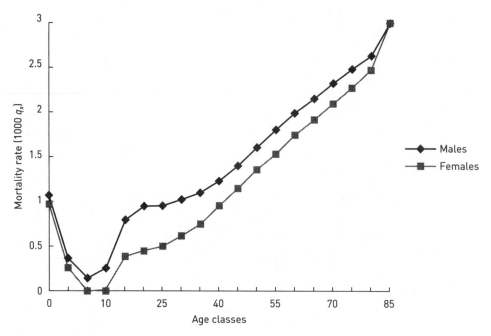

Figure 4.11 Mortality curves for United States human population, based on 1986 vital statistics (Anonymous 1988).

expectancy of an individual of a given age, x? The survivorship table allows us to directly read the average life span at birth. We simply look for the age at which l_x equals 0.50. However, average expectation of life is age-specific. Obviously, if life expectancy at birth for a human male is 75 years, a male who has survived to be 65 years old should, on average, live well beyond 75. In fact recent data suggest that a 65-year-old human male can look forward to an average of at least 15 more years of life.

Age-specific life expectancy is measured by taking the area under the survivorship curve beyond an age x, and dividing by the number or proportion of survivors of that age, x. In theory, using integral calculus, expectation of life, e_x to the last age class (w) is found using the formula:

$$e_x = \frac{\int_x^w l_x d_x}{l_x} \tag{4.4}$$

In practice we use discrete age classes. The survivorship between ages x and $x + 1$, if the age interval is reasonably small, is estimated as follows:

$$L_x = \frac{l_x + l_{x+1}}{2} \tag{4.5}$$

Table 4.4 Calculation of average age-specific life expectancy.

Age, x	l_x	L_x	T_x	Expectation of life, e_x
0	1.000	0.850	2.180	2.18
1	0.700	0.600	1.330	1.90
2	0.500	0.400	0.730	1.46
3	0.300	0.200	0.330	1.10
4	0.100	0.075	0.130	1.30
5	0.050	0.035	0.055	1.10
6	0.020	0.015	0.020	1.00
7	0.010	0.005	0.005	0.50
8	0	0	0	0
		$\Sigma = 2.18$		

L_x is the mean survivorship for any particular age interval, and assumes that, on average, an organism dies halfway between two age classes. The area under the survivorship curve for an individual of a given age, x, to the age, w, at which the oldest individual dies, is estimated as:

$$T_x = \sum_{x}^{w} L_x = \sum_{x}^{w} \frac{l_x + l_{x+1}}{2} \tag{4.6}$$

Therefore expectation of life is estimated by Equation 4.7:

$$e_x = \frac{T_x}{l_x} \tag{4.7}$$

In the hypothetical example in Table 4.4 we see that at birth the expectation of life is two years ($l_x = 0.50$). But the average two-year-old individual can expect to live another 1.46 years.

4.6 Net reproductive rate, generation time, and the intrinsic rate of increase

Once we have the basic life table, we are in a position to calculate the same types of growth-rate statistics we discussed in the first two chapters. The first of these is the net reproductive rate, R_0. This is an equivalent to the R (Eqn. 1.4) we developed for populations with discrete generations. The net reproductive rate represents the increase in the population **per generation**, and is defined as the **mean number of female offspring produced per female in the population per generation**. This value is found by incorporating both the fertility and survivorship functions of the life table. For each age class the product of $l_x \times m_x$ is found. This product is the contribution a particular age class is making toward population growth per generation. The net reproductive rate for the population as a whole is the sum of these products for all age classes:

$$R_0 = \sum l_x m_x \qquad\qquad (4.8)$$

In Table 4.3 the calculated net reproductive rate for the gray squirrel population is 1.119. This means that the average female squirrel replaces herself with 1.119 female squirrels per generation. As in the case of the net reproductive rate for non-overlapping populations, an $R_0 > 1$ means that the population, according to the life table, has the potential to increase every generation. The opposite is also true: an $R_0 < 1$ means that the population is decreasing every generation.

Although the net reproductive rate is an important statistic, we usually want to know the growth rate per year (or some other defined period). When we compare growth rates among different types of populations, the usual currency is r, the intrinsic rate of increase, or the finite rate of increase (λ), since both are measured for a specific unit of time. The intrinsic rate of increase can be extracted from life history data using an equation developed by Euler, although some authors give credit to Lotka (see Mertz 1970, or Case 2000 for its derivation). It is most often known as the Euler equation; but in any event, it is considered to be a "characteristic equation" of demography (Dingle 1990).

$$\sum l_x m_x e^{-rx} = 1 \qquad\qquad (4.9)$$

This equation is useful because it allows us to determine the intrinsic rate of increase from the life table. However, since r is an exponent in a summation, it cannot be explicitly solved for if there are more than two age classes. Values of r must be estimated and tried in the Euler equation until a value is found that satisfies it. However, Laughlin (1965) and May (1976a) showed that there exists an excellent approximation for r. Assuming a stable age distribution, the approximation is based on the following:

If G = generation time, we can write: $\dfrac{N_G}{N_0} = R_0$.

It is also true (Eqn. 1.8) that $\dfrac{N_G}{N_0} = e^{rG}$.

Therefore, we can set $R_0 = e^{rG}$.

Taking natural logs of both sides of the equation gives us $\ln R_0 = rG$ and therefore:

$$r = \frac{\ln R_0}{G} \qquad\qquad (4.10)$$

This tells us that the intrinsic rate of increase can be found by dividing the natural log of the increase per generation by the generation time. We now have an approximation for r, but we must calculate G, the mean generation time. Mean generation time is actually a somewhat slippery concept, and can be defined in various ways. Here we will use the definition, **the mean age of the mothers at the time of their daughter's birth**. This is the same definition as, "the average interval between reproductive onset in two successive generations" (Dingle 1990). Generation time is estimated according to the following equations, in which the age, x, is weighted by its realized fecundity, $l_x m_x$. In the second equation, discrete age intervals are used:

$$G = \int x l_x m_x d_x \qquad (4.11a)$$

$$G = \sum x l_x m_x \qquad (4.11b)$$

These equations, however, can only be used if the population is not growing. In order to account for the number of offspring being produced per individual female, the right side of the equation must be divided by the net reproductive rate, R_0. Therefore we use Equation 4.12 in estimating G:

$$G = \frac{\sum x l_x m_x}{R_0} \qquad (4.12)$$

Once we have approximated the value of G, r can be estimated using Equation 4.10. Note that all approximations of r gained using Equation 4.10 must be verified by the Euler equation! Since Equation 4.10 simply approximates r, the value of r must be verified, or the approximation adjusted using Equation 4.9, the Euler equation. The way r is estimated and confirmed is illustrated in Example 4.1.

4.7 Age structure and the stable age distribution

In the next sections, we are ready to begin examining the interactions between the age distribution of a population and its life table. As stated previously, the actual age distribution of a population has potentially dramatic effects on population growth in the short term. The age distribution of a population is defined as the proportions of the population belonging to various age categories at a given point in time. The proportion belonging to a given age category, x, is calculated by dividing the number of individuals in that age category by the total population size, N, producing c_x:

$$c_x = \frac{n_x}{N} = \frac{n_x}{\sum n_x} \qquad (4.13)$$

c_x is the proportion of the population belonging to an age category, x, and n_x equals the number of individuals in that age category.

Whenever survivorship and fertility remain constant for long enough, a population will converge on a particular age distribution, known as the **stable age distribution**, which is unique for each combination of survivorship and fertility. Once this stable age distribution is achieved, the age distribution no longer changes unless and until survivorship or fertility change in the life table. Furthermore, the population will grow or decline at the steady rate, λ (unless the r-value $= 0$, in which case the population is unchanging and $\lambda = 1$), and each age class will change at the same rate as the population as a whole. If a population has a stable age distribution, λ is easy to calculate, since $N_{t+1}/N_t = \lambda$. Since $r = \ln \lambda$ (Eqn. 1.13), it is also simple to calculate r.

The stable age distribution itself can be calculated from the survivorship column of the life table. In order to predict the stable age distribution, however, it is also necessary to know the value of r as well as the survivorship function, l_x. Since the Euler equation requires the knowledge of fertility (m_x), we actually must know both survivorship and fertility. The formula for predicting the stable age distribution is as follows:

$$c_x = \frac{e^{-rx}l_x}{\sum e^{-rx}l_x}$$ (4.14)

4.8 Projecting population growth in age-structured populations

Examine Table 4.5. The basic information on survivorship and fertility by age class would have been gathered through fieldwork. In order to project population growth into the future, we also must know the actual number of individuals belonging to each age class, again based on data we have obtained in the field. Once these data are available we can do a year-by-year projection not only of the population size as a whole, but also of the expected number of individuals in each age class. We must assume, however, that the survivorship and fertility functions do not change.

In Table 4.6 we have done some basic calculations that will tell us generally what to expect from this population. The net reproductive rate tells us that we expect this population to grow every generation ($R_0 > 1$). In the ensuing projection we will need the age-specific probability of surviving to the next age class (p_x); accordingly we have devoted a

Table 4.5 Hypothetical life table for a population. This table will be used to illustrate a simple population projection (Table 4.6).

Age, x	l_x	m_x	p_x	q_x	$l_x m_x$
0	1.00	0	0.50	0.50	0
1	0.50	2.0	0.40	0.60	1.0
2	0.20	1.0	0.50	0.50	0.2
3	0.10	1.0	0	1.00	0.1
4	0	0	–	–	–
Sums		GRR = 4.0			$R_0 = 1.3$

Table 4.6 Projected population growth based on life history from Table 4.5 and starting with 200 individuals in age class zero (newborn).

Age, x	n_x at $t=0$	c_x at $t=0$	n_x at $t=1$	c_x at $t=1$	n_x at $t=2$	c_x at $t=2$	n_x at $t=3$	c_x at $t=3$	Calculated stable age distribution
0	200	1.00	200	0.67	240	0.63	300	0.625	0.625
1	0	0	100	0.33	100	0.26	120	0.250	0.258
2	0	0	0	0	40	0.11	40	0.083	0.082
3	0	0	0	0	0	0	20	0.042	0.035
4	0	0	0	0	0	0	0	0	0
Sums	200	1.00	300	1.00	380	1.00	480	1.00	1.00

column in Table 4.5 to p_x as well as to its opposite (q_x), the age-specific probability of dying in the next age class. The last column allows us to calculate R_0.

We begin this projection by assuming that we have founded this population with 200 newborn females at time $t = 0$ (Table 4.6). We have completed the first two columns based on that assumption (n_x = the number in an age category, c_x = the proportion of the population in that age category). To project the population, the number of individuals in each of the non-zero age categories is found by multiplying the number of individuals in the appropriate age category by the suitable p_x value. For example, to find the number of one-year-old individuals at time $t = 1$, we multiply the number of newborn individuals at time $t = 0$ by p_0. The number of two-year-olds at time $t = 1$ is found by multiplying the number of one-year-olds at time $t = 0$ by p_1, and so on. In this case, to find the number of one-year-old females at time $t = 1$, we simply multiply 200 by p_0 (0.50), giving us 100. Since we lack one-, two-, or three-year-old individuals at time $t = 0$, there can be no two-, three-, or four-year-old individuals at time $t = 1$.

To find the number of newborn individuals at time $t = 1$, multiply the appropriate m_x values by the number of individuals in each age class at time $t = 1$. In this example, to find the number of newborn individuals we multiply the number of one-year-olds by the appropriate m_x value. In this case, $m_1 = 2.0$; that is, each of the 100 females that has survived has 2.0 female offspring, on average. Therefore there are 200 newborn females at time $t = 1$. The total population is 300 and our c_x values are based on that number (Table 4.6). To find the number of one-year-old females at time $t = 2$, we repeat the procedure outlined above, giving us 100 again. The number of two-year-old females at time $t = 2$ is found by multiplying p_1 (0.40) by 100, giving us 40 two-year-old individuals. The number of newborn individuals is found by multiplying 100 one-year-old females by 2.0 and 40 two year old females by 1.0. That is, multiply by the appropriate m_x values. The number of newborn individuals is therefore 240 and the total population is 380. Confirm the values for $t = 3$. The last column in Table 4.6 is the calculated **stable age distribution**, based on Equation 4.14. Notice that after only three time periods the population proportions (c_x) have moved to within a few tenths of the stable age distribution, even though we started at time $t = 0$ with only one age class (200, $x = 0$).

In the above exercise, the finite rate of increase can be calculated for each of the time periods t to $t + 1$ as N_{t+1}/N_t. The results are as follows:

λ from $t = 0$ to $t = 1$: 300/200 = 1.50
λ from $t = 1$ to $t = 2$: 380/300 = 1.27
λ from $t = 2$ to $t = 3$: 480/380 = 1.26
λ from $t = 3$ to $t = 4$: 586/480 = 1.22

Notice how quickly λ moves to 1.22. The calculated r from the Euler equation for this life table is 0.206 and therefore the predicted value of λ at the stable age (e^r) distribution is 1.23. Exercises like this show us three things: (i) in the short term the growth of a population is heavily influenced by its age distribution; (ii) nevertheless, a population with a short generation time can move rapidly to its stable age distribution; and (iii) the finite rate of increase is influenced by the actual age distribution, but it settles in at the predicted value from the life history table (= e^r) as the population reaches the stable age distribution.

In Example 4.1, find the necessary information to project the population. As you do that, test yourself to see if you understand how to calculate generation time, G, the intrinsic rate of increase, and the stable age distribution (SAD) from the life history table.

Example 4.1

Find GRR, R_0, and G. Estimate r and then find its true value with the Euler equation. Verify the predicted stable age distribution (SAD). Project this population as described above. After reading the next section, use the Leslie matrix to project the population.

Age	l_x	m_x	P_x	q_x	$l_x \times m_x$	$x \times l_x \times m_x$	Euler based on $r = 0.152$	Euler based on $r = 0.154$	$l_x \times e^{-rx}$	c_x of SAD
0	1.00	0	0.250	0.750	0.00	0.00	0	0	1.000	0.733
1	0.25	0	0.400	0.600	0.00	0.00	0	0	0.214	0.157
2	0.10	7.0	0.800	0.200	0.70	1.40	0.517	0.514	0.073	0.054
3	0.08	7.5	0.500	0.500	0.60	1.80	0.380	0.378	0.050	0.037
4	0.04	5.0	0.250	0.750	0.20	0.80	0.109	0.108	0.022	0.016
5	0.01	0	0.000	1.000	0.00	0	0	0	0.005	0.003
6	0	0	–	–	0.00	0	0	0	0.000	0.000
Σ		GRR = 19.5			$R_0 = 1.50$	4.00	1.006	1.000	1.364	1.000

$$G = \frac{4.00}{1.50} = 2.67$$

Estimated value of $r = 0.152$

Predicted value of $\lambda = e^r = 1.17$

Population projection

Age	n_x at $t = 0$	c_x at $t = 0$	n_x at $t = 1$	c_x at $t = 1$	n_x at $t = 2$	c_x at $t = 2$	n_x at $t = 3$	c_x at $t = 3$	n_x at $t = 4$	c_x at $t = 4$
0	250	0.714	318	0.743	359	0.729	420.6	0.733	492.10	0.733
1	60	0.171	62.5	0.146	79.5	0.162	89.75	0.156	105.15	0.157
2	20	0.057	24	0.056	25	0.051	31.8	0.055	35.90	0.054
3	12	0.034	16	0.037	19.2	0.039	20	0.035	25.44	0.038
4	6	0.017	6	0.014	8	0.016	9.6	0.017	10.00	0.015
5	2	0.006	1.5	0.004	1.5	0.003	2	0.003	2.40	0.003
6	0	0.000	0	0.000	0	0.000	0	0.000	0	0
	N = 350	1.000	N = 428	1.000	N = 492.2	1.000	N = 573.75	1.000	N = 670.99	1.000

$$\lambda = \frac{428}{350} = 1.22$$

$$\lambda = \frac{492}{428} = 1.15$$

$$\lambda = \frac{574}{492} = 1.17$$

$$\lambda = \frac{671}{574} = 1.17$$

4.9 The Leslie or population projection matrix

This process of projecting the population one age class and one year at a time, as done above, is time-consuming and tedious. Leslie (1945) showed that populations could easily be projected through the use of matrix algebra. If you are not familiar with matrix

Table 4.7 General matrix format for projecting a population with five age classes for one time period ($t = 0$ to $t = 1$).

Age classes	Matrix					$t = 0$	$t = 1$
0	p_0m_1	p_1m_2	p_2m_3	p_3m_4	0	n_0	n_0
1	p_0	0	0	0	0	n_1	n_1
2	0	p_1	0	0	0	n_2	n_2
3	0	0	p_2	0	0	n_3	n_3
4	0	0	0	p_3	0	n_4	n_4

(with $\times$ before the $t=0$ column and $=$ between the $t=0$ and $t=1$ columns)

For age class 0 the number of individuals is based on: $p_0m_1n_0 + p_1m_2n_1 + p_2m_3n_2 + p_3m_4n_3 + 0$.
For age class 1 the number of individuals is calculated as: $p_0n_0 + 0 + 0 + 0 + 0$.
For age class 2 the number of individuals is calculated as: $0 + p_1n_1 + 0 + 0 + 0$.
For age class 3 the number of individuals is calculated as: $0 + 0 + p_2n_2 + 0 + 0$.
For age class 4 the number of individuals is calculated as: $0 + 0 + 0 + p_3n_3 + 0$.

algebra see the primer in Appendix 2. The matrix approach allows quick calculations of changes in the age structure and total population size as well as a quick method for finding λ when there is a stable age distribution. The survivorship and fertility columns are placed in matrix form $|A|$. The population itself is considered a column vector which, when multiplied by the matrix, produces a new column vector representing the population at time $t + 1$:

$$N_{t+1} = |A|N_t \tag{4.15}$$

The format for the matrix is as shown in Table 4.7. The p_x values (the probabilities of surviving from age x to age $x + 1$) appear in the matrix in the off diagonal. The first row consists of the products $p_x \times m_{x+1}$. The matrix must be a square matrix, with the final column consisting of zeros. In this example, with five age classes, we have a 5×5 matrix. Given the rules of matrix multiplication, the product of the matrix times the column vector (representing the population by age classes at $t = 0$) results in a column vector at time $t = 1$.

If we use the life table from Table 4.5, the resultant matrix is as shown in Table 4.8. If we then multiply this matrix by the column vector for time $t = 2$ from Table 4.6, we can calculate the column vector for time $t = 3$. The result is identical to the projection we did above in Table 4.6.

4.10 A second version of the Leslie matrix

To this point, in constructing our life tables we have assumed that the year begins with the reproductive season. For example, if we are developing a life table for white-crowned sparrows we might assess the number of eggs that have hatched and assign the hatchlings to age class zero. However, in some studies this approach is not practical and the study begins with counts of animals that have completed at least one year of life. The count is done before the production of newborn individuals, **but there is no count of the number of newborn individuals.** Assume it is possible, however, to estimate both fertility and

Table 4.8 Matrix projection based on life-history data found in Table 4.5. Column vectors are based on time periods $t = 2$ and $t = 3$ in Table 4.6

Age classes	Matrix					Column vectors	
						$t = 2$	$t = 3$
0	$0.5 \times 2 = 1.0$	$0.4 \times 1 = 0.4$	$0.5 \times 1 = 0.5$	0		240	300
1	0.5	0	0	0		100	120
2	0	0.4	0	0	$\times$	40 $=$	40
3	0	0	0.5	0		0	20
Total	0	0	0	0		380	480

Age, x	Calculations for column vector, time $t = 3$	Resultant column vector for time $t = 3$
0	$(1.0*240) + (0.40*100) + (0.50*40)$	300
1	$0.50*240$	120
2	$0.40*100$	40
3	$0.50*40$	20
Total		480

survivorship for age classes from year 1 onward. For example, see Table 4.9, which is based on Table 4.5. The life table simply begins with age class 1. Note that in this version of the matrix all of the fertility values in the first row are simply multiplied by p_0, since in order to find the number of one-year-old individuals we are multiplying the total number of newborn by the probability of surviving to age class 1.

We end up with the same result by either method, but we do not have exact data on age class zero in this second method. Interestingly enough, since age-class-zero individuals do not reproduce, the identical finite rate of increase (λ) can be derived from either method. And since the second method requires a smaller matrix, calculations can be vastly simplified, as shown in the next section.

Table 4.9 A population projection matrix, based on Table 4.5, in which the year begins prior to reproduction and no count is made of the zero-year age class. The population is projected from time $t = 2$ to 3 as in Section 4.9.

				$t = 2$	$t = 3$					$t = 2$	$t = 3$
$m_1 p_0$	$m_2 p_0$	$m_3 p_0$	$m_4 p_0$	n_1	n_1	1.0	0.5	0.5	0	100	120
p_1	0	0	0	n_2	n_2	0.4	0	0	0	40	40
0	p_2	0	0	n_3 $\times$	$= n_3$ $\Rightarrow$	0	0.5	0	0	0 $=$	20
0	0	p_3	0	n_4	n_4	0	0	0	0	0	0

Table 4.10 Simplified life table for the gray squirrel.

Age or stage class	m_x	p_x
Y = young adult	1.28	0.25
A = mature adult	2.28	0.80

4.11 The Lefkovitch modification of the Leslie matrix

Lefkovitch (1965) noticed that for many organisms the yearly fertility and survivorship functions remain relatively constant once adulthood is reached. Instead of specific age classes, he proposed using "stage classes" based on life stages such as juvenile, young adult, adult, etc. Recall our discussion in the first part of this chapter of the necessity to use such a method for plant populations. Now review Table 4.3 in section 4.3 above for the North Carolina gray squirrel population. Notice that by using method two (Section 4.10 above) the gray squirrel life table can be simplified to two stages as shown in Table 4.10. To do this we have eliminated some of the variability in year-to-year adult survivorship, using 0.80 for p_A (adult survivorship). Further, we have estimated survivorship in the first year of life as $p_Y = 0.25$.

Lefkovitch showed that in spite of this lumping the growth rate λ is conserved. In setting up the projection matrix based on this lumping, we must realize that adults from many of the age classes that we have placed together are simply recycled back into the same stage class from which they came for many years. That is, a three-year-old that survives is placed back in the "mature adult" stage class. Therefore, when placing p_A (mature adult survivorship) into the matrix, it ends up in the bottom right-hand corner of this 2 × 2 matrix.

Using Y = "young adult" and A = "Mature adult" stages, the gray squirrel matrix can be simplified as follows:

$$\begin{vmatrix} m_Y p_0 & m_A p_0 \\ p_A & p_A \end{vmatrix} = \begin{vmatrix} 1.28 \times 0.25 & 2.28 \times 0.25 \\ 0.80 & 0.80 \end{vmatrix} = \begin{vmatrix} 0.325 & 0.57 \\ 0.80 & 0.80 \end{vmatrix}$$

Let us now project this population and determine its growth rate, λ. We will start with 50 "young adults" and 90 "mature adults" (Table 4.11). The yearly calculated value of $\lambda = \dfrac{N_{t+1}}{N_t} = 1.28$ (180/140, 230/180, 294/230, etc.). By contrast, if you were to use all eight

Table 4.11 Projection of the gray squirrel population using the 2 × 2 simplified matrix.

Stage class	Matrix		t = 0		t = 1		t = 2		t = 3
Y	0.325	0.570	50	=	68	=	86	=	110
A	0.80	0.80	90		112		144		184
			$\Sigma = 140$		$\Sigma = 180$		$\Sigma = 230$		$\Sigma = 294$

age classes, including age class zero, the λ-value turns out to be 1.26 (you can prove this to yourself using Table 4.3 and projecting the population). Given that we assumed one p_x for all adult age classes, and given the uncertainties of actual survivorship and fertility data we might gather in the field, these two estimates of the finite rate of increase are adequately close. As Lefkovitch (1965) emphasized, the reduced matrix has the same λ as its larger counterpart using all age classes.

4.12 Dominant latent roots and the characteristic equation

Following the rules of matrix multiplication, if the matrix does not change (meaning that l_x and m_x remain constant), we can write:

$$N_1 = |A|N_0$$

$$N_2 = |A|N_1 = |A||A|N_0$$

$$N_3 = |A|N_2 = |A||A||A|N_0$$

This generalizes to:

$$N_t = |A|^t N_0 \tag{4.16}$$

Therefore a population can be projected to any time in the future. We can also project a population backwards in time. The advantage of this is that we can examine properties a population might have had in the past, assuming the present life table. The backwards projection requires the "identity matrix." The identity matrix is equivalent to the number one in algebra and is such that one can write the following:

$$I|A| = |A|I = |A| \tag{4.17}$$

Each number has an inverse in algebra (except zero) such that $(x) \times (1/x) = 1$. The inverse of a matrix is such that $|A||A|^{-1} = I$ (the identify matrix). If the inverse of $|A|$ is $|B|$ then $|A||B| = I$.

For example, if $|A| = \begin{vmatrix} 2 & 5 \\ 3 & 8 \end{vmatrix}$, then $|B| = \begin{vmatrix} 8 & -5 \\ -3 & 2 \end{vmatrix}$. The result of multiplying $|A||B|$ is the two-by-two identity matrix: $\begin{vmatrix} 1 & 0 \\ 0 & 1 \end{vmatrix}$ (See Appendix 2).

The inverse of a matrix only exists when the matrix is square and when the matrix has a "determinant." A determinant for a square matrix is a particular scalar number (see Appendix 2) that is easy to calculate for a 2×2 matrix, but becomes increasingly complicated for larger matrices. The projection matrix has an inverse if $[A][B] = I$. We can then use [B] to project the matrix backwards from time $= 0$ to $t = -1, -2, -3$, etc:

$$N_t = |B|^t N_0 \tag{4.18}$$

This allows us to compare the actual population of 1955, for example, with the potential population in that year, based on present-day survivorship and fertility values.

As discussed earlier, when a population has a stable age distribution, each age group grows as the same rate as the population as a whole; that is, at the finite rate of increase, λ. Thus: $n_{x(t+1)} = n_{x(t)}\lambda$ for all age classes (x). And $N_{t+1} = N_t\lambda$. In matrix form, we have:

$$N_{t+1} = \lambda N_t = \lambda(I\ N_t) = \lambda I(N_t)$$

Since $N_{t+1} = |A|Nt = \lambda I\ Nt$, we can write:

$$|A| - \lambda I = 0 \tag{4.19a}$$

Zero is a column vector consisting of all zeros.

Equation 4.19a is known as the **characteristic equation** for a matrix. It only exists for square matrices, and λ is known as the "latent root," "the characteristic root" or the **Eigenvalue** of the characteristic equation. When the matrix $|A|$ is of the order n (that is, a four-by-four matrix is of the order 4), the characteristic equation is a polynomial of degree n, and has n solutions. But for the Leslie matrix, there is only one positive root (or **dominant latent root**). This dominant latent root $= \lambda$ and Leslie has shown that $\lambda = e^r$. Thus matrix algebra can be used to solve for r from basic life history data.

As explained in detail by Case (2000), one way of solving for λ is to find the determinant of the expression $|A| - \lambda I$ and setting it equal to zero. In other words, instead of projecting the matrix to find λ as we did above, we can find λ if we solve the expression:

$$\det(|A| - \lambda I) = 0 \tag{4.19b}$$

For more information on determinants and basic matrix operations, see Appendix 2. For a 2×2 matrix, we start by multiplying λ by the identity matrix, giving us:

$$\begin{vmatrix} 1 & 0 \\ 0 & 1 \end{vmatrix} \times \lambda = \begin{vmatrix} \lambda & 0 \\ 0 & \lambda \end{vmatrix}$$

Now we subtract this from our 2×2 matrix (see rules of subtraction, Appendix 2), yielding:

$$\begin{vmatrix} a_{11} & a_{12} \\ a_{21} & a_{22} \end{vmatrix} - \begin{vmatrix} \lambda & 0 \\ 0 & \lambda \end{vmatrix} = \begin{vmatrix} a_{11} - \lambda & a_{12} \\ a_{21} & a_{22} - \lambda \end{vmatrix}$$

The determinant of a simple 2×2 matrix is found by taking the difference between the cross products, resulting in the following:

$$(a_{11} - \lambda)(a_{22} - \lambda) - (a_{21})(a_{12}) = \lambda^2 - \lambda(a_{11} + a_{22}) + (a_{11}a_{22}) - (a_{21}a_{12}) = 0 \tag{4.20}$$

Solving for λ looks difficult, and finding the determinant for more complex matrices is extremely tedious without computer software. As discussed above, thanks to Lefkovitch (1965), many matrices can be simplified to this 2×2 form.

Now let's return to the 2×2 matrix for the gray squirrel population we found in the previous section:

$$\begin{vmatrix} 0.325 & 0.57 \\ 0.80 & 0.80 \end{vmatrix}$$

Solving for λ using the method outlined above (Eqn. 4.20) we have:

$$(0.325 - \lambda)(0.80 - \lambda) - (0.80 \times 0.57) = 0.$$

This gives us:

$$\lambda^2 - 1.125\lambda + 0.26 - 0.456 = \lambda^2 - 1.125\lambda - 0.196 = 0$$

We can solve this equation using the formula for the solution of a quadratic equation ($ax^2 + bx + c = 0$, recall your high school algebra!):

$$\lambda = \frac{-b \pm \sqrt{b^2 - 4ac}}{2a}$$

$$\lambda = \left(1.125 \pm \sqrt{1.266 + 0.784}\right)\Big/2 = \frac{1.125 \pm 1.432}{2} = 1.28$$

Although there is another solution, the positive or dominant root for λ is 1.28, which is the same value we found above by projecting population growth and finding λ after evaluating N_{t+1}/N_t.

4.13 Reproductive value

As biologists and conservation biologists evaluate life histories, they are often interested in which age classes contribute most heavily to present and/or future population growth. For this analysis they calculate a parameter known as the reproductive value. One application of the reproductive value focuses on whether natural selection can regulate events late in the life span of an organism when reproductive value is very low. Behavioral biologists have suggested that in dominance hierarchies, individuals with the greatest reproductive potential or value will be supported by their mothers or others in the population (Alberts and Altmann 2003). Conservation biologists attempting to evaluate what intervention strategy will give them the biggest bang for their buck (in terms of long-term survival of the population) may use reproductive value in determining which age classes are likely to produce the desired increase in growth rate.

To evaluate reproductive potential, simply examining the fertility column is usually misleading. For example, if one-year-old females have, on average, two female offspring, but two-year-old females have four female offspring, one might conclude that the two-year-old individuals contribute more to population growth. However, suppose that the survivorship value for $l_1 = 0.1$ and for $l_2 = 0.01$. The products of $l_x m_x$ tell us that one-year-old females produce 0.2 females and two-year-olds only 0.04 females per generation. Evidently an evaluation of reproductive potential of a given age class must take into account both survivorship and fertility. Furthermore, the value of a given female depends not on

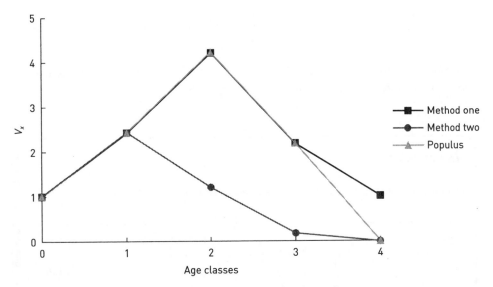

Figure 4.12 Three methods of computing reproductive value.

just her present reproduction, but the potential for future reproduction. For example, in many species females become more fecund and more adept at successfully raising offspring as they mature, leading to an increase in m_x with age, followed by a decrease as they senesce. On the other hand, in mammals that suckle their young for more than a year, reproductive rates may be suppressed in the year following a birth, leading to oscillations in birth rates by age class across the population. For example, in Amboseli baboons $m_{10} = 0.816$, $m_{11} = 0.649$, $m_{12} = 0.818$, and $m_{13} = 0.554$ (Alberts and Altmann 2003).

Reproductive value is a number that measures the relative reproductive potential of an individual of a given age. It can also be thought of as the weighted average of present and future reproduction by a female or male of age x. It is the relative value of a daughter born i time-units in the future when the population size will be N_{t+i}. Reproductive value is scaled so that the value for the first age class equals 1.0 (Fig. 4.12). There are several different ways of computing reproductive value (Lanciani 1998). In Example 4.2 I present the traditional equation for reproductive value as method one (Lanciani 1998). A second method produces reproductive values that are closer to the original concept of R.A. Fisher (1930) and to that derived from the matrix-algebra approach (Caswell 1989). Conceptually, the second value is defined as "the present value of future offspring" of a female of age x, and ignores reproduction by the present age class (method two). In other words, values from age $x + 1$ to the end of life are used. Reproductive value can also be found in the Populus computer simulations developed by University of Minnesota ecologist Don Alstad (2001). In method one, as in Populus, values from age x to the end of life are used.

The differences between the two approaches can be gleaned by working through Example 4.2. In the formula for reproductive value, the survivorship functions l_x, m_x, and r are as usual. In Equation 4.21, the numerator is simply the Euler equation. In method two the summation is from the age $x + 1$ to the end of life. The summation in the numerator is from the age class in question (x) to the end of the life span (z).

$$V_x = \frac{\sum_x^z e^{-rx} l_x m_x}{e^{-rx} l_x}$$

(4.21)

Sample calculations are found in Example 4.2. Note that you should be able to confirm that $r = 0.371$ and $\lambda = 1.449$, assuming a stable age distribution.

Example 4.2

To calculate reproductive value, we need the following information:

Age, x	l_x	m_x	$e^{-rx} \times l_x \times m_x$	$e^{-rx} \times l_x$
0	1.00	0	0	1.000
1	0.60	0	0	0.414
2	0.50	3.0	0.714	0.238
3	0.40	2.0	0.263	0.132
4	0.10	1.0	0.023	0.023
5	0	0	0	0
Sum			1.000	

Calculation of reproductive value, V_x:

Age, x	Method one	Method two
0	$(0 + 0 + 0.714 + 0.263 + 0.023)/1.000$ = **1.000**	$(0 + 0.714 + 0.263 + 0.023)/1.000$ = **1.000**
1	$(0 + 0.714 + 0.263 + 0.023)/0.414$ = **2.42**	$(0.714 + 0.263 + 0.023)/0.414$ = **2.42**
2	$(0.714 + 0.263 + 0.023)/0.238 =$ **4.20**	$(0.263 + 0.023)/0.238 =$ **1.20**
3	$(0.263 + 0.023)/0.132 =$ **2.17**	$0.023/0.132 =$ **0.17**
4	$0.023/0.023 =$ **1.000**	$0/0.023 =$ **0**

All methods produce the typical triangular shape when reproductive value, V_x, is graphed against age (Fig. 4.12). See Fig. 4.8 of Alberts and Altmann (2003) for an example of triangular reproductive values of male and female Amboseli baboons. If you run Populus and bring up reproductive value for this life table it produces a hybrid. The values for method 1 are identical for ages 0–3. However, instead of assigning a value of 1.00 to the last age class, it drops it to zero, as in method 2.

4.14 Conclusions: sensitivity analysis

We have learned that age- or stage-structured growth is common in most plant and animal populations, but the details of both survivorship and fertility differ greatly across species. For example, natural populations display many variations in survivorship, although a type III survivorship schedule is most common. Most of the "charismatic megafauna" of interest to conservation biologists and the general public shows age structured growth. Thus the techniques outlined in this chapter are necessary before we can make predictions about the potential for growth and recovery of an endangered population. We must understand the effects that age-specific survivorship and fertility have on the behavior of a population before we are in a position to implement a management plan that would actually be effective in promoting its long-term survival.

For example, examine Table 4.12. These data were gathered by Schmutz *et al.* (1997) on a population of the emperor goose (*Chen canagica*) (analysis from Morris and Doak 2002). Given this information, is this population growing or declining? What aspects of its life history are most important to its long-term growth rate? To answer these questions, we will use the matrix format from Section 4.10. It turns out that the long-term λ for this population is 0.989.

We now ask, "How sensitive is population growth (or extinction risk) to particular demographic changes?" Specifically, will a particular change in survivorship or fertility have a large impact on the growth rate, λ? Using the above matrix, we have substituted survivorship values from 0.136 to 1.00 for both hatchlings (S_0) and older birds ($S_{\geq1}$), and fertility rates of 0.136–2.000 for two-year-old (F_2) and three-year-old and older birds ($F_{\geq3}$). Figure 4.13 summarizes this analysis. Obviously survival, especially that of the older birds, has the greatest effect on the growth rate (λ). Increases in fertility have virtually no effect. A related point is that, since adult survival is so critical to population growth, errors in our estimates would have a very large impact on our conclusions about this population. As it stands now, from this matrix our estimate of λ is 0.989 < 1.000, and we expect this population to slowly decrease. A small change in our estimation of survivorship, however, would lead us to believe that this population is stable or growing. For example, a

Table 4.12 Survivorship and fertility of an emperor goose population (Schmutz *et al.* 1997).

Survival of hatchlings = S_0	Survival of one-year-old and birds = $S_{\geq1}$	Fertility of two-year-old birds = F_2	Fertility of three-year-old and older birds = $F_{\geq3}$
0.136	0.893	0.639	0.894

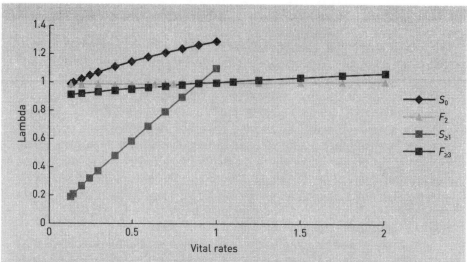

Figure 4.13 The effect of changing vital rates on the value of lambda in an emperor goose (*Chen canagica*) population. An increase in survivorship of adults (one-year-old and older birds) has the greatest effect. Increases in fertility have a negligible effect on λ.

change in adult survival from 0.893 to 0.905 changes lambda from 0.989 to 1.001. A change in juvenile survivorship from 0.136 to 0.155 changes the expected lambda to 1.000.

An examination of this sort is known as a perturbation or "sensitivity analysis." Although Fig. 4.13 is convincing, conservation biologists have sought to summarize the kind of analysis we have done above into a single number that would summarize the sensitivity of lambda to particular vital rates. The most common basic measure of sensitivity is the slope of the tangent taken on the curve of lambda as a function of each vital rate. Problems with this approach include the possibility of nonlinearity in the relationship between lambda and a particular vital (survivorship or fertility) rate. A second issue is the scaling of sensitivity values. Obviously survivorship scales on a strict 0.0–1.0 scale, while reproduction can scale to very large numbers (number of acorns produced by an oak tree). These comparisons can be made more meaningful by examining the proportional change in lambda as a proportion of change in the vital rates. These calculations result in a measure known as **elasticity**. Elasticity, then, is a standardized sensitivity that measures the effects of proportional changes in vital rates. That is, elasticities tell us the effect of perturbations in vital rates that are all of the same relative magnitude. Elasticities are standardized to sum to 100%.

For the emperor goose population discussed above the vast majority (92% of elasticity) of sensitivity was in the survival of the one-year-old and older birds (Morris and Doak 2002). An analysis of the Amboseli baboon population by Alberts and Altmann (2003) led to similar conclusions. That is, fertility represented just 9% of the total elasticity for both males and females. Survival of the pre-reproductive age classes accounted for 37% of the total elasticity for females and 62% for males. Details on the calculation of both sensitivity and elasticity values can be found in Morris and Doak (2002) or Alberts and Altmann (2003).

What should be done to promote the long-term survival of these two populations, or of other populations described in the first paragraph of this chapter? What evolutionary forces have led to a particular life history in the first place? These are just two of many questions for which there are no easy answers. Still, the analyses outlined in this chapter should have given you the tools necessary to at least begin to address these issues.

5

Metapopulation ecology

- Metapopulations and spatial ecology
- MacArthur and Wilson and the equilibrium theory
- The Levins or classical metapopulation
- Extinction in metapopulations
- Metapopulation dynamics of two local populations
- Source–sink metapopulations and the rescue effect
- Non-equilibrium and patchy metapopulations
- Spatially realistic models
- Assumptions and evidence for the existence of metapopulations in nature

5.1 Introduction

On August 27, 1883 Krakatau, an island about the size of Manhattan located between Sumatra and Java, underwent a series of volcanic eruptions releasing as much energy as 100 megatons of TNT (Wilson 1992). Magma, ash, and rock flew 5 km into the air and fell back into the sea, creating a tsunami 40 m in height, washing away villages in Java and Sumatra, killing 40,000 people. Waves were still a meter high when they came ashore in Sri Lanka. A total of over 18 cubic kilometers of rock and ash was thrown into the air with dust and sulfuric acid aerosol reaching 50 km into the stratosphere, where their effects were seen as brilliant sunsets for several years thereafter. All of this airborne material produced "darkness at noon" in areas near the former Krakatau.

Only the southern end of Krakatau remained. This island, which became known as Rakata, was covered by pumice 40 m thick. The pumice had been heated to between 300 and 850 °C, and all living things had been destroyed; Rakata was a sterile island. Yet living things soon began colonizing this lifeless rock. Nine months after the explosion a visitor found a small spider spinning its web. In the fall of 1884, a year after the eruption, biologists found a few shoots of grass. By 1886 there were 15 species of grasses and shrubs; by 1897 there were 49 species; and in 1928 300 species of plants were found. In 1919 there were patches of forest; by 1929 most of the island was forested, forcing the grasses into small pockets (Wilson 1992).

What Wilson described, as outlined in the preceding paragraphs, sounds like a typical successional sequence, proceeding from a community of colonizing species to a mature or "climax" forest. What we want to emphasize, as we consider the topic of metapopulations, is the two processes at work on Rakata. Obviously one of those processes is **colonization**. New species continually arrive on the island from the nearby mainlands. The other process is **local extinction**. Many species that arrived on this island, and were recorded early in the twentieth century, are no longer present. Again, this may not be surprising to students of succession. So-called climax species are supposed to outcompete and eliminate earlier successional species. But is that what happened? At least one animal species that we would associate with the more mature community, the reticulated python (*Python reticulatus*), was present as early as 1933, but was gone by the 1980s. The bird community is perhaps more to the point. In 1908, 13 species of birds were recorded on Krakatau; by 1920 there were 31 species; and in 1933, 30 species were found. Wilson (1992) believed that an "equilibrium" number for Krakatau was approximately 30 species of birds, and that number had been reached by about 36 years. More important, however, is that the actual composition of the bird community has not remained stable. During the interval between the 1920 and 1933 surveys, five species of birds went extinct on Krakatau, to be replaced by four new species (MacArthur and Wilson 1967). For example, the sooty-headed bulbul (*Pycnonotus aurigaster*) and the long-tailed shrike (*Lanius schach*) had become extinct on Rakata between 1920 and 1933.

The history of Krakatau illustrates two major points: (i) local populations are continuously subject to the twin processes of colonization and extinction; and (ii) communities are continuously changing. Even when the number of species in the community is static, the composition of the community is not.

5.2 Metapopulations and spatial ecology

Many of the population models we have examined, particularly the deterministic models, have underlying assumptions that natural populations are numerous, widespread, and occupy contiguous habitats. The reality is that these assumptions may never have been true for certain populations, and that the natural world is now increasingly fragmented due to human activities. Wildlife populations are now more likely to be small, restricted in distribution, and increasingly isolated from each other. Partially as a reaction to these realities and partially because of increasingly sophisticated theoretical developments, ecologists have begun stressing the importance of the spatial context in populations, communities, and ecosystems. What we can broadly call **spatial ecology** is the progressive introduction of spatial variation and complexity into ecological analysis, including changes in spatial patterns over time. Krakatau is an example of both temporal and spatial complexity. Not only have its communities and populations changed over time, but the community found on present-day Krakatau is still very different from that of the nearby mainland forests of Java and Sumatra. The local community of Krakatau remains distinct from those of the mainland (Wilson 1992).

Spatial ecology is distinguished by two different approaches: (i) **landscape ecology** and (ii) **metapopulation ecology**.

Landscape ecology usually focuses on a larger geographic scale than traditional ecological studies; it was founded largely by community- or ecosystem-oriented ecologists, geographers, and landscape planners. Landscape ecology explicitly recognizes the heterogeneity or "patchiness" of the environment both spatially and over time. It provides a large-scale perspective that describes the physical structures of patchy environments as well as the movements of both individuals and resources among them. Generally, landscape ecologists focus at the community, as opposed to the population, level. Furthermore, landscapes have a more complex structure than usually allowed in simple metapopulation models, with habitat suitability being on a continuous scale, rather than simply "suitable" or "unsuitable" (Hanski 1999). Landscape ecologists do not usually work on population dynamics (Turner *et al.* 2001), but much of their work is relevant to metapopulations. For example, both landscape and metapopulation models often attempt to incorporate the roles of edge habitats, movements of individuals between patches via habitat corridors, spatial location of the patches, habitat fragmentation, landscape disturbance, and spatial and temporal variation in the quality of the habitat.

The metapopulation approach begins by stressing that local populations are influenced by immigration/emigration and extinction, as well as by birth and death processes. Until the 1960s, the idea that populations might routinely go locally extinct was rarely discussed in the literature. However, the population geneticist Sewall Wright (1940), as well as ecologists such as Andrewartha and Birch (1954), introduced the notions that populations are connected by migration and that local extinctions might be commonplace (Hanski 1999). The importance of immigration and emigration to the long-term persistence of a local population, however, was first emphasized by Levins (1970), who coined the term **metapopulation**. For Levins a metapopulation was a population consisting of many local populations. He asserted that all local populations have a finite probability of extinction, and long-term survival of a species was at the regional or metapopulation level (Hanski 1999). Beginning in the 1990s, as it became obvious that the natural world was becoming increasingly fragmented, the metapopulation approach became standard in the world of conservation biology. An understanding of metapopulations, the probabilities of local extinctions in different-sized natural reserves, and the rates of immigration and emigration between these preserves, became one of the fastest-growing research areas in population, community, landscape, and conservation biology. As currently defined, **metapopulations** are **regional assemblages** of plant and animal species, with the long-term survival of the species depending on a shifting balance between **local extinctions** and **re-colonizations** in the patchwork of fragmented landscapes.

Whereas in landscape ecology we begin with the assumption of complex environmental heterogeneity, in a simple metapopulation analysis, the landscape, from the perspective of a given species, is assumed to consist of discrete patches of suitable habitat, surrounded by unsuitable areas. All of these patches are of the same quality and size, and while these patches are isolated from each other there is no special recognition given to how far apart they are or any measure of ease of movement from one patch to another (connectivity). Nor are local population dynamics emphasized. The concern is simply whether a patch is occupied or not, the extinction rate on patches, and the overall colonization rate among patches. From this simple approach, more complex spatially realistic models have been developed to recognize differences in habitat quality and size, local population dynamics, and differences in connectivity among local patches.

Spatial ecology includes a variety of approaches, including so-called lattice models. In patch or lattice models, the habitats occupied by local populations are represented in continuous space or made up of a series of spatially subdivided cells or "lattice" segments.

These models include explicit spatial locations for the habitats. In patch models the habitats have just two states, occupied or empty, and no effort is made to estimate the size of the population in the occupied patches. Cells change states according to simple rules, and stochastic extinctions and colonizations take place. Many of these simplifying rules can be relaxed in a spatially explicit variation of this approach known as the incidence function model, examined later in this chapter (Hanski 1999).

The spatial distribution of a species is based on environmental patchiness. However, the recognized patterns of spatial distribution (**clumped**, **random** or **regular**) only describe the current spatial pattern; they do not address the underlying causation of the pattern, or the long-term persistence of the population at that site.

The term "metapopulation" has been used for any spatially structured population and "metapopulation dynamics" has been used to refer to any population dynamics involving spatial patterns (Hanski 1998). Furthermore, as Harrison (1994) has pointed out, several different types of metapopulations exist: classical metapopulations, mainland–island metapopulations, non-equilibrium metapopulations, and patchy metapopulations. We will explore these later in this chapter.

As pointed out by Tilman and Kareiva (1997), we must recognize that an individual organism only interacts with its local environment and with the competitors and/or predators present in that local environment. Such realities have been largely ignored by ecologists in the past, particularly when deriving theoretical models for competition or predation. Both classical and modern studies of competition and predator–prey interactions can be easily integrated into a simple metapopulation (spatial) context. For example, envision the environment as a series of patches and apply the competitive-exclusion principle (to be elaborated in a later chapter). If we have a superior competitor that drives other species extinct on a given patch, the inferior competitors are only driven locally extinct. They can remain in the region as long as they can have a higher dispersal rate than extinction rate and there are empty habitat patches. Similarly, a predator may drive a prey species locally extinct, but the prey population remains regionally present as long as it continues to colonize empty habitat patches faster than its predator. Although this is an over-simplification of models presented later, consider the following two examples.

Huffaker (1958) worked with orange mites and their predators in the laboratory. He found that coexistence of the two species was impossible on any given orange. Here one orange represents a small homogeneous habitat or patch. Through the addition of spatial complexity, however, as well as barriers to limit the rates of movements between patches, the orange mite and its predator coexisted in the laboratory for many months. A higher dispersal rate and environmental complexity allowed the prey species to remain regionally present, even though it was continually driven locally extinct on a given orange once the predator arrived. Notice the parallels to epidemiological and host–parasite interactions. An unoccupied, but suitable, habitat patch is the equivalent of an individual susceptible to a parasite, yet not infected. An infected individual is the equivalent of an occupied habitat patch.

Spatial complexity can also have important effects on competitive interactions. For example, spatial complexity can help explain the coexistence of more species than expected based on the theory that the number of coexisting species should not exceed the number of limiting resources (Hutchinson 1961). As pointed out by Lehman and Tilman (1997), usually there is a trade-off, often expressed in terms of energetic investments, between **competitive ability** and **colonization ability**. While a superior competitor may take over a given site, if the less competitive species is a better colonizer, it may simply escape to a new site before competitive exclusion can occur. For example, Hubbell *et al.* (1999) proposed that high tree diversity on Barro Colorado Island in Panama is due, at least in part, to the low dispersal ability in competitively dominant species. Through a combination of low local abundance, low dispersal, and chance events, many plant species are absent from the local neighborhood in which a tree is located. Many sites are colonized by "default" species that were not the best competitor for the site. For example, an individual tree sapling competes with only 6.3 neighbors on average. Thus, plants compete only with those individuals sufficiently nearby to shade them or whose roots overlap with theirs in the soil. In order to "win" locally, a tree must only compete with those species that have "shown up" in the local neighborhood. Inferior competitors "win" by default. Because winners are only the best competitors that happen to have colonized a specific site, this process can lead to an almost unlimited diversity (Tilman 1994, Hubbell *et al.* 1999).

While landscape ecology and metapopulation ecology have started at different scales and with different assumptions and traditions, these two branches both ask similar questions. Many landscape ecology courses include sections on metapopulations. An important future task will be the reconciliation of these two approaches and the establishment of common methodologies and principles.

5.3 MacArthur and Wilson and the equilibrium theory

Spatial ecology has its roots in the MacArthur and Wilson equilibrium (or dynamic) theory of island biogeography. MacArthur and Wilson (1963, 1967) brought a quantitative theoretical framework to the study of biogeography. Even before Darwin carried out his pioneering work on the Galapagos, islands and island examples have been of great importance in biology, and islands have been analyzed as natural laboratories and experimental systems. They are small, contained ecosystems in which certain species found in continental ecosystems may be missing. The lessons learned from examining islands can also be applied to those continental areas that are comparable to islands. That is, streams, lakes, tidal pools, caves, and mountaintops can be thought of as habitat islands in a "terrestrial sea." The approach of island biogeography has also been applied to host animals as habitat patches for parasites. Finally, as noted above, the natural world is increasingly fragmented, surrounded by roads, agricultural crops, shopping malls, industrial sites, and urban development. As conservation biologists became increasingly aware that wildlife preserves were essentially islands, a set of rules for the design of natural areas was inferred from the MacArthur and Wilson theory (Diamond 1975, Terborgh 1975, Wilson and Willis 1975, Willis 1984).

The basic principles derived from the MacArthur and Wilson theory are:

1 There is a relationship between habitat island area and the number of species found there (the species–area curve);
2 local extinction is a normal, common occurrence, particularly on small islands with small populations;
3 local diversity is based on an interplay between colonization from a "mainland" source of species and local extinction, resulting in an "equilibrium" number of species;

4 island size and distance from the source of species will affect the "equilibrium" number of species. That is, large islands that are close to the mainland will have more species than small islands far from the mainland.

The relationship between number of species on an island and the area of the island is one of the cornerstones of island biogeography theory. The species–area relationship has been discussed since the nineteenth century, and MacArthur and Wilson (1967) proposed that the number of species on an island could be approximated by the equation:

$$S = CA^z$$

where S = the number of species on the island, A = the area of the island, C = a constant (the y-intercept, see below), and z = a constant which remains fairly consistent within a taxonomic group and/or the types of islands being considered.

The above equation can be log-transformed as follows:

$$\text{Log } S = \text{Log } C + z \text{ Log } A \qquad\qquad (5.1)$$

This is an equation for a straight line with a slope $= z$, with log C as the y-intercept. Thus, if data are gathered on the area of islands of different sizes and on the number of species on each island, a regression of the log-transformed data will produce a linear equation with slope z. The slope is relatively consistent within a taxonomic group but also depends on the type of island system. That is, the z-value depends on whether we are dealing with true oceanic islands, recently isolated islands ("land-bridge" islands), or habitat islands. According to MacArthur and Wilson (1967), z-values range from 0.20 to 0.40 for oceanic islands, 0.1 to 0.25 for arbitrary portions of the mainland, and greater than 0.26 for habitat islands (Gould 1979, Quinn and Harrison 1988). Preston (1962) showed that a z-value 0.26 is expected when the log of species abundance versus the number of species has a normal distribution.

Gould (1979) pointed out that a slope of 0.25 is extremely common for species–area curves. What is of interest are those z-values differing significantly from 0.25. When we simply sample larger and larger areas of habitats not isolated from each other, the z-values are theorized to be smaller than the expected 0.25. When small areas are sampled they include a number of transient species passing through the area, raising the number of species. The result is a smaller-than-expected rise in the number of species with increasingly large sample areas. Thus, ants from non-isolated continental areas in New Guinea (Wilson 1961) have a z-value of 0.17, mammals from the Sierra Nevada in California have a z-value of 0.12 (Brown 1971b), and birds from the Great Basin of the USA a z-value of 0.17 (Brown 1978). By contrast, larger-than-expected z-values arise when islands contain great habitat diversity, with semi-isolated unique biota encountered as sample areas are increased. Examples include terrestrial invertebrates found in caves ($z = 0.72$, Culver et al. 1973), mites on cushion plants ($z = 0.42$–0.69, Tepedino and Stanton 1976), and mammals on isolated mountaintops ($z = 0.43$, Brown 1971b, and $z = 0.33$, Brown 1978). Lawrey (1991, 1992) has suggested that pollution, by reducing interspecific competition, produces larger-than-expected z-values for lichen species on rocks of differing sizes. Whereas z-values varied from 0.16 to 0.21 for six undisturbed sites, a site disturbed by air pollution near the Capital Beltway in Maryland yielded a species–area curve with a z-value of 0.28.

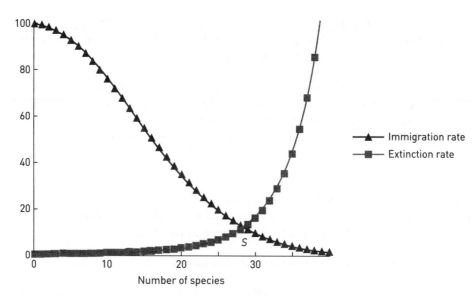

Figure 5.1 Immigration and extinction curves from the island biogeography model of MacArthur and Wilson (1967). *S* is the equilibrium number of species, where the two curves intersect.

Some scientists have asserted that as islands get larger the topography becomes more complex, there are more habitats, and therefore we have more species. In their study of red mangrove islands, however, Simberloff and Wilson (1969, 1970) found that species number increased with island size alone and was unrelated to habitat diversity.

The number of species found on an island, according to MacArthur and Wilson, was due to two contrasting processes of (i) **immigration** and (ii) **extinction**. Extinction was envisioned as a normal, locally common event, while new species were added through immigration form the mainland. Diversity was the result of the equilibrium between immigration and extinction. Furthermore, the theory indicated that once the "equilibrium" number of species was reached, the only constant was the **number of species** in the community, not the identity of the species involved (Fig. 5.1). Since extinction is a locally common process, there should be a regular "turnover" in the species found on the island.

The expected number of species on an island is affected not only by the area of the island, but also by the distance of the island from the source of species. Immigration rates are lower on smaller islands and on islands further from the "mainland" source of species. By contrast, immigration rates are higher on larger islands and on islands closer to the "mainland." Extinction rates are expected to be higher on small islands, since average population sizes are smaller (Wilson 1992).

The rate at which new immigrant species establish themselves on the island falls as the number of species on the island increases. As more species become established on the island, fewer individual immigrants will belong to a species not already present; moreover it will be harder for a new species to successfully colonize due to competition with the already established species. Species with high dispersal rates are those that arrive quickly, while those with lower dispersal rates arrive more slowly. Because of the proposed colonization–competition trade-off, the species with lower dispersal rates are likely more competitively dominant.

The extinction curve rises as more species arrive on the island. The more species that are present, the more that can become extinct. But again, as more species are present, competition increases and the average population size per species declines, leading to an increased probability of extinction. Finally, if succession proceeds to a "climax" stage, the community will be saturated with species. At equilibrium, the number of species will be constant on the island, though some new species will continue to arrive while others will go extinct.

The MacArthur and Wilson equilibrium theory captured the imagination of ecologists, conservation biologists, and biogeographers, making it the leading paradigm for the spatial dynamics of species during the 1980s. It shares much of the conceptual framework of metapopulation biology. Both view nature as subdivided into discrete fragments of suitable habitat; both view local populations as subject to stochastic processes and prone to extinction; and both stress the importance of movements of individuals between habitats (or islands). A key difference is that island biogeography stressed the community property of diversity rather than focusing on the dynamics of individual populations. Furthermore, island theory was developed to explain patterns at large spatial scales as opposed to fragmentation of landscapes at small scales (Hanski 2002).

The MacArthur and Wilson model is now categorized as a **mainland–island metapopulation**. There is a constant source of species, the mainland. The mainland population is seen as permanent, with no chance of extinction. Furthermore, dispersal is one-way. Species move from the mainland to the island; the reverse is not significant. Finally, no movement from one island to another is included in this type of metapopulation.

5.4 The Levins or classical metapopulation

According to the Levins model, metapopulation persistence is due to a stochastic balance between local extinction and re-colonization of empty habitat patches. The rate of change in occupied habitat patches is a function of colonization rates (c) and extinction rates (ε) as shown in Equation 5.2 (Levins 1969). P is the proportion of patches occupied by the population under consideration.

$$\frac{dP}{dt} = cP(1 - P) - \varepsilon P \tag{5.2}$$

As described by Hanski (2001), if we define P' as the number of habitat fragments occupied by the species (rather than the proportion), and define T as the total number of habitat patches available, the equation can be modified as follows:

$$\frac{dP'}{dt} = cP'(T - P') - \varepsilon P \tag{5.3}$$

Both of these models are deterministic descriptions of the rate of change of metapopulation size, even though the models are based on stochastic events. Assumptions include:

1 The local populations are identical and have the same behavior;
2 extinctions occur independently in different patches and therefore local dynamics are asynchronous;

3 colonization spreads across the entire patch network and all patches are equally likely to be "encountered;"
4 furthermore, all patches are equally connected to all other patches.

In the Levins model we are not concerned with population dynamics within each population. We do not attempt to assess the number of individuals in each patch; we simply record a patch as occupied or not occupied. For this reason we also do not assess the size or quality of the patches.

The equilibrium value of P can be obtained by setting $dP/dt = 0$. This produces the expected proportion of patches to be occupied and amounts to a carrying-capacity term such as is found in the logistic equation.

$$0 = cP(1 - P) - \varepsilon P = P(c - cP - \varepsilon)$$

Since $P = 0$ is not an interesting solution, we have:

$$0 = c - cP - \varepsilon, \quad \text{and} \quad \varepsilon = c - cP$$

The equilibrium value of P, defined as $\hat{P}$ and found by solving the above for P, is shown in Equation 5.4:

$$\hat{P} = \frac{c - \varepsilon}{c} = 1 - \frac{\varepsilon}{c} \tag{5.4}$$

The implication here is that **colonization must be greater than extinction or the equilibrium proportion of patches occupied will be zero**, and the colonization rate must be greater than the extinction rate for persistence of the metapopulation. If we consider colonization a "birth" event and extinction a "death" event (thereby using $c - \varepsilon$ as the equivalent of the growth rate, r, in the logistic equation) and we use $1 - \varepsilon/c$ to represent a "carrying capacity" term (equivalent to K in the logistic equation) as mentioned above, we can model metapopulation dynamics as a modification of the logistic (Equation 5.5). No matter what the starting patch frequency is (assuming $1 \geq P > 0$), over time it moves to the expected value based on $\hat{P} = 1 - \varepsilon/c$ (see Fig. 5.2).

$$\frac{dP}{dt} = (c - \varepsilon)P\left(1 - \frac{P}{1 - \varepsilon/c}\right) \tag{5.5}$$

This simple model has helped ecologists develop insights into the consequences of habitat destruction and fragmentation. For example, imagine a fragmented landscape in which a fraction of the habitat patches is destroyed. The extinction rate is not affected, but the colonization rate is. This is because there are fewer local populations and fewer empty patches. If the patch connectivity is reduced, it can be modeled by reducing the value of c. Habitat destruction can therefore lead to a reduction in the proportion of patches that are occupied. Alternatively, if no patches are destroyed but they are reduced in area, this would result in lower average population sizes, which would increase the extinction rate. Simultaneously, colonization rate would be reduced due to the smaller population sizes in the occupied patches. The net result again is a reduction in the fraction of occupied patches (Fig. 5.3).

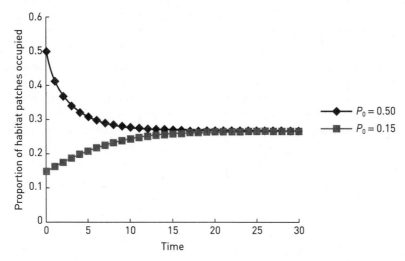

Figure 5.2 Expected proportion of habitat patches occupied, based on Equation 5.5. In this example $P_0 = 0.50$ and 0.15, the colonization rate $c = 0.75$, and the extinction rate $\varepsilon = 0.55$. The expected proportion of patches occupied at equilibrium $= 1 - \varepsilon/c = 0.27$.

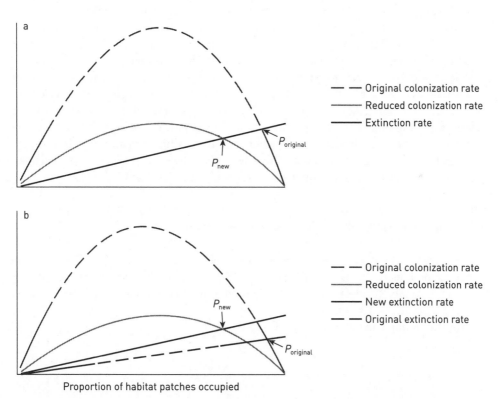

Proportion of habitat patches occupied

Figure 5.3 (a) The effect on patch occupancy of a lowered colonization rate due to a reduction in the number of habitat patches. (b) Expected changes in patch occupancy with lower colonization rate and increased extinction rate. Colonization rate is reduced by loss of habitat number; extinction rate is increased by reduction in patch area. Adapted from Hanski (1999).

Table 5.1 Potential causes of local and metapopulation extinctions (Hanski 1998).

	Local extinction	Metapopulation extinction
Stochastic processes	(a) Demographic	(a) Extinction–colonization interaction
	(b) Environmental	(b) Regional processes
Extrinsic causes	Habitat loss	Habitat loss and fragmentation

5.5 Extinction in metapopulations

In metapopulation dynamics, as well as in the MacArthur and Wilson theory, the extinction of a local population is not uncommon. But we are more interested here in the extinction of the metapopulation. Table 5.1 (Hanski 1998) summarizes the potential causes of both local and metapopulation extinctions.

We need only comment on the comparison of stochastic processes in local versus metapopulation extinctions. One of the assumptions for long-term metapopulation persistence is that the expected number of new populations generated by one existing population during its lifetime must be greater than one. That is, the replacement rate must be greater than one, as is true for a local population to persist. In a small metapopulation, however, all local populations may go extinct by chance. This is known as "extinction–colonization stochasticity" (Hanski 1998). This is an exact analogue to demographic, stochastic extinction of a local population (Chapter 1). This may happen even if the replacement rate is greater than one in both local populations and metapopulations. Regional stochasticity is due to processes such as large-scale weather patterns, which produce synchrony among the independent local populations. This effectively reduces the number of independent populations and makes metapopulation-level persistence less likely.

5.6 Metapopulation dynamics of two local populations

Recall that in the discrete-time population model, when r was set > 2.69 (Fig. 2.19) the population underwent chaotic behavior (May 1974, 1976b). However, if two such populations are connected to each other by migration a number of interesting and unexpected changes occur (Hanski 1999).

In this example the Ricker model, is used:

$$N_{t+1} = N_t\, e^{r\left(1 - \frac{N_t}{K}\right)}$$

(5.6)

In populations one and two, shown in Fig. 5.4, $r = 3$ and $K = 3$. Population one is initiated with one individual ($N_0 = 1$), while for population two $N_0 = 2$. Each population and the metapopulation (Fig. 5.5) behave chaotically. The metapopulation is simply the sum of populations one and two. At time = 49 the two populations are connected by allowing 30% of the individuals to emigrate. The emigrants are divided equally between the

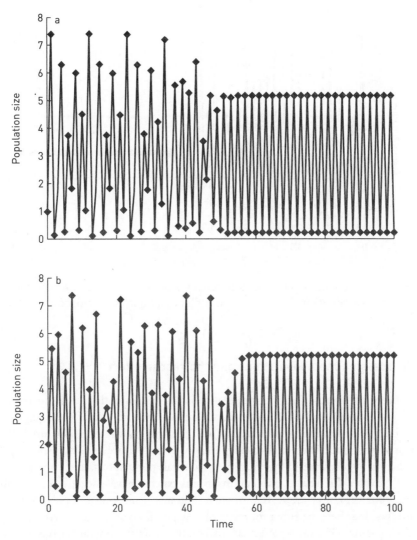

Figure 5.4 Two local populations. (a) Population one: $N_0 = 1$, $r = 3$, $K = 3$.
(b) Population two: $N_0 = 2$, $r = 3$, $K = 3$. The two populations are connected
at $t = 49$.

two populations. The migrations calm the chaotic behavior, and by time = 55 the two
populations (Fig. 5.4) have moved to a two-point limit cycle. The two populations go through
cycles out of phase with each other, and the **metapopulation** (Fig. 5.5) is **completely stabilized** (Hanski 1999).

Gyllenberg *et al.* (1993) have confirmed that migration can help stabilize local population dynamics, although some mortality must occur during migration to have a stabilizing effect. Similarly, migration from a permanent (mainland) population or from a
population with a low growth rate also has a stabilizing effect. Movement of individuals
between local populations has, at least theoretically, a stabilizing effect on the local
populations themselves as well as on the metapopulation. We are well advised, however,
not to push this theoretical point too far in field populations (Hanski 1999).

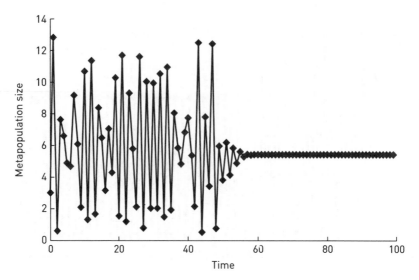

Figure 5.5 Metapopulation of populations one and two from Fig. 5.4. The two populations are connected at $t = 49$. The metapopulation is stabilized at $t = 55$.

5.7 Source–sink metapopulations and the rescue effect

The concept of source–sink metapopulations was put into the literature by Pulliam (1988), and is based on the fact that habitats are not uniform but differ in quality. High-quality patches produce large populations with positive growth rates, and are likely to be a source of emigrants. These high-quality areas, where $r > 0$, are known as **source** patches. Other habitat patches are of low quality, have small populations, and consistently have negative growth rates. That is, populations in **sink** patches have a negative r in the absence of immigration (Hanski 1999). If and when migrants from source populations arrive at sink patches, they become either founders of new populations, or new members of established populations. The **rescue effect** is based on the idea that emigrants from source areas regularly supplement these small, extinction-prone populations. If the expected size of the small population is increased through this supplementation it becomes less prone to, or is **rescued** from, extinction (Brown and Kodric-Brown 1977). In a true sink habitat a population would decline to extinction if cut off from its source population. A **pseudo-sink** population is one in which the population would decline to a lower equilibrium, but not go extinct, if cut off from its source population (Watkinson and Sutherland 1995). In practice it is difficult to distinguish between a true sink and a pseudo-sink population in the field.

The MacArthur and Wilson (1967) island biogeography theory, a mainland–island metapopulation, has much in common with source–sink metapopulations in that mainlands are large compared to islands and, in theory, not prone to extinction, while islands are small and there is always a finite probability of extinction. The mainland is the source, while the islands are, collectively, sink populations. While MacArthur and Wilson (1967) emphasized that extinction is a normal process in a community, they envisioned the mainland as a more or less uniform patch when compared to the island populations, and that the mainland would function as a source indefinitely. Moreover, while extinction was assumed to occur on islands, the island populations were also assumed to have positive growth rates in a normal year (Elmhagen and Angerbjörn 2001).

One of the best illustrations of the source–sink concept comes from the study by Hubbell and Foster (1986a, 1986b) of a 50-hectare plot of tropical moist forest on Barro Colorado Island in Panama. They mapped over 238,000 individual trees and shrubs of 314 species over a 13-year period. They found that at least one-third of the rare species were not self-maintaining populations. They were not reproducing effectively and their presence in the plot was a result of immigration from outside of the 50-hectare plot.

Sink populations can be important to the long-term survival of a source population. Assume that the source population is prone to chaotic behavior, or shows great fluctuations in size because of disease or sensitivity to environmental fluctuations such as drought or fire. In this case the source population itself could be rescued if connected to a sink (Gyllenberg *et al.* 1993).

5.8 Non-equilibrium and patchy metapopulations

In examining the literature on metapopulation studies, Harrison (1991, 1994) found that the term has also been used to describe two other kinds of situations. In a **non-equilibrium metapopulation** the rate of extinction among the populations exceeds the colonization rate. Without some change in the dynamics of the system, including perhaps restoration of functioning habitat patches, the ultimate fate of such a metapopulation is extinction.

By contrast, in a **patchy population** the migration rate is so high that the so-called subpopulations function effectively as one single population. Hanski (1999) has asserted that patchy populations should be excluded from metapopulation theory since they lack distinct breeding subpopulations. The classical metapopulation of Levins is found between the extremes of patchy and non-equilibrium metapopulations (Elmhagen and Angerbjörn 2001).

5.9 Spatially realistic models

The Levins model assumed that local populations were identical; extinctions occurred independently in different patches; all patches were equally likely to be found; and patches were equally connected to each other. He did not try to describe how individuals move from one population to another, nor did he allow for differences in patch size or quality, in spite of the importance assigned to island size by MacArthur and Wilson.

Most metapopulation models also assume unconditional emigration. That is, there are no consequences to the source population from losing individuals. This seems to be a reasonable assumption in most populations since emigrants are often pictured as "extra" individuals when a population has reached or exceeded its carrying capacity. In many species non-breeding males venture away from the family group looking for an empty territory or an existing group that they may join in order to become breeders. Females of butterflies and other insects emigrate in order to find newly available host plants for egg deposition. Male and female dung beetles emigrate when the dung pile they were born

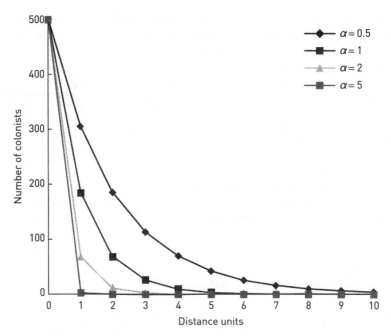

Figure 5.6 Number of colonists arriving at different distances, based on a negative exponential model, $C_i = \beta e^{-\alpha d_i}$ and using different α values. C_i = the number of colonists arriving at a distance d_i from the source population. In all cases $\beta = 500$.

into has become depleted. If the mortality rate of dispersers, however, is much higher than that of individuals who remain "home," a very high population dispersal rate (large percentage dispersing) can lead to local, and sometimes metapopulation, extinction (Hanski 2002, J. Mickelberg, personal communication).

The rate and the scale of re-colonization of an empty habitat depend on the shape of the dispersal curve of the population. A simple model of animal movement is based on a random walk, using a coefficient of diffusion (D) and a normal distribution for movements, the variance of which increases with time ($2Dt$) (Okubo 1980). Studies of animal emigration, however, indicate that more individuals move very short and very long distances than predicted by random walks (Johnson and Gaines 1990). In metapopulation models emigration distances are usually modeled using a negative exponential function, which is a reasonable approximation of reality (Hanski 2002) (Eqn. 5.7, Fig. 5.6). Two characteristics determine the dispersal efficiency of a species. One is the number of individuals dispersing (here equal to β) and the dispersal ability of each individual (α). Species differ a great deal in the amount of energy invested in reproduction each year. Obviously the more energy invested in reproduction, the greater the number of dispersal units. But for a given reproductive effort, species also differ in whether they produce a smaller number of large offspring or a larger number of small offspring. The smaller the dispersing unit, the greater distance it is likely to travel. Many highly dispersed organisms are so small that the wind can carry them hundreds or even thousands of miles. On the other hand, acorns only move as far as gravity or squirrels will take them. In Equation 5.7 α is directly associated with a greater colonizing ability per unit of dispersal, while β is associated with the number of colonists produced per individual from the source population. C_i is the colonization probability per unit time for population i and d_i is the distance from the source population.

$$C_i = \beta\, e^{(-\alpha d_i)} \tag{5.7}$$

where α and β are site- and species-specific parameters.

What is most important to a metapopulation is the proportion of individuals that move long distances. The exact shape of the distribution of migration distances among emigrants is therefore of great importance, but it is very hard to estimate in most populations (Hanski 2002).

We will discuss one approach that has been proposed to make metapopulation models more realistic, the **incidence function model** (IFM) championed by Hanski and his colleagues (Hanski 1994a, 1994b, 1999, Hanski and Gilpin 1997). The IFM is a **stochastic patch model** in which the population in each patch has one of two states, presence or absence. The IFM includes: (i) a finite number of habitat patches; (ii) patches of different sizes (sometimes including differences in quality and shape); and (iii) each patch having a unique spatial coordinate so that interactions among patches are localized in space. Since habitat patches are simply occupied or not, there is usually no estimation of population sizes or dynamics within patches. The major virtue of the IFM is that it is constructed so that parameters can be estimated from field data. This allows the application of this model to real populations.

The IFM begins with the assumption that for an empty habitat patch, i, there is a constant probability, C_i, of re-colonization per unit time. If a patch is occupied, there is a constant probability, E_i, of extinction per unit time. One event, either colonization or extinction, is allowed per time period. The long-term probability of the patch being occupied is called the "incidence" or J_i (Eqn. 5.8). The incidence function model is based on discrete time intervals, and is a stochastic rather than a deterministic model (Hanski 2001).

$$J_i = \frac{C_i}{C_i + E_i} \tag{5.8}$$

There are a number of difficulties if we are modeling a true metapopulation with no "mainland" source of species. With no external mainland, metapopulation extinction is the only true steady state (Hanski 1999). However, a metapopulation may theoretically persist for very long periods of time.

Recall that in the rescue effect the probability of extinction on a habitat patch is reduced through immigration of individuals from other patches. In order to allow for the rescue effect, Hanski modified the probability of extinction between times t and $t+1$ by substituting $(1 - C_i)E_i$ for E_i. This modified equation is:

$$J_i = \frac{C_i}{C_i + E_i - C_i E_i} \tag{5.9}$$

Hanski (1999) then derived a relationship between extinction probability, E_i, and the size of the patch area, A_i, using the basic reasoning of the species–area curve. That is, extinction probability depends on population size, which is a function of patch area. The general relationship is as expressed below (Eqn. 5.10) in which e and X are estimated from the data. The value e is a parameter related to the probability of extinction per unit time in a patch of a given size. The parameter X is a measure of environmental stochasticity (Hanski 1999).

$$E_i = \frac{e}{A^X}$$

(5.10)

Colonization probability, C_i, is a function of the number of immigrants, M_i. In a simple mainland–island metapopulation, the colonization probability is a function of distance (d_i) from the mainland, which was expressed above as Equation 5.7.

For a true metapopulation, however, M_i is the sum of individuals arriving from all of the surrounding habitat patches. M_i can be written as a summation for all patches:

$$M_i = \beta S_i = \beta \sum_{j \neq i} e^{(-\alpha d_{ij}) p_i A_j}$$

(5.11)

In this equation d_{ij} represents the distance between patches i and j; p_j is 0 for an unoccupied patch and 1 for an occupied patch; and A_j is the size of the patch. The summation term is represented by S_i, which becomes a measure of **patch isolation** or, to put it positively, patch **connectivity**. If population sizes (N_j) are known for each patch, S_i can be written as:

$$S_i = \sum_{j \neq i} e^{(-\alpha d_{ij}) N_j}$$

(5.12)

As above, the term α describes how fast the immigration rate from patch j declines with distance. Hanski suggests that this term can be found through mark–recapture data.

If interactions among immigrants are negligible, C_i increases exponentially with M_i. However, there is often a sigmoid relationship between the number of immigrants and successful re-colonization by a given species. Therefore C_i can be written as:

$$C_i = \frac{M_i^2}{M_i^2 + y^2}$$

(5.13)

where y is a parameter fitted from the data. In the sections below the terms y and β are combined simply as y (Hanski 1999).

Once equations were developed for the dependence of extinction on patch size and colonization on patch connectivity, Hanski combined them into the usual form of the incidence function model:

$$J_i = \frac{1}{1 + \dfrac{ey}{S_i^2 A_i^X}} = \left(1 + \frac{ey}{S_i^2 A_i^X}\right)^{-1}$$

(5.14)

As before, J_i is the probability that a patch, i, is occupied; y is a parameter related to successful immigration; A_i represents the area of the patch i; X is the rate of change of extinction per unit time with increasing patch size (a measure of environmental stochasticity); and S_i describes the connectivity between patches, that is, the effect of distance on immigration rate. Equation 5.14 can be rewritten as:

$$J_i = \frac{1}{1 + e^{\ln(ey)-2\ln S_i - X \ln A_i}} \tag{5.15}$$

If we manipulate and take natural logs of both sides of this equation we come up with a linear relationship between the expected patch occupancy (J_i) and the two independent variables, connectivity (S_i) and size (A_i) of the habitat patches:

$$\ln\left(\frac{J_i}{1 - J_i}\right) = -\ln(ey) + 2 \ln S_i + X \ln A_i \tag{5.16}$$

Hanski (1999) has described how to estimate all of these parameters from field data, and has applied the model to simulate metapopulation dynamics in butterflies (Hanski et al. 1995, Wahlberg et al. 1996), the American pika (*Ochotona princeps*) (Moilanen et al. 1998), and a number of other species. The basic information needed in order to apply this model in the field is simply the area of each habitat patch and the inter-patch distances (d_{ij}). Subsequently, model parameters must be estimated. First α is estimated from mark–recapture data or estimated from patch-occupancy data. The parameters y and e are fitted to empirical data using nonlinear regression techniques. The value of X is fitted from the data and can be modified to include the rescue effect.

In Table 5.2, from Kindvall (2000), are the results of fitting the incidence function model to field data gathered from a fragmented population of the bush cricket *Metrioptera bicolor*. Kindvall (1995) found that using occupancy data from a single year did not result in realistic predictions about the metapopulation. When parameters for the IFM were estimated from patch occupancy over a five-year period, better results were obtained.

Table 5.2 Results of fitting the incidence function model to occupancy data of the bush cricket *Metrioptera bicolor* for two areas of Sweden during the period 1989–94. *P* is mean proportion of available habitats actually occupied from 1990 to 1994. Predicted mean proportion of patches occupied is based on 100 replicates. Adapted from Kindvall (2000).

Parameters	Western area	Eastern area
Number of available patches	66	50
Actual *P*	0.82	0.71
Predicted *P*	0.80	0.62
Multiplier of *X* for the rescue effect	0.05	0.001
α	2.0	6.0
X	0.876	0.514
y	7.278	2.571
e	0.072	0.029

Table 5.3 The predicted and observed annual extinction and colonization rates for three species of shrews (*Sorex*) on small islands. Parameters for this mainland–island incidence function model were from 68 islands and applied to a different set of 17 islands. X and **e** are annual extinction parameters. C = annual colonization rate; E = annual extinction rate. Adapted from Peltonen and Hanski (1991), Hanski (1993).

Species	Body size (g)	Parameter estimates		Predicted		Observed	
		X	e	C	E	C	E
S. araneus	9	2.30	0.20	0.26	0.04	0.20	0.04
S. caecutiens	5	0.91	0.53	0.03	0.28	0.05	0.33
S. minutus	3	0.46	0.73	0.18	0.53	0.13	0.46

Kindvall (2000) compared the incidence function model to three other spatially realistic models, including a logistic regression model (Sjögren-Gulve and Ray 1996), and found the logistic regression model performed best in its ability to predict regional occupancy, local occupancy, and the number of colonizations and extinctions. Nevertheless he found the IFM did a reasonable job of predicting the actual outcomes for the bush cricket in Sweden.

The application of the IFM to mainland–island populations can be simpler since there are fewer parameters to estimate. One example is a study on the occurrence of small mammal populations on islands in lakes and in the sea (Peltonen and Hanski 1991, Hanski 1993). The IFM was based on the occurrence of three species of shrew (*Sorex*) on 68 islands. Using parameters estimated from this study, Hanski (1993) predicted the annual colonization and extinction probabilities on 17 additional islands. The observed and the predicted rates were well matched (Table 5.3). In this study only the extinction parameters (X and e) were estimated since colonization was from the mainland and was assumed not to differ among islands (Hanski 1999). The value X is inversely related to the strength of environmental stochasticity: a large value of X means weaker environmental stochasticity. The value of X is directly correlated with body mass in the shrews described in Table 5.3, as well as in birds from four different areas (Cook and Hanski 1995). What this implies is that species with larger mass (such as *Sorex araneus*) are less affected by environmental stochasticity as compared with species with a smaller mass (Hanski 1999). Table 5.3 also suggests that only *Sorex araneus* has a long-term metapopulation survivorship, since it is the only species with a $C > E$.

5.10 Minimum viable metapopulation size

As discussed in Chapter 1, conservation biologists introduced the concept of minimum viable population size (Soule 1980), although this approach has been replaced by various population-viability analyses. The MVP size was intended to estimate the minimum number of individuals necessary for a population to have a specific probability of surviving for a fixed period of time. When applied to metapopulations the analogous concept would be defined as the minimum number of local populations necessary for the long-term persistence of the metapopulation. Gurney and Nisbet (1978) and Nisbet and Gurney (1982) developed a stochastic version of the Levins model with a finite number of habitat patches and local populations. In the analysis of their results they defined long-term persistence of the metapopulation, T_M, as at least 100 times the expected time of local extinction, T_L.

If $\hat{P}$ is the fraction of occupied patches at equilibrium, and H is the total number of habitat patches, Gurney and Nisbet (1978) found that the product of $\hat{P}\sqrt{H}$ must be greater than 3:

$$\hat{P}\sqrt{H} > 3 \qquad\qquad (5.17)$$

For example, if there are 50 habitat patches, this equation says that colonization and extinction rates must be such that $\hat{P} > 0.42$ for a metapopulation to persist more than 100 times T_L. This relationship does not take into account the size and quality of the habitat patches, but does demonstrate that long-term metapopulation persistence benefits from a large number of habitat patches.

5.11 Assumptions and evidence for the existence of metapopulations in nature

The different types of metapopulations described above (mainland–island, classical, source–sink, patchy, and non-equilibrium) are all variations on the same themes. Local extinctions are commonplace, there is an equilibrium involving colonization and extinction rates, and so on. They differ in the levels of detail, whether they allow for patches to be of different quality, whether they allow for differing levels of connectivity between patches, and whether they include local population dynamics.

What are the general assumptions we are making in all of these models? How can we demonstrate that the long-term persistence of a species in a landscape is due to metapopulation, rather than local population, processes? One difficulty is that long-term population data on patch occupancy are hard to gather. For this reason, many studies of population persistence in a fragmented landscape, meeting the criteria for a metapopulation, come from short-lived, easily monitored organisms. Accordingly, there has been an emphasis on populations of butterflies such as the Glanville fritillary (*Melitaea cinxia*) and the bay checkerspot (*Euphydryas editha*) (Hanski 1999, Ehrlich and Hanski 2004). On the other hand, metapopulation theory was famously, though perhaps inappropriately, used in designing a conservation plan for the northern spotted owl (*Strix occidentalis*) (Boyce 2002), and is being implemented in the management of both black and golden lion tamarin (*Leontopithecus chrysomelas* and *L. rosalia*) populations in South America (J. Mickelberg, personal communication).

Assumptions that metapopulation dynamics are decisive to the structure of the regional population include (Hanski and Kuussaari 1995, Hanski 1999):

1 The species has local breeding populations in relatively discrete habitat patches. This condition stresses that the population is spatially structured and therefore most individuals interact with others only in the local habitat patch.

2 No single local population is large enough to have a longer expected lifetime than the expected lifetime of the metapopulation itself. This excludes mainland–island populations.

3 Empty habitat patches are common. In the Glanville fritillary butterfly study in Finland, for example, 70% of approximately 1600 habitat patches have been empty at a given time (Hanski *et al.* 1995).

4 The habitat patches are not too isolated to prevent re-colonization. Long-distance movements may be facilitated by habitat corridors or other mechanisms.

5 Local dynamics are sufficiently asynchronous to make simultaneous extinction of all local populations unlikely. With complete synchrony, the metapopulation only lasts as long as the local population with the lowest chance of extinction. The greater the asynchrony, the longer the metapopulation is likely to last. In a recent review of the literature, Elmhagen and Angerbjörn (2001) found eight studies (four insect species and four small mammal species) in which asynchrony of population dynamics among patches was confirmed.

6 Population turnover, local extinctions, and the establishment of new populations form the basis for metapopulation dynamics, and metapopulations persist despite population turnover. Elmhagen and Angerbjörn (2001), in a review of the literature, found 22 studies confirming turnover. See supportive data below.

7 Population size or density is significantly affected by migration. This is the basis for source–sink populations and the rescue effect (Pulliam 1988, 1996).

8 Population density, colonization rate, and extinction rates are all affected by patch size and isolation.

9 Metapopulations can affect competitive, predator–prey, and parasite–host interactions. These ideas were discussed at the beginning of this chapter and will be elaborated in later chapters.

Reviews of metapopulation studies by Elmhagen and Angerbjörn (2001) and Harrison (1991, 1994) found that many of these criteria are frequently not met in the published metapopulation literature. Nevertheless, there is an extensive literature, particularly from studies of butterflies, supporting most of these assumptions (Ehrlich and Hanski 2004).

Supportive field studies

1 Boycott (1930) studied freshwater mollusk populations in 84 ponds in England. Over a 10-year period he recorded 64 extinctions and 93 colonizations of 18 species.

2 In their study of arthropod populations on red mangrove islands, Simberloff and Wilson (1969, 1970) removed all arthropod species by fumigation. After re-colonization of the mangroves by the arthropods, they found an equilibrium of between 20 and 40 species per island, depending on island size. But the turnover rate of species was approximately 2% of the species pool per day.

3 As described at the beginning of this chapter, after the explosion on Krakatau the number of bird species on Rakata seemed to reach equilibrium at 30 by 1934. Yet a consistent turnover has continued.

4 Hanski *et al.* (2004) described an experiment with the Glanville fritillary in which 10 local populations, derived from 72 larval groups, were transported from the Finnish island of Åland to the island of Sottungia in August 1991. Sottungia is a 4 km by 2 km island containing 20 small meadows suitable for this butterfly, although there were none at the time of the introduction. An

Table 5.4 The number of surviving local populations, the number of extinctions, and the number of colonizations per year for the metapopulation of *Melitaea cinxia* on Sottungia island in the Baltic Sea off the coast of Finland. Extinctions + colonizations = turnover events. Larvae were transported in August 1991. Adapted from Hanski *et al.* (2004).

Year	Number of local populations	Number of extinctions	Number of colonizations	Total number of turnover events
1991	10	–	–	–
1992	5	5	0	5
1993	5	1	1	2
1994	6	2	3	5
1995	3	3	0	3
1996	6	0	3	3
1997	10	0	4	4
1998	14	0	4	4
1999	2	12	0	12
2000	11	0	9	9

examination of Table 5.4 shows that this metapopulation has persisted, in spite of the fact that none of the original 10 populations lasted for the entire 11 years of the study. These data confirm the persistence of the metapopulation as a stochastic balance between local extinctions and re-colonizations of available habitat patches. Note that we are only keeping track of presence or absence of the population in a habitat; we are not making an assessment of local population numbers or local population dynamics, and there is no assessment of habitat quality.

5 Long-term work on the same species of butterfly on the Åland archipelago (Nieminen *et al.* 2004) over an eight-year period found: (i) the number of extinct populations varied from 131 to 234 per year; (ii) the number of colonizations ranged from 97 to 230 per year; (iii) the total number of extant populations varied from 303 to 496 per year; and (iv) the number of empty patches varied from 749 to 3507 per year.

6 Crone *et al.* (2001) examined the six-year data gathered by Pokki (1981) on vole (*Microtus agrestis*) populations found on the Tvärminne archipelago in Finland. Extinction and re-colonization of local island subpopulations were common, and a spatially explicit model such as the incidence function model provided reasonable predictions of the structure and function of this metapopulation. However, contrary to assumptions, the mainland did not prove to be a significant source of dispersing animals; in fact, an important source of immigrants to the larger islands came from tiny, ephemeral populations found on small islands. While these small populations were unlikely to persist, they were an important source of immigrants to larger islands on the archipelago. Crone *et al.* also found that the parameters fitted to the incidence function models varied dramatically from year to year. Most importantly, this study suggests that the "rescue effect" can be turned around such that a small ephemeral population may help stabilize a larger, more permanent population.

5.12 Conclusions

Metapopulation biology and spatial ecology have provided a new framework for both population and conservation biology. Spatial locations of populations and the interactions among local populations have as great an effect as the traditional parameters of birth and death rates, age structures, and interspecific interactions. Although a metapopulation was originally just a "population of populations" (Levins 1970), many of Levins' simplifying assumptions have been relaxed in modern models. As we have seen, the persistence of the metapopulation is highly influenced by: (i) the number of patches; (ii) the size and quality of the patches; and (iii) the connectivity between the patches. Spatially explicit models attempt to encompass these variables.

We have not specifically discussed the topic of corridors between patches. The usefulness of corridors to metapopulation persistence has been widely discussed. The general idea is that, since movements between populations have the potential to stabilize both the local population and the meta-population, a corridor between habitat patches would increase connectivity and facilitate these movements. Yet in spite of intuitive appeal and theoretical support, the benefits of corridors remain a controversial topic in metapopulation and conservation biology. Although corridors have not yet been proven irrefutably to be beneficial to wild populations, Laurance and Laurance (2003) concluded that the preponderance of available evidence is positive. They recommended that corridors be regarded as beneficial in fragmented landscapes unless specific local evidence suggests otherwise.

The metapopulation approach has challenged the dogma that populations only exist in locations where they are optimally adapted. Rather, we know that local populations go extinct on even the best-quality habitats, and that so-called sink populations hang on in areas of marginal habitat. Further-more, most natural populations are small enough to be subject to stochastic extinctions. Metapopulations, in which populations in different patches have independent growth and decline dynamics, may therefore be necessary for the long-term persistence of the regional population (Hanski 1999). Foppen *et al.* (2000) have even demonstrated that sink populations can be essential to the preservation of the populations found in larger patches.

Within a decade of the publication of MacArthur and Wilson's theory of island biogeography it dominated conservation biology to the extent that "rules" of refuge design were based on it. In the 1990s there occurred a shift, and now metapopulation theory has replaced island biogeography as the major theoretical basis for conservation biology, although its applications to specific situations should not be undertaken lightly (Doak and Mills 1994). Metapopulation theory may be replaced by some other paradigm in the future. But the overall message is clear. Spatial dynamics matter.

6

Life-history strategies

- Diversity of life histories
- Power laws
- The metabolic theory of ecology
- The pioneering work of Cole and Lewontin
- The MacArthur and Wilson *r*- and *K*-selection theory
- Cost of reproduction, allocation of energy, and clutch sizes
- Predation and life histories
- The Grime model of life histories for plants

6.1 Introduction

To the uninitiated, nothing could be worse than accompanying a bunch of "birders" on a field trip. They keep stopping, peering through their binoculars, whispering to each other, and motioning you to keep quiet. Why are they so fascinated with birds? Aren't they all pretty much the same?

Of course not. Even the most naive non-biologist knows that birds come in an amazing variety of colors and sizes; amateur birders are legion. What we are interested in exploring in this chapter, however, is the variety and potential adaptive value of life histories found in all groups of organisms. Since ornithologists such as David Lack have contributed so much to our understanding of life histories, we begin by using birds to illustrate the complexity and diversity of life histories. These accounts are mostly based on Janzen (1983).

1 The groove-billed ani (*Crotophaga sulcirostris*) is a common and conspicuous bird found in the lowlands and mid-elevations of Central America. Females are about 65 g in mass, but lay extremely large 11 g eggs. Since each female may deposit 4–8 eggs in the nest, the combined total mass of her eggs may exceed her body weight. What is more extraordinary, however, is that this species engages in a communal breeding system. The birds live and breed in a group ranging from two to eight adults, with an equal number of males and females. The group defends a common territory, year-round. A single nest is

constructed and all females deposit their eggs in it, forming a communal clutch. All members of the group contribute to incubation and feeding of the nestlings. Anis are highly social, roosting and sleeping in close contact with each other while engaging in mutual grooming. However, Vehrencamp (1977, 1978) found that there are specific costs and benefits to the individuals participating in this group endeavor. For example, there are individual differences in the number of eggs that get into the nest, in the amount of time and effort put into incubation, and in the care of the nestlings. Furthermore the eggs and offspring of the dominant females and males benefit the most. Dominant females lay their eggs last and actually remove eggs laid by other females from the nest. These dominant females then behave like brood parasites in that they actually put less effort into incubation and feeding than do the subordinate females. On the other hand, so-called alpha males, who have the most eggs in the nest (and the most to lose), perform a large share of the incubation. What is the advantage of communal nesting, especially for the subordinate birds? How do the dominant females get away with dumping the subordinates' eggs while they do less of the work?

2 The northern jacana (*Jacana spinosa*) is found from Costa Rica northward in Central America wherever there is floating aquatic vegetation. Jacanas have reversed the usual roles of the genders. Males build the nests and incubate and care for the young. Females lay one egg a day for four days in a typical clutch. Females are able to lay a second clutch of eggs elsewhere within 7 to 10 days, if necessary. The eggs are quite small (7.9 g) as compared with the average weight of the females (160.9 g). Males are smaller (mean weight of 91.4 g) than females. The mating system is polyandrous. Each male defends a small territory while each female defends a territory containing one to four males. Once chicks reach 12–16 weeks of age the females often provide a second clutch for the males to care for. The ratio of males to females varies seasonally and from place to place, but is often skewed in favor of the males. For example, the long-term average at Turrialba, Costa Rica was 2.3 males per female (Jenni 1983). Jacanas have a very high reproductive potential, but the hatching and fledging survivorship rates are very low.

3 The frigatebird (*Fregata magnificens*) is a large (800–1700 g) seabird with a life history that is unusual because of its low reproductive potential. Both sexes do not become mature until 5–8 years of age. Females breed only every other year and lay one egg in a clutch. The egg takes 55 days for incubation and the nestlings grow very slowly. They are fed primarily by the females for as long as 14 months. Given a 50 : 50 sex ratio, a new female is produced, on average, only every four years! The potential *r*-value for this species is extremely low, but by contrast survivorship of adults is very high. The life span is 40 or more years. What selective pressures resulted in a life history so radically different from that of most bird species?

4 Brown pelicans (*Pelecanus occidentalis*) are one of the best-known birds in the western hemisphere. They are found on both the Atlantic and Pacific coasts from North Carolina to Brazil and from British Columbia to Chile. Breeding colonies may contain as many as 500 pairs. An adult brown pelican weighs between 2 and 5 kg; it takes 3–5 years to attain adult plumage. Males and females share chick-raising duties equally, and the normal clutch size is three eggs. Incubation takes 30 days and the nestlings need 10–12 weeks to fledge.

Schreiber and McCoy visited a pelican colony four times during the breeding season of 1979 on Isla Guayabo in Costa Rica. Of 430 nests surveyed, most had three eggs, but the average was 2.42. By their fourth visit the number of surviving fledglings was 506, which was an average 1.18 per nest. The brown pelican is much larger than the frigatebird, has a much higher reproductive potential, but also has a lower survival rate.

5 Oropendolas (*Zarhynchus wagleri*), which are related to blackbirds and orioles, nest in colonies. Males weigh twice as much as females (212 versus 110 g) and they have been shown to take twice the energy to fledge as opposed to a female. As a result, male mortality among chicks is much higher during times of food scarcity. The sex ratio at colonies is normally 5 : 1 in favor of females. In Costa Rica, nesting begins with the dry season (December) and three complete breeding cycles are possible before the beginning of the rainy season in May. The normal clutch size is two, but breeding success is very low. The average number of chicks fledged per nest is 0.40. On the other hand, survivorship of adults is very high. Adults have been recorded living beyond the age of 26 in the field. By contrast to frigatebirds, which also have very long adult life spans, this species has a much higher reproductive potential.

So what have we learned about life histories from these birds? Nesting ranges from communal to colonial to pair-wise. Breeding systems vary from communal to polyandrous to simple pair bonds. Fecundity varies from one egg every other year to as many as eight in one clutch. Survivorship of the chicks is as low as only 0.40 per nest, but adult survivorship is as high as 40 years. What accounts for all this variation in life histories? Under what conditions do we find high versus low fecundity and/or survivorship? These are questions we want to attack in this chapter.

Another set of questions we wish to address concerns the relationship between the body size of an organism and its reproductive potential. Although body mass does not determine all aspects of life history, it is a powerful influence. For example, Fig. 6.1 is based on data for 24 species of mammals found in Costa Rica. The log of the length of the pre-reproductive period was graphed against the log of adult body mass. The obvious conclusion is that there is a higher likelihood of delay in reproductive maturity in the larger animals. Similarly, Fig. 6.2 demonstrates that animals with larger mass also have a longer interval between births. Litter size and total reproductive output per year were negatively associated with body mass, though the relationships were weak in this set of data.

Basic life-history equations were available early in the twentieth century (Lotka 1925). But a serious comparison of life-history parameters across a large number of species did not begin until 1954. In that year Frederick Smith and L.C. Cole published important papers that have become the foundation of life-history analysis.

Smith (1954) surveyed the literature and published what was known about r, R_0, and G (generation time) at that time. Using the well-known relationship:

$$N_t = N_0 \, e^{rt}$$

and setting t = the generation time, G, we have:

$$N_G = N_0 \, e^{rG} \quad \text{and} \quad \frac{N_G}{N_0} = e^{rG}$$

Taking an equation from Chapter 1 we have:

$$N_G = N_0 R_0 \quad \text{and} \quad \frac{N_G}{N_0} = R_0$$

The result is:

$$R_0 = e^{rG} \tag{6.1}$$

or

$$\ln R_0 = rG \tag{6.2}$$

If two of these variables are known, the third can be determined, as least for a population with a stable age distribution. However, **there is no necessary relationship among all three of these variables.**

Smith (1954) and a number of more recent surveys of life histories have shown that:

1 r is inversely related to generation time, G. r-values respond very strongly to generation time, as we will show later in this chapter.
2 Generation time is directly related to size. Since generation time is highly influenced by the length of the pre-reproductive period and by intervals between births, Figs 6.1 and 6.2 illustrate the relationship between these parameters and size (mass).

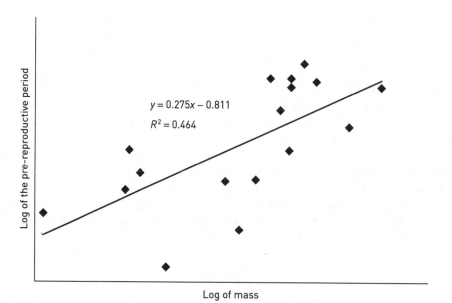

Figure 6.1 Log length of the pre-reproductive period versus log of mass in 24 species of Costa Rican mammals. Data from Janzen (1983). Note that the slope is close to the predicted value of 0.25. Linear regression is significantly positive.

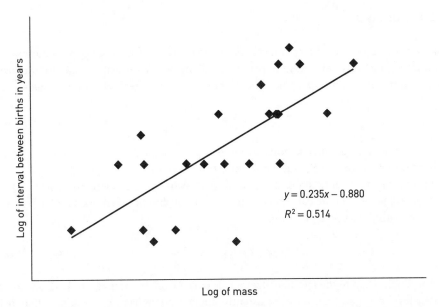

Figure 6.2 Log interval between births in years as a function of the log of mass based on 24 species of Costa Rican mammals. Data from Janzen (1983). Note that the value of the slope is close to the predicted value of 0.25. Linear regression is significantly positive.

Points 1 and 2 lead to point 3:

3 r is inversely related to size (mass);
4 life span and growth rate are negatively associated (Stearns 1992);
5 high individual growth rates are positively correlated with r;
6 growth rate is inversely related to body mass.

In summary, small size is associated with fast growth rates, high r-values and short generation times, while organisms with large mass have slower growth rates, small r-values and long generation times (Smith 1954, Enquist *et al.* 1999).

Smith also speculated that r is related to the harshness of an animal's environment, and that large organisms live in a more benign world. That is, small environmental changes are less catastrophic for large organisms. This is a point made by Hanski (1999; see Chapter 5). He speculated that environmental stochasticity was reduced for larger organisms. By contrast, organisms with high r-values have the ability to recover more quickly from events that decimate their populations. Furthermore, over evolutionary time, the fossil record shows many lines of organisms continually increasing in size until they go extinct. This is known as "Cope's rule" (Southwood 1976).

The general questions are: Under what conditions is it advantageous to have a high r-value, combined with small size? Under what conditions will evolution select for large size, with its presumed lower r-value? Is it inevitable that large size must be combined with a low r-value?

6.2 Power laws

The relationship between size (mass) and a wide variety of metabolic, physiological, and ecological functions (everything from skeletal muscle contraction rate to incubation period to maximum life span) have been shown to follow general **power** or **scaling laws**:

$$Y = Y_0 M^b \qquad (6.3a)$$

M is mass; b is an allometric constant; Y is a physiological rate or some other variable dependent on mass; and Y_0 is the normalization constant.

Taking the log of both sides of Equation 6.3a leads to Equation 6.3b, which shows us that on a log–log scale we can expect a linear relationship between mass and a dependent variable, with the slope equal to b.

$$Log\ Y = \log Y_0 + b \log M \qquad (6.3b)$$

Amazingly, despite the fact that size ranges over 21 orders of magnitude (from bacteria to whales), these scaling laws consistently apply, and it has been asserted that b is a multiple of 1/4 (Brown *et al.* 2004). For example, heart rate versus mass scales as −0.25, life span versus mass as 0.25, and the length of both mammalian aortas and tree trunks versus mass as 0.25 (West *et al.* 2000). These are known as allometric relationships because the scaling factor, b, is not equal to one. If $b = 1$ the relationship would be isometric and plot as a straight line on both an arithmetic and a logarithmic axis. However, in these allometric relationships b is not equal to one. Kleiber (1932) showed long ago that the relationship between mass and basal metabolic rate followed an allometric relationship with $b = 3/4$. Figure 6.3 shows us that, in fact, the 3/4 relationship between metabolic rate and mass appears to apply to everything from unicellular organisms to both poikilothermic and homeothermic vertebrates. More recently West *et al.* (2000) claimed that $M^{3/4}$ also "coincides with the respiratory rate . . . of mammalian mitochondria . . . [and] . . . even with that of the molecular respiratory complex and terminal oxidase molecular units within mitochondrial membranes!"

The 3/4-power scaling relationship for animals was, and still is, known as Kleiber's law. But a different scaling power was rooted in the botanical literature based on geometric or Euclidian principles. The basic idea is that "resources are acquired by surfaces and used by volumes" (Horn 2004). For example, a cell acquires nutrients across a two-dimensional surface area, but must distribute those resources throughout its three-dimensional volume. For a forest, the area of ground that a tree covers scales as the square of the diameter of ground area covered or shaded (canopy "footprint"). The volume scales according to volume or the cube of this same diameter. The number of individuals that can fit into a given area is assumed to be the reciprocal of the area covered. Plant volume and biomass were therefore proposed to be proportional to negative 3/2 (Equation 6.4a) of population density.

This is the basis, for example, of the "geometric model" of self-thinning. Agronomists and plant ecologists had proposed that mortality during the intraspecific self-thinning process followed a −3/2 power law (Yoda *et al.* 1963). This law asserted that the mean weight per plant increased faster than density decreased. During the thinning process, if density (N) and mean weight per plant ($\overline{w}$) are points on a graph, over time these points will

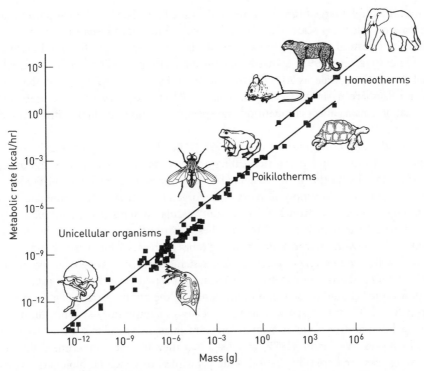

Figure 6.3 Allometric scaling of metabolic rate for organisms ranging from unicellular Protista to homeothermic mammals. The regression lines all have a slope of 0.75. After Hemmingsen (1960); reprinted with permission from the Steno Diabetes Center.

produce a linear relationship in which the slope will be −3/2 (Eqns 6.4a and 6.4b). Total mass continues to increase until the constant final yield (*C*) is reached.

$$\bar{w} = CN^{-3/2} \tag{6.4a}$$

$$Log\ \bar{w} = Log\ C - (3/2)\ Log\ N \tag{6.4b}$$

The −3/2 power law, however, began to be questioned late in the twentieth century (Lonsdale 1990), since thinning exponents were shown to be highly variable among species. In a recent series of papers (West *et al.* 1997, Enquist *et al.* 1998) a different model was proposed. This model assumes that growth continues until it is limited by a resource, and that resources are acquired through branching, linear, or "fractal" networks that distribute resources throughout the organism. In plants, Enquist *et al.* (1998) used whole-plant xylem transport as an estimate of resource use. They found that resource use scales as the 3/4 power of mass, the same exponent that Kleiber found. West *et al.* (1997) and Enquist *et al.* (1998) therefore asserted that the real exponent in Equations 6.4a and 6.4b should be −4/3, not −3/2.

Again, the quarter-scaling rule has many applications in both animal and plant populations. Calder (1984) showed that maximum life span, reproductive maturity, incubation time for birds, and a large number of metabolic functions scaled to the 1/4 power against body mass on a log–log scale. For example a regression line through ln(density) versus ln(mass) in terrestrial mammals has a slope of −0.77 ($b \approx -3/4$) (see fig. 6 of Brown *et al.* 2004). Enquist *et al.* (1998) showed that the population densities of both plant and animal populations scale as $M^{-0.75}$. From such similarities, they suggested that plants and animals share a similar scaling law that "reflects how resource requirements of individual organisms affect competition and spacing among individuals within ecological communities."

West *et al.* (1997) and Enquist *et al.* (1998, 2000) have proposed the existence of a common mechanism underlying the scaling or power laws in both plants and animals. They assert that living things are sustained by the transport of materials through linear networks that branch to supply all parts of the organism. These networks include mammalian blood vessels and bronchial trees, plant vascular systems (xylem and phloem), as well as tracheal tubes in insects. Most of these distribution systems can be described as a branching network in which the size of the tubes decreases. This "vessel-bundle" structure is characteristic not only of plant vascular systems but also of vertebrate and invertebrate circulatory systems. These networks vary in the properties of the tubes, in the type of fluid transported (liquid to gas), and in the nature of the pump. A pulsating compression pump is found in cardiovascular systems, a bellows in respiratory systems, diffusion in insect tracheae, and osmotic and vapor pressure in plant vascular systems. In spite of these differences, West *et al.* (1997) proposed that they all follow the same scaling laws. The scaling laws are based on the following principles, to which all biological networks should adhere. (i) The system must fill the volume of the organisms so as to distribute essential nutrients to all of its cells. (ii) The terminal branches of these networks should be roughly the same size, regardless of body size. Indeed, capillary size does not vary much among animals, and the same is true of petioles or terminal xylem in plants. (iii) Supply networks should be so efficient that fluids move through them with a minimal loss of energy. Models of fluids flowing through networks lead to the conclusion that the summed cross-sectional areas of the "daughter" branches at each level should be equal to that of the "parent" branch. Such an area-preserving pattern is found in trees and in artery–arteriole branches. Enquist *et al.* (2000) have modeled the physical structures of the networks and substituted them into equations describing fluid volume and rate of flow in a system. The end result is a formula for metabolic rate versus mass with a scaling exponent of 0.75. In other words, their model successfully confirms the 3/4-power scaling law.

In plants, Enquist *et al.* (1998, 2000) predicted that if the log of average mass per plant is graphed against density, the slope would be −0.75. However, since the independent variable is density, and mass is the dependent variable, the expected slope is the inverse of −3/4, or −4/3. Enquist *et al.* (1998) then gathered data on plants ranging from *Lemma* to *Sequoia* and showed that the slope is as predicted, −4/3 with an $R^2 = 0.963$ on the regression model. The slope of −3/2, predicted by the geometric model, was not found.

Enquist *et al.* (1998) concluded that both animals and plants share a common set of allometric relationships involving body mass, growth rates, life spans, and densities. Like all grand theories, there are likely to be complications and exceptions. Darveau *et al.* (2002) raised doubts about any "single cause" explanation for scaling of basal metabolic rates, since metabolic rates are controlled by a number of steps and not by any single rate-controlling factor. They believe that the scaling constant b in Equations 6.3a and 6.3b is based on the sum of multiple metabolic factors, not on a limit based on any one metabolic rate. As Agutter and Wheatley (2004) point out, not everyone is convinced that a uniform scaling constant exists, and many other models of metabolic scaling have been proposed. Nevertheless, the group lead by Brown, Enquist, and West have gone so far as to propose a "metabolic theory of ecology."

6.3 The metabolic theory of ecology

The metabolic theory is based on the 1/4-power law. Brown *et al.* (2004) begin by linking basic metabolic processes to body mass and temperature. Since metabolism determines the rate of acquisition of energy and nutrients by organisms, it determines the rate of resource use and sets constraints on the allocation of resources to growth, reproduction, and other components of fitness.

Given Kleiber's finding that metabolic rate scales according to $b = 3/4$, we can write an equation for whole-organism metabolic rate (I) as a function of mass (M) and a normalization constant, I_0. Again, the justification for the 1/4-power scaling is based on West *et al.* (1997, 1999a, 1999b), in which "whole-organism metabolic rate is limited by rates of uptake of resources across surfaces and rates of distribution of materials through branching networks. The fractal-like designs of these surfaces and networks cause their properties to scale as 1/4 powers of body mass or volume, rather than 1/3 powers that would be expected from Euclidean geometric scaling" (Brown *et al.* 2004).

$$I = I_0 M^{3/4} \tag{6.5}$$

Next, Brown *et al.* (2004) introduce the effect of temperature on biological processes. Biological activity increases exponentially with temperature, and a general equation for the kinetics of this process is described in Equation 6.6. In this equation E is the activation energy, k is the Boltzmann constant, and T is absolute temperature in degrees Kelvin. The Boltzmann factor describes how temperature affects reaction rate by changing the proportion of molecules with sufficient kinetic energy.

$$e^{-E/kT} \tag{6.6}$$

What interests us here is that nearly all biological rates are temperature-dependent, including population growth rates, development time, and life span (Brown *et al.* 2004).

The effect of mass and temperature on metabolic rates can be combined into one equation as shown below (Eqn. 6.7a), where i_0 is a normalization constant independent of both body size and temperature. After taking the natural log of both sides of Equation 6.7a we have 6.7b:

$$I = i_0 M^{3/4} e^{-E/kT} \tag{6.7a}$$

$$\ln(IM^{-3/4}) = -E(1/kT) + \ln(i_0) \tag{6.7b}$$

Equations 6.7a and 6.7b essentially predict that mass-corrected whole-organism metabolic rate is a linear function of the inverse of absolute temperature ($1/kT$). Metabolic rate, in turn, determines the rate of resource acquisition from the environment. Brown *et al.*

(2004) then showed how these metabolic relationships can determine life-history characteristics such as population growth rate, development rate, age at maturity and life span. For example, based on a variety of data, they showed that maximum growth rate (r_{max}) is dependent on both temperature and mass. The slope ($b = -0.23$) seems to confirm the $-1/4$ allometric relationship between mass and r_{max} that has been previously reported (Slobodkin 1962, Blueweiss et al. 1978).

The "metabolic theory of ecology" has already generated a great deal of commentary and controversy (see volume 85, issue 7 of Ecology), and it is highly likely to become a significant theme for ecological research in the twenty-first century.

One question we posed in the introduction ("is it inevitable that large size must be combined with a low r-value?") has been answered in the affirmative. But other questions remain. Under what conditions is it advantageous to have a high r-value, combined with small size? Under what conditions will evolution select for large size and/or high adult survivorship combined with a low r-value? How do we explain all of those mating systems that have been evolved by the birds we looked at in the introduction?

6.4 Cole and Lewontin

Cole (1954) reviewed the basic equations of population ecology and noted that it was easy to visualize situations where natural selection would tend to increase the value of r (after all, fitness is measured in terms of differential production of offspring). However, there exist many life histories in which the value of r is quite low. For example, California condors (Gymnogyps californianus), like frigatebirds, lay one egg every two years, and one female egg once every four years. What ecological or evolutionary conditions lead to these life histories?

Cole broke down life histories into components. He analyzed each separately in order to determine which were most crucial to the r-value and therefore to the life history of the organism. In his simple models he analyzed three of them. What is remarkable about Cole's work is that he did it without the aid of modern calculators or computers. In 1954 computers were available to few individuals, and hand calculators, as we know them, did not exist.

Cole recognized that since $r = b - d$, a species may increase r by reducing mortality. However, he chose to limit his investigations to factors that might influence b. In most of his models, therefore, he assumes no mortality! Next he broke his investigation of reproduction down into three areas:

1 What are the advantages of "repeated breeding" or **iteroparity**, as opposed to "one-time breeding" or **semelparity**? This basic dichotomy in life histories had neither been widely discussed nor previously investigated. Cole appears to have invented these two terms.
2 What is the effect of the length of the pre-reproductive period on r?
3 How does litter size (B) or clutch size (defined as the number of offspring produced at one time) affect r?

Cole made several simplifying assumptions, which he asserted did not seriously alter his general conclusions. Furthermore, he wanted to show the maximum gains a species might realize from altering a life-history pattern. These assumptions were:

1 No mortality (indefinite survival);
2 the litter size does not change with age;
3 iteroparous species continue to breed indefinitely.

It is unfortunate that the first part of the paper became more famous than the remainder. What became known as "Cole's result" is really due to his first assumption, which was corrected in the literature by Charnov and Schaffer (1973). "Cole's result" centers on his discussion of semelparity versus iteroparity. In the following analysis Cole is trying to determine what the advantage of iteroparity is in terms of its effect on r.

Consider an annual plant or animal. If it matures in late summer and dies in the fall at the time of reproduction, then, $B = \lambda = e^r$. For a semelparous organism we have:

$$r = \ln B \qquad\qquad (6.8)$$

where B = the litter size in terms of the number of female offspring.

By how much would r be increased if the organism lived for another year? First of all, if the litter size is only 1, since $\ln 1 = 0$, no growth would be possible and the species would have to be iteroparous for the population to grow.

Cole next developed an equation involving α = the age at first reproduction, ω = the age at last reproduction, and B = litter size.

Starting with the Euler equation $(1 = \Sigma(l_x m_x\, e^{-rx}))$, assume that: $l_x = 1$ for all ages, and m equals b and is a constant for all ages after α. Then the number of litters beginning at age α and including age ω is $n = (\omega - \alpha) + 1$. Cole showed that the Euler equation could be approximated under these conditions by:

$$1 = e^{-r} + B\, e^{-r\alpha} - B\, e^{-r(n+\alpha)} \qquad\qquad (6.9)$$

Now consider the extreme case of $\alpha = 1$ with n infinitely large. That is, reproduction begins in the first year after birth, and there is an infinite life span. The final term becomes zero, and we have:

$$1 = e^{-r} + B\, e^{-r} = e^{-r}(1 + B)$$

Taking natural logs:

$$0 = -r + \ln(1 + B)$$

which results, for an iteroparous organism, in:

$$r = \ln(1 + B) \qquad\qquad (6.10)$$

Compare this with $r = \ln B$ for the semelparous organism.

This is "Cole's result." "The absolute gain in intrinsic population growth which could be achieved by changing to the perennial habit is equivalent to adding one individual to the average litter size" (Cole 1954, p. 118).

In other words, a litter or clutch size of 101 in an annual organism would serve the same purpose as a clutch size of 100 forever in a perennial. Therefore, why would organisms evolve mechanisms for adults to survive winter if the same result is obtained through adding one individual to the litter size?

The answer, as was shown by Charnov and Shaffer (1973), is that there is great difference for most species between **juvenile** and **adult survivorship**. Organisms with high juvenile mortality, small litter sizes, or long pre-reproductive periods (slow maturity) have a great deal to gain by repeated reproduction (iteroparity). Similarly, if organisms live in an unpredictable environment, such that reproduction can fail completely in some years, selection would strongly favor adult survivorship and repeated reproduction. This is sometimes called a "bet hedging" life history. Recall the life histories of the frigatebirds, pelicans, and oropendolas described in the introduction to this chapter.

"Cole's result" is based on his assumption of complete survivorship of all life stages. If, on the other hand, juvenile survivorship is superior to adult survivorship, then semelparity should be favored. Examples include dragonflies, mayflies, and periodical cicadas (*Magicicada septendecim*).

Cole (1954) then analyzed the relative gains on r of (i) age of first reproduction, (ii) litter size, and (iii) iteroparity. As was hinted at in the introduction to this chapter, for mammals found in Costa Rica, r is most sensitive to changes in the pre-reproduction period (maturation time) (Figs. 6.1 and 6.2). The r-value is somewhat sensitive to litter size, and is least influenced by changes from semelparity to iteroparity (but see above).

For example, examine Table 6.1. Presented are four life histories. In each case the survivorship function is held constant. The fertility columns reflect differences either in the maturation time, or in the amount of reproduction which is "front loaded" to an earlier age class. In scenario one, reproduction begins at age 2 and the majority of reproduction takes place in this age class. The result is an r of 0.102 and a generation time of only 2.2 years. In scenario two the gross reproductive rate (GRR) is greater than that of scenario one, while R_0 remains at 1.25. The pre-reproductive period is also the same as

Table 6.1 Four fertility scenarios demonstrating the importance of early reproduction on the value of r. l_x, survivorship; m_x, fertility; GRR, gross reproductive rate; R_0, net reproductive rate; G, generation time; r, intrinsic rate of increase; λ, finite rate of increase.

Age	l_x	Scenario one m_x	Scenario two m_x	Scenario three m_x	Scenario four m_x
0	1.00	0	0	0	0
1	0.100	0	0	0	0
2	0.050	20	10	0	0
3	0.025	10	30	40	50
4	0.010	0	0	0	0
5	0	0	0	0	0
		GRR = 30	GRR = 40	GRR = 40	GRR = 50
		$R_0 = 1.25$	$R_0 = 1.25$	$R_0 = 1.00$	$R_0 = 1.25$
		$G = 2.2$	$G = 2.6$	$G = 3.0$	$G = 3.0$
		$r = 0.102$	$r = 0.086$	$r = 0.000$	$r = 0.074$
		$\lambda = 1.11$	$\lambda = 1.09$	$\lambda = 1.00$	$\lambda = 1.08$

Table 6.2 Four population scenarios for a hypothetical human population.

Age	l_x	Scenario one m_x	Scenario two m_x	Scenario three m_x	Scenario four m_x
0	1.00	0	0	0	0
1	0.991	0	0	0	0
5	0.989	0	0	0	0
10	0.988	0	0	0	0
15	0.987	1.000	0	0	0
20	0.985	0.500	1.000	0	0
25	0.982	0	0.500	1.000	0
30	0.979	0	0	0.500	1.000
35	0.975	0	0	0	0.500
40	0.970	0	0	0	0
45	0.961	0	0	0	0
		$R_0 = 1.48$	$R_0 = 1.48$	$R_0 = 1.47$	$R_0 = 1.47$
		$G = 16.6$	$G = 21.6$	$G = 25.6$	$G = 31.6$
		$r = 0.091$	$r = 0.073$	$r = 0.061$	$r = 0.052$
		$\lambda = 1.095$	$\lambda = 1.076$	$\lambda = 1.063$	$\lambda = 1.053$

in scenario one. However, because reproduction is mostly "back loaded" onto age class 3, the generation time is 2.6 years and the resulting r-value is only 0.086. In scenario three, by increasing the pre-reproductive period to age three, even though we increase the GRR to 40, the net reproductive rate falls to 1.00 and $r = 0$. In scenario four, we see that, because reproduction does not start until age 3, we must raise the GRR to 50 in order for the population to grow. This produces an R_0 of 1.25, but since generation time is 3 years, the r-value is only 0.074.

In Table 6.2 we apply Cole's ideas to a human population. The survivorship column is based on 1985 Vital Statistics from the United States (Peters and Larkin 1989). The four fertility scenarios are based on a population in which the average female has a gross reproductive rate of 1.5 females. In scenario one, the average female has 1.0 female offspring between the ages of 15 and 20, the remainder between 20 and 25. In each of the other scenarios, the females delay reproduction by five years, ten years, and finally fifteen years. The result is that when reproduction begins at the age of 15, the r-value is 0.091, as compared to an r-value of 0.052 when reproduction begins at 30. Generation time is obviously almost doubled when reproduction begins at 30 as compared to 15 years. If a population had a stable age distribution, it would grow from 1000 to 2500 in 10 years if $r = 0.091$, whereas it grows to only 1600 if $r = 0.052$.

R. C. Lewontin (1965) further developed the themes of Cole in his paper, "Selection for colonizing ability." He was concerned with the problem of the establishment of a population in a new geographic or ecological space. Lewontin was able to examine more subtle changes in the life cycle, and illustrated how changes in fecundity, longevity, and length of pre-reproductive period affect r. Using more sophisticated mathematical techniques, he was able to relax some of Cole's simplifying assumptions.

In his analysis, Lewontin summarized l_x and m_x into a function he called V_x. $V_x = l_x m_x$. He removed Cole's assumptions concerning infinite life span and constant fecundity with

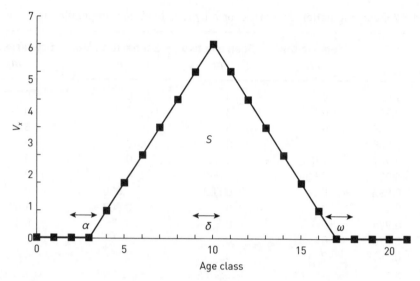

Figure 6.4 Summary of the Lewontin analysis of the effect on r of changes in life-history parameters. V_x = the product of l_x and m_x. α = age at which reproduction begins. δ = the age of maximum reproduction. ω = age at which reproduction ends. S = area under the triangle, which equals the total lifetime reproductive output. In the Lewontin analysis, the area S is held constant, while the other parameters are manipulated.

age. He then showed that a common form of V_x is a basic triangle with V_x on the y-axis and age on the x-axis (Fig. 6.4). He identified four important parameters, α (age at which reproduction begins), ω (age at which reproduction ends), δ (the age of maximum reproductive output), and S (the total reproductive output, which equals the area of the triangle).

The integral form of the Euler equation was modified as shown below. Since the function V_x has the shape of a triangle, Lewontin was able to solve the integral and devised expressions relating r to α, ω, δ, and S.

$$1 = \int l_x m_x \, e^{-rx} dx$$

$$1 = \int V_x dx/e^{rx} \tag{6.11}$$

Lewontin manipulated the triangle sizes and shapes by varying the values along the x-axis. His findings can be summarized by the following examples. In each case he invest-igated how basic life-history changes affect the value of r.

1 Rigid translation. This means moving the entire triangle along the x-axis (varying by age classes), leaving S constant. In this case, α, δ, and ω are moved simultaneously to younger (to the left along the axis), or older (to the right) age categories.

2 Change the age to sexual maturity (pre-reproductive period) only. This means moving α to earlier or later ages without changing δ, ω, or S.

3 Change the age of maximum reproduction only. That is, move δ to earlier or later ages while all else remains constant.

4 Change the age of last reproduction, ω, leaving all else constant.

Lewontin's results can be summarized as follows. Let us assume $r = 0.300$ and we wish to know what life-history changes will increase r to 0.330. Lewontin showed that doubling the lifetime fecundity from 780 to 1350 would produce such a change in r, all else being held constant. However, holding fecundity constant at 780, a similar increase in r was achieved through one of the following:

1 Rigid translation of the life history. That is, by simultaneously decreasing α, δ, and ω by 1.55 days, the value of r increased from 0.300 to 0.330. This is a change of approximately 13%, over the total span of the life history.

2 If only α, the pre-reproductive period, is changed, it takes a decrease of 2.20 days (approximately an 18% change), to produce the desired value of r.

3 If only δ is decreased, it must be shortened by 5.55 days (a 24% change).

4 If the only change is in ω, it must be decreased by 21.0 days (a 38% change).

In other words, a simultaneous reduction of 13% in the three measured aspects of the fertility schedule was roughly equivalent to a 100% change in fecundity. Furthermore, reducing the single parameter, α, by 18% resulted in the increase of r from 0.300 to 0.330. As was proposed by Cole, a reduction in the pre-reproductive period is the single aspect of the life history with the greatest influence on the value of r.

Lewontin pointed out that development time or rate is closely regulated by natural selection and varies little among geographic races, whereas less important factors such as fecundity and longevity show substantial variation.

6.5 The theory of r- and K-selection

We discussed MacArthur and Wilson's theory of island biogeography extensively in the last chapter. Another application of their findings was the theory of r- and K-selection as applied to life histories. MacArthur and Wilson (1967) asserted that for "colonizing species" the ability to grow rapidly and to disperse was the major component of fitness. They theorized that new, unpopulated environments would be colonized first by these "r-selected" species. On the other hand, once an environment, either an island or the mainland, was completely colonized, it became densely populated (crowded or saturated), with all populations near their carrying capacities (K). They reasoned that such populations were exposed to "K-selection." The major component of fitness in a crowded environment was the ability, not to grow quickly, but to survive under these highly competitive conditions.

According to the theory, in an environment with no crowding (an r-selecting environment), those genotypes which harvested the most resources and produced the most offspring in the shortest period of time, even if they were not efficient at using their resources, would be favored. In the crowded, K-selecting environment, those genotypes

Table 6.3 Correlates of *r*- and *K*-selection. Adapted from Pianka (1970).

Trait	*r*-selection	*K*-selection
Climate	Variable or unpredictable	Predictable, less variable
Mortality	Density-independent	Density-dependent
Survivorship	Type III	Type I or II
Population size	Variable, below *K*,	Fairly constant; at or near *K*,
Recolonization frequency	Re-colonization common	Re-colonization uncommon
Competitive ability	Usually poor	Usually keen
Investment in defense	Little energetic investment	Great energetic investment
Parental care	Minimal	Usually great
Length of life	Short	Long
Stage of succession	Early	Late
Rate of development	Rapid	Slow
r_{max}	High	Low
Pre-reproductive period	Short	Long
Body size	Small	Large
Number of offspring	Many	Few
Size of offspring	Small	Large
Dispersal ability	Excellent	Fair to poor

which replaced themselves at low resource levels would survive and reproduce. Evolution was said to favor efficiency in these *K*-selecting environments.

Thus *r*-selected species would be favored in areas where there has been a disturbance of the established community. *K*-selected species would be favored in areas, such as mature communities, where competition is high. It is natural to think of *r*- and *K*-selection in terms of ecological succession, in which colonization and competition are assumed to be the primary forces determining the appearance of plant communities over time. *r*-selected species appear early in succession, to be eventually replaced by *K*-selected species in the later stages.

Biologists were led to consider what aspects of life histories and of the physical environment would be correlated with *r*- or *K*-selection. In 1970 Pianka published a paper entitled "On *r*- and *K*-selection," which laid out general expectations of *r*- and *K*-selection for a number of life-history traits. Table 6.3 is adapted from that paper with modifications and embellishments.

In evaluating the theory of *r*- and *K*-selection, we should recognize that it is based on the assumption that life histories have evolved as a response to the twin selective pressures of competition and colonizing or dispersal ability. As described in section 6.9, life histories that appear to be *r*-selected may have evolved in response to predation or to the uncertainties of the physical environment (Wilbur *et al.* 1974). Caution must therefore be used in making assumptions about the selective pressures that have shaped a particular life history. Nevertheless, the terms *r*- and *K*-selection are ingrained in the ecological lexicon and have retained a certain utility as handy shorthand for a particular suite of life-history traits.

In a review of the *r*- and *K*-selection theory, Reznick *et al.* (2002) gave it credit for stimulating a great variety of empirical and theoretical work on life-history evolution. An important aspect of the theory was its focus on density-dependent selection as an important force in the shaping of life histories. But *r*- and *K*-selection was a starting point. Now it is clear that age-specific mortality, resource limitation, predation, environmental variation, metabolic rate, and density-independent factors all must be included in models of life-history evolution.

6.6 Cost of reproduction and allocation of energy

The theory of *r*- and *K*-selection illustrated that as life-history traits evolve, trade-offs have taken place among growth, maintenance, and reproduction. Another way of approaching life-history evolution is to examine the allocation of energy devoted to reproduction over a life span. This is known as **reproductive effort**. Energetic trade-offs are found in all aspects of a life history. When the parent organism devotes energy to reproduction it has less energy to devote to growth and maintenance. The energy devoted to reproduction can be partitioned in various ways: large versus small offspring, small versus large litter size, the amount of energy devoted to parental care, etc.

Cole (1954) described the dichotomy of semelparous versus iteroparous reproduction. In semelparity, reproduction is channeled into one major reproductive effort. Most insects, many invertebrates, fish, and many plants (annuals, biennials, and some bamboos) have this life cycle. A semelparous life history in a disturbed habitat offers no mystery. More interesting are organisms which are long-lived, yet have semelparous reproduction. For example, mayflies often spend several years as larvae, and periodical cicadas live 13 or 17 years below ground as juveniles, but the adult phases are only a few hours or days in mayflies and a few weeks in periodical cicadas. Hawaiian silverswords (*Argyroxiphium* spp.) live 7 to 30 years before flowering once and dying. Some bamboo species delay flowering for 100–120 years before producing a massive seed crop and dying. Janzen (1976) and others have proposed that this life history allows escape of highly vulnerable juvenile stages through predator satiation.

Iteroparity is common among most vertebrates and perennial plants. Iteroparous species, nevertheless, are extremely diverse in their life histories: (i) short versus long pre-reproductive periods; (ii) annual versus periodical reproduction; (iii) small versus large amount of reproductive effort; and (iv) many small offspring versus a few large offspring.

Since energy devoted to reproduction is not available for growth and maintenance, reproduction itself has a "cost" in terms of increased mortality or decreased growth of the adult organism. Thus, an individual that reproduces in a given year often has reduced survivorship and/or may reproduce at a lower rate in the near future. In addition, there is usually a limit to the number of offspring that an adult or pair of adults can successfully produce without causing harm to themselves, their survivorship, their future reproduction, or the survivorship of this year's brood.

Therefore, an organism may have greater evolutionary fitness over the long term if it postpones reproduction or limits the allocation of energy to current reproduction. This will be true if the energetic allocation to growth and maintenance (as opposed to reproduction) produces a sufficient gain in future reproduction to compensate for losses in the present. According to this argument, an organism may have greater evolutionary fitness over the long term if it postpones (or limits the allocation of energy to) reproduction in the current year.

Field studies documenting the effect of reproduction on growth, survival, and future reproduction are not abundant. Primack and Hall (1990) showed that reproduction in pink lady's slipper orchids (*Cypripedium acaule*) is limited by bee pollination, but seed production could be greatly increased through hand pollination. As seed production increased, the cost of reproduction took effect in the third and fourth years. Hand-pollinated plants had lower growth and flowering rates than controls. For example, an average-sized hand-pollinated plant lost 10–13% of leaf area and had a 5–16% lower flowering rate as compared to control plants. In red deer (American elk) (*Cervus elaphus*) (Clutton-Brock 1984, Clutton-Brock *et al.* 1982, 1989) and in lizards (Tinkle 1969) it has been shown that great reproductive effort leads to declining fecundity and reduced survivorship. In American bison (*Bison bison*) sons suckle longer than daughters (up to 15 months). Cows that have produced sons breed later and are more likely to be barren in the next year as compared with cows that had females in the previous year (Wolff 1988).

Experimental field studies are often done with birds, in which the clutch size can be manipulated. Reid (1987) studied the optimal brood size of the glaucous-winged gull (*Larus glaucescens*) in which the natural range of chicks per nest is 1–3. Reid added and subtracted eggs, producing broods of from one to seven chicks per nest. He found that when more than three chicks were present in the nest, adult survivorship declined significantly.

A study by Beissinger (1990) on snail kites (*Rostrhamus sociabilis*) at two field sites is also instructive. Brood size was manipulated with similar results. The normal brood range is 1–3 eggs. With only one egg in a nest, since one parent can raise one young success-fully, both females and males often deserted the nest and tried to start another one. Desertion rates were almost 100% with one egg per nest, with females deserting twice as often as males. With two young per nest, desertion rate was about 50%. With three young in the nest there was no desertion. All broods with one or two young were 100% successful at fledging. However, the kites had difficulty raising three young and were unable to raise four young. Desertion is an adaptive response that allows the kites to adjust their parental investment to the number of chicks present in a nest.

In the Venezuela field site, Beissinger found that no parents raised four young, and only 40% of nests with three young raised them all. At control nests no broods of three young occurred naturally (205 nests). In Florida, however, 22% of the control nests fledged three young. The investigators found that total food delivery rates increased up to three young per nest, but there was no further increase with four per nest. Thus, natural selection closely controls the clutch size and the amount of parental effort per nest in snail kites.

6.7 Clutch size

To produce viable offspring, a certain amount of energy must be devoted to each young. Lack (1954, 1966) hypothesized that clutch size in birds evolved through natural selection to correspond to the number of young that could be fed. Although this idea is appealing, there are a number of exceptions and complications. For example, in 35 out of 60 stud-ies, an increase in brood size resulted in an increase in the number of young fledged, without affecting parental survival (Lessells 1991). However, very often the juveniles had lower survivorship and fertility than in normal broods. In other studies, birds with larger clutches bred later the next year.

Although the Lack hypothesis is generally supported over the long run, there are many cases where clutch size is modified through brood reduction after egg laying. Studies of red-winged blackbirds (*Agelaius phoeniceus*) (Forbes *et al.* 1997), for example, confirm the "insurance hypothesis," which states that hatching asynchrony creates "marginal" offspring that serve as replacements for failed earlier-hatched "core" offspring. Forbes *et al.* showed that "marginal" offspring had very high mortality rates in control nests, but survived well if broods were experimentally or naturally reduced. In asynchronous hatching the older siblings beg more vigorously for food and/or the older siblings may simply kill the weaker sibling. If food shortages arise, the late-hatched young usually die of starvation. In such cases, the parents, who cannot predict food resources, can use brood reduction to adjust reproduction to actual food availability.

6.8 Latitudinal gradients in clutch size

Lack (1966), Cody (1966), and Ashmole (1963) all developed hypotheses to explain observed latitudinal gradients in clutch sizes. Most groups of birds show a gradient in which clutch sizes increase from the tropics to higher latitudes (see Table 6.4). Lack proposed that particular clutch sizes are adaptations to food supplies, and that increasing day length allows more foraging time in the higher latitudes. Cody developed an allocation-of-energy argument. He believed that both competition and predation were more intense in the tropics, and that tropical birds invested more energy both in competitive interactions and in predator avoidance. Therefore they had less energy to allocate to reproduction than their temperate counterparts. Ashmole believed that clutch size varies in proportion to the seasonal variation in resources. In birds, population density in the higher latitudes is regulated by winter mortality, when resources in the temperate zone are scarce. Higher winter mortality produces a high ratio of available food per breeding adult. This results in large clutch sizes. Migratory birds can also be thought of as annual colonizers, whose reproductive styles tend to be *r*-selected. In addition, plant productivity per day is very high in the short growing season of the higher latitudes. Thus, it is not too surprising that clutch sizes are high. Limited data from plants and insects reinforce the general trend of higher reproductive effort at higher latitudes.

Martin *et al.* (2000) monitored 1331 nests in Argentina and 7284 nests in Arizona. They found that clutch sizes of passerine birds in Argentina averaged 2.41, similar to averages from the humid tropics. As expected, the Arizona site averaged 4.61 eggs per clutch.

Table 6.4 Clutch size in passerine birds in relation to habitat. Data from Lack (1968), as summarized by Southwood *et al.* (1974).

Habitat	Number of species	Average clutch size
Tropical forest	82	2.3
Tropical grass and shrub lands	260	2.7
Tropical deserts	21	3.9
Middle Europe	88	5.6

However, although predation rates were significant at their sites, Martin *et al.* concluded that nest predation does not explain the smaller clutch sizes at the Argentina site. Furthermore, they concluded, contrary to Ashmole's theory, that clutch size was not related to food delivery rates, and food limitation is not greater in the South American site. Martin *et al.* (2000) favor the idea that adult mortality is reduced in the tropics. This leads to a life history involving a reduction in reproductive effort and/or an allocation of a greater amount of energy to a smaller number of offspring.

Another interesting pattern is that many island populations have lower clutch sizes than their mainland counterparts. Litter sizes among mammals also follow this pattern (Fons *et al.* 1997). Islands often have fewer predators, so both juvenile and adult survivorship is higher. With fewer predators and fewer competing species, densities of the island populations will be higher. Since adults live longer it is likely to be hard for young of the year to find vacant territories, and selection appears to favor lowering the number of young thorough a lower clutch size. For example, on Santa Cruz Island scrub jays (*Aphelocoma insularis*) have a clutch size of 3.71, which is significantly lower than that of the mainland (4.34). Adult survivorship on the island is 0.935 as compared to 0.833 per year on the mainland (Atwood *et al.* 1990). Johnston *et al.* (1997) have shown that tropical passerine birds have higher adult survivorship than do comparable temperate-zone species. This provides additional support for the theory that reproductive effort is based on age-specific mortality patterns.

6.9 Predation and its effects on life-history characteristics

Predation also plays a role in the shaping of life histories. Studies by Reznick and Endler (1982) and Reznick *et al.* (1997) on guppies (*Poecilia reticulata*) have confirmed the effect of predation on life histories. Reznick and Endler (1982) came to the following conclusions:

1 When a predator prefers mature fish, the guppies devote a high percentage of their body weight to reproduction, there are short inter-brood intervals, and they mature at a small size.
2 When a predator prefers immature stages, or no predators are present, the guppies devote a low percentage of body weight to reproduction, there are long inter-brood intervals, and they mature at a large size.

In an experimental test of their predictions, Reznick *et al.* (1997) compared the life histories of two populations of guppies living in upstream versus downstream habitats. The upstream habitats lacked several species of fish, many of which prey upon adult guppies, which were present downstream. The upstream, low-predation, population had later sexual maturity, lower fecundity, less frequent litters, and larger offspring as compared with the high-predation, downstream population. In other words, lack of predation produced the suite of traits listed under point 2 above. If we knew nothing about predation, we might conclude that such traits were consistent with the *K*-selection syndrome, and that these life histories were shaped by competitive interactions. In a transplant experiment, Reznick *et al.* (1997) moved guppies from the high-predation to the low-predation habitat. After 11 years the transplanted guppies had evolved life histories similar to the resident populations.

In a similar study, Crowl and Covich (1990) showed that freshwater snails (*Physella virgata virgata*) also shift their life-history characteristics when exposed to their major predator, the crayfish *Orconectes virilis*. When exposed to water-borne cues released when crayfish fed on other members of the population, the snails exhibited rapid growth and delayed reproduction until they reached a size of 10 mm after eight months. In the absence of the cue, snails typically grew to about 4 mm in 3.5 months and then began reproduction. Since crayfish selectively prey on the smallest individuals, juvenile mortality is high. By delaying the onset of reproduction and growing to a larger size, snails decrease mortality from size-specific predation.

In both of these studies, when predators concentrated their efforts on immature or smaller stages, the life history of the prey species shifted to a "*K*-selected" syndrome. Obviously caution must be used when evaluating the potential causes of a particular life history.

6.10 Bet hedging

When the environment is unpredictable, selection may favor the concentration of reproduction early in life. Alternatively, it may favor spreading reproduction over several seasons or years. If recruitment of offspring is unpredictable from year to year, selection should favor adult survival at the expense of present fecundity (Stearns 1992). When Sandercock and Jaramillo (2002) examined survivorship rates of six species of wintering sparrows in California they found that the life histories were consistent with a bet-hedging hypothesis.

6.11 The Grime general model for three evolutionary strategies in plants

Grime (1977) developed a life-history model specifically for plants. In this model Grime categorized factors that limit plant biomass or productivity into:

1 **Stress**. Stress is any condition that restricts plant production. Examples would be shortage of light, water, or nutrients, or low temperatures.
2 **Disturbance**. A disturbance is partial or total loss of plant biomass arising from the activities of herbivores, pathogens, man, or from environmental factors such as wind, frost, desiccation, soil erosion, or fire.

Grime then put together a chart (Table 6.5) in which stress and disturbance are placed into categories by intensity (low or high), and the result is a 2×2 table. Each of the four possible interactions is a possible life-history strategy, but he claims that there is no viable strategy for high stress combined with high disturbance. This leaves three ecological life-history strategies for plants.

A **Competitive strategy**. Competition is the tendency of neighboring plants to use the same quantum of light, ion of mineral nutrient, molecule of water, or volume of space. A wide range of studies has shown that while competitive ability is important in productive conditions, it declines in importance in

Table 6.5 Grime model for three basic life histories for plants.

Intensity of disturbance	Intensity of stress	
	Low	High
Low	Competitive strategy	Stress-tolerant strategy
High	Ruderal strategy	No viable strategy

    unproductive habitats. During periods of high stress, characteristics related
to rapid potential growth become disadvantageous.

B Stress-tolerant strategy. This strategy is best represented in unproductive
habitats. Examples are: (i) low-temperature Arctic or alpine areas (here, important life forms are low evergreen shrubs, small perennial herbs, bryophytes,
and lichens); (ii) arid habitats (plants with small hard leaves, succulents,
or other water-conserving mechanisms are favored); (iii) shaded habitats
(shade tolerance is important); (iv) nutrient-deficient habitats (plants with slow
growth rates and mechanisms for conserving leaves and other plants parts
have been selected for).

C Ruderal strategy (= *r*-selected strategy of MacArthur and Wilson). This
strategy is common in severely disturbed, but potentially productive habitats.
In this case, rapid colonization, rapid growth, and high reproductive rates are
favored.

    During succession, Grime sees the plant community changing over time from ruderal
to competitive to stress-tolerant. However, a highly productive habitat may stay in the competitive mode. Grime displays the three strategies on an equilateral triangle (Fig. 6.5) with

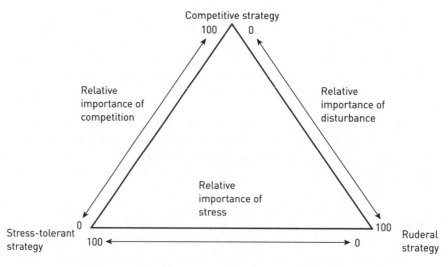

Figure 6.5 The Grime model for three evolutionary strategies in plants.

each corner representing maximum importance of one strategy or another. The apex is the competitive strategy, the right corner is ruderal, and the left corner is stress-tolerant. Life forms are organized along these three axes. Annual herbs are basically ruderals, lichens are usually stress-tolerant, trees and shrubs are a combination of stress-tolerant and competitive, and bryophytes are a combination of stress-tolerant and ruderal (Grime 1977).

6.12 Conclusions

An amazing diversity of life histories can be found in almost any ecological community. Since at least 1954 ecologists have been attempting to understand and explain this diversity. A wide range of organisms in a wide variety of environments have shown us that there exists a basic dichotomy between the reproductive rate and the size or life span of an organism. The metabolic theory of ecology emphasizes that mass and temperature heavily influence life-history traits, producing an inverse relationship between mass and r. Another dichotomy is between dispersal ability and competitive ability, which is emphasized in the theory of r- and K-selection.

Early theorists, such as Cole and Lewontin, emphasized life-history characteristics that would be most effective in increasing r. As a result they focused their attention on generation time and the fertility aspects of the life table. In many environments, however, we have found that the major adaptive syndrome revolves around reduction in adult mortality. Tropical environments, islands, and unpredictable survivorship of juvenile stages all have selected for long-lived adults with low mortality rates and iteroparous reproduction. The Grime stress-tolerant strategy for plants can be viewed as a set of adaptations for environments in which resources are limited, and recruitment of juveniles is uncertain. The result is mature plants that are long-lived and iteroparous, exactly what we normally find in climax stages of vegetation. It will always be difficult to distinguish among selective factors that might have produced an observed life history: a stochastic physical environment, predation, competition, mating system, or some other factor we haven't thought of yet. The amazing variety of life histories will keep ecologists and evolutionary biologists busy for a long time.

Part II
Interspecific interactions

At this point we are ready to consider interactions among populations of different species. These interactions may be at the same trophic level (interspecific competition) or between different trophic levels (predator–prey, parasite–host, parasitoid–host, and plant–herbivore interactions). Some of these interactions are considered "symbiotic." A symbiosis is the intimate biotic association of phylogenetically unrelated species, and is thought to develop as a consequence of coevolution. While symbiosis is sometimes considered exotic or rare in nature, it is actually a rather common phenomenon. For example, the eukaryotic cell appears to be a coevolved symbiotic complex involving organelles such as mitochondria and chloroplasts that were originally free-living organisms. Lichens consist of a symbiotic complex of algae living inside fungi, and the roots of higher plants have symbiotic associations with fungi (mycorrhizae). Moreover, the roots of leguminous plants have nodules within which bacteria (*Rhizobium* and related genera) live. Corals have a symbiotic relationship with algae known as zooxanthellae, and many species of flowering plants have complex relationships with ants.

Types of interactions

The following are symbolic representations of interactions between two species. In each case the symbol indicates whether the interaction is positive (+), negative (−), or neither (0) for each of the two species involved.

- −/− Competition, which is considered to be a reciprocally negative interaction for both of the species involved.
- +/− A biological association in which one individual benefits while the other is harmed. Such interactions include predator–prey, herbivore–plant, and parasite–host relationships.
- +/0 Commensalism. One partner benefits from the interaction, while the other experiences no particular benefit or harm. Examples include: (i) seeds with barbs that stick to the fur of animals and are thereby dispersed; (ii) organisms (mites, bacteria, etc.) that live on the skin of animals but do no harm to the host; (iii) a bird's nest in a tree; or (iv) a cattle egret (*Bubulcus ibis*) which feeds more efficiently in the company of a cow.

+/+ Mutualism. Both partners benefit from the interaction. Examples include: (i) "cleaners" and their hosts such as oxpeckers and ungulates, or cleaner shrimp and fish; (ii) pollinator–plant interactions; (iii) frugivore–plant relationships; (iv) a variety of ant–plant interactions; and (v) relationships between higher plants and fungi (mycorrhizae) or bacterial (*Rhizobium*) nitrogen fixers.

We do not usually consider the above mutualistic associations to be "symbiotic" unless there is a definite "living together" of the associates. Thus the algae and fungi that make up lichens are symbiotic, as are *Rhizobium* bacteria that live in the roots of legumes. But plant–pollinator and bird–seed-dispersal interactions are not symbiotic.

Predator–prey, parasite–host, and plant–herbivore interactions

These interactions are of fundamental importance in that they have effects at the population, the community, and the ecosystem level:

1 At the population level, predation, herbivory, and parasitism often control or help regulate animal and plant populations. Predators and parasites may also change the demographic structure of the prey/host population and thereby drive population cycles (Dobson and Hudson 1992).
2 At the community level, a predator, parasite, or herbivore can decimate a community, eliminating certain species entirely, and fundamentally changing the community (Fritts 1988, Rodda *et al.* 1992). Alternatively, it may have a positive effect on community diversity by allowing the coexistence of less competitively dominant species (Paine 1966).
3 At the ecosystem level, predator–prey, plant–herbivore, and parasite–host interactions provide means of energy flow from one trophic level to another.

Terminology

Terms used in discussing interactions among species are explicitly defined below.

Facultative. Term applied to an interaction or association among species that is not required for the survival of the individuals involved. A parasite or an herbivorous insect often has several hosts, and a mutualist may be able to survive without its partner.
Obligatory. Term applied to an interaction or association that is required for the survival of the organisms involved. In this case, the specialized pollinator, parasite, herbivore, or mutualist cannot survive without the other species with which it is associated.
Carnivorous predators kill living prey and usually consume them immediately.
Parasitoids, on the other hand (examples are certain wasp species), lay eggs on their prey. The parasitoid larva that hatches from the egg eventually devours the prey species (caterpillars, for example) from the inside. The prey

may be paralyzed or may continue feeding for some time as it is being consumed. The host (prey) is eventually killed by the time the wasp larva pupates.

Parasites do not usually kill their hosts, but harvest energy from the hosts over a period of time. During this time the parasites, reproduce copiously.

Herbivores are similar to parasites in that they consume plant parts, but usually not entire plants, and they do not usually kill the plants they feed upon.

Seed predation, on the other hand, is an example of true predation if the seed embryo is consumed. On the other hand, **fruit eating** is usually an example of mutualism. The fruit is consumed, but individual seeds are dispersed without the embryo being killed. **Scatter hoarding**, such as a squirrel–acorn interaction, is a fascinating combination of seed predation and seed dispersal.

Types of parasites

Parasites that live outside the host are termed **ectoparasites**, while those that live inside the host are **endoparasites**. Parasites also differ with regard to: (i) the level of harm caused to the host (virulence); (ii) their ability to move from one host to another (transmissibility); and (iii) the degree to which they are restricted to a given type of host species (specialization or host-specificity).

The table overleaf summarizes ecological characteristics of parasite–host, herbivore–plant, and predator–prey relationships. In the next few chapters, we will first discuss interspecific competition before moving on to mutualism, followed by parasite–host, predator–prey, and herbivore–plant relationships.

Comparisons of ecological characteristics along a spectrum from parasite to predator. Modified from Dobson (1982) and Toft (1986).

Organism type:	Microparasite	Macroparasite	Parasitoid	Small herbivore	Large herbivore	Solitary predator	Pack predator
Examples	Viruses/bacteria	Tick	"Parasitic" wasp	Insect	Howling monkey or ungulate	Lynx or mountain lion	Wolf
Body size	Much smaller than host	Smaller than host	Adult stages are similar in size	Smaller than host	Larger than host when feeding on grasses or herbs; smaller if host is a shrub or tree	May be larger or smaller than prey	Individuals are usually smaller than prey
Intrinsic rate of increase, r, as compared to host	Much larger than host	Larger than host	Slightly smaller than host	Larger than host	Comparable to or smaller than host	Usually smaller than prey	Comparable to or smaller than prey
Effect of interaction on the host individual	No effect to seriously deleterious	Mildly to seriously deleterious in vertebrate host; can be highly virulent in intermediate host	Eventually fatal	Little effect to seriously deleterious	Little effect to very deleterious	Immediately fatal	Immediately fatal

7

Interspecific competition

7.1 Introduction

The ecological niche

Our understanding of interspecific competition and the development of the concept of the niche have shared an interesting history. The term ecological niche can be traced both to the American ecologist Joseph Grinnell and to the British ecologist Charles Elton. In his book *Animal Ecology*, Elton (1927) defined niche in the context of trophic position and feeding habits of an animal. Elton viewed the niche as a subdivision within the traditional trophic grouping of herbivore, carnivore, etc. Grinnell, however, described the niche in terms of the potential distribution of a species over habitat types, and in his paper on the feeding habits of the California thrasher he connected competitive exclusion to the term niche: "It is, of course, axiomatic that no two species regularly established in a single fauna have precisely the same niche relationships" (Grinnell 1917, p. 433). However, because various authors used the term niche differently, and because there was no formal definition of the niche, many ecologists used the term informally. For example, in his ecology text Krebs (1994) defined niche as "the role or 'profession' of an organism in the environment; its activities and relationships in the community." In their ecology text, Begon *et al.* (1986) defined the niche more formally as "the limits, for all important environmental features, within which individuals of a species can survive, grow and reproduce."

Ricklefs (1997) used both the informal and the formal definition of the niche: "the ecological role of a species in the community" (informal) and "the ranges of many conditions and resource qualities within which the organism or species persists, often conceived as a multidimensional space" (formal).

The second, formal, definition is based on the concept, devised by G.E. Hutchinson (1957), of the niche as an N-dimensional hypervolume. In Hutchinson's definition, important environmental features such as temperature, pH, nutrient availability, food types and/or sizes are depicted as niche dimensions. For each dimension the limits are identified. For example, what are the minimum and maximum temperatures within which the species can survive? Or, what are the smallest and largest food particles upon which a species can feed? The range within which the species can survive represents the fundamental niche for the species, for that niche dimension. All of these niche dimensions combined represent the overall **fundamental niche** of the species. If we take three niche dimensions into account, the niche can be represented as a three-dimensional volume. If we consider N dimensions, the resultant niche was described by Hutchinson as an **N-dimensional hypervolume**.

The fundamental niche is the largest ecological niche that an organism or species can occupy. It is based mostly on interactions with the physical environment and is always in the absence of competition. The **realized niche**, on the other hand, is that portion of the fundamental niche that is occupied after interactions with other species. That is, the niche after competition. The realized niche must be part of, but smaller than, the fundamental niche.

7.2 Interspecific competition: early experiments and the competitive exclusion principle

Early in the twentieth century, Tansley (1917) experimentally demonstrated the potential power of interspecific competition in shaping ecological communities. Tansley had observed that closely related plant species living in the same region were often found in different habitats or different soil types. For his experiment he selected two species of an herbaceous perennial, bedstraw, in the genus *Galium* (Rubiaceae). One species, *G. saxatile*, is normally found on peaty, acidic soils, while the second species, *G. sylvestre*, is an inhabitant of limestone soils. Tansley obtained soils from both areas, planted each species singly in each soil type and then placed the two species together in each soil. He found that each species, when planted alone, was able to survive in both soils. Therefore the fundamental niche for both species includes both acidic, peat-rich soil and limestone soil. However, growth and germination were best on the soil where the *Galium* species was normally found. When grown together on limestone soil, *G. sylvestre* overgrew and outcompeted *G. saxatile*. The opposite was true in the acidic peat soil. At this early date, Tansley had established that competitive exclusion could be demonstrated, and that the results differed by environment.

The work by Tansley, however, was not developed further until the publication by Gause (1934) of *The Struggle for Existence*. Through a series of experiments with yeast (Gause 1932) and protozoans, Gause found that competitive exclusion is observed most often between two closely related species (two species in the same genus, for example), when grown in a simple, constant environment. For example, see Fig. 7.1. Gause prepared organic extracts

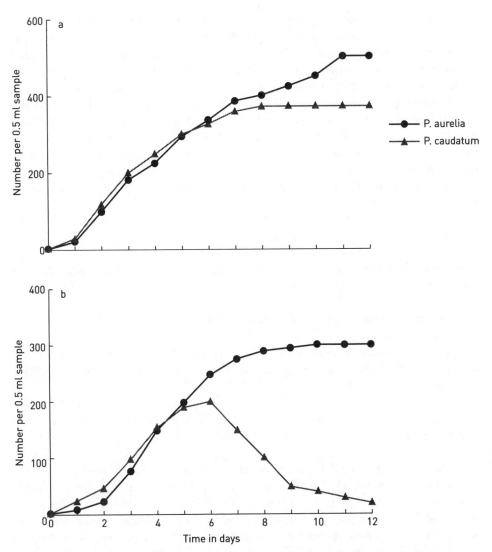

Figure 7.1 Population dynamics of *Paramecium aurelia* and *P. caudatum*: (a) when grown separately; (b) when grown together. Adapted from Gause (1934).

and introduced bacteria as food. When either *Paramecium caudatum* or *P. aurelia* was introduced alone, each flourished and grew logistically, leveling off at a carrying capacity (Fig. 7.1a). When placed together, however, *P. caudatum* diminished and eventually went extinct, while *P. aurelia* grew to a steady level (Fig. 7.1b). There are two lessons from this experiment. First, two closely related species were unable to coexist in the simple test-tube environment. Second, even though we declare *P. aurelia* the "winner," notice that its steady state of approximately 300 per 0.5 ml sample (Fig. 7.1b) is less than the carrying capacity of 500 when this species was grown alone (Fig. 7.1a). Recall the definition of competition as a reciprocally negative interaction, meaning that competition has a negative effect, even on the winners.

Gause's laboratory work inspired many others who worked with yeast, grain beetles, fruit flies, and other organisms easily grown in the laboratory. Crombie (1945, 1946, 1947), Thomas Park (1948) and others did some particularly interesting work with grain beetles. Different species were grown together in vials or other simple environments, usually resulting in competitive exclusion. Crombie, however, showed that the species excluded could change depending on the temperature under which the experiment was run. And when he added glass tubing as a refuge, he found that a grain beetle in the genus *Oryzaephilus* was able to coexist with a related species in the genus *Tribolium*. Without the glass tubing, *Tribolium* drove *Oryzaephilus* extinct. All of this work suggested that closely related species, whose niches are very similar, are unlikely to coexist in a simple environment. From his research with *Paramecium*, Gause (1934) proposed what became known as Gause's theorem or principle:

A *Two species cannot coexist unless they are doing things differently.*

This was eventually rephrased such that competition and the niche concept became integrated.

B *No two species can occupy the same ecological niche.*

Based on such results, Hardin (1960), three years after the publication of Hutchinson's definition of the niche as an *N*-dimensional hypervolume, proposed the **competitive exclusion principle**:

Species which are complete competitors, that is, whose niches overlap completely, cannot coexist indefinitely.

When comparing the fundamental niches of competing species it becomes obvious that their niches overlap on many, if not all, dimensions. On the other hand, if we look at enough niche dimensions, since all species are genetically differentiated from one another, each species will have a unique niche. Therefore, complete niche overlap between two species is virtually impossible. Yet **competitive exclusion** does occur. A question we might pose is: how closely can niches overlap before competitive exclusion occurs? The other side of that question is: under what conditions do potential competitors coexist?

7.3 The Lotka–Volterra competition equations

The first important attempts to model interspecific competition were devised by Lotka (1925) and Volterra (1926). They based their models on the logistic equation, which is, of course, a model of intraspecific competition.

In modeling interspecific competition, Lotka and Volterra assumed that the growth rate of each species would be decreased as the population of its competitors increased. Thus the impact of species 2 on the growth rate of species 1 is expressed as a modification of the logistic equation. Simultaneously, the number of individuals of species 1 modifies the growth of species 2. To model competition between two species, Lotka and Volterra wrote two simultaneous equations, one for each species. Each equation is based on the logistic, but includes a new term, the competition coefficient (α_{ij}), which describes the effect of one species on another. In the equations which follow:

N_1 = the number of individuals of species one;
N_2 = the number of individuals of species two;
r_1 = the intrinsic rate of increase of species one;
r_2 = the intrinsic rate of increase of species two;
K_1 = the carrying capacity of species one;
K_2 = the carrying capacity of species two;
α_{12} = the competition coefficient: effect of species two on species one;
α_{21} = the effect of species one on species two;
t = time.

$$dN_1/dt = r_1 N_1 \left[\frac{K_1 - N_1 - \alpha_{12}N_2}{K_1} \right] \tag{7.1}$$

$$dN_2/dt = r_2 N_2 \left[\frac{K_2 - N_2 - \alpha_{21}N_1}{K_2} \right] \tag{7.2}$$

The competition coefficients try to model the impact of adding one individual of species i on species j. The value of the competition coefficient is usually between 0 and 1, for the following reasons. (i) A competition coefficient of zero would mean that there is no competition between the two species. If that were the case, there is no reason to try to model this interaction. (ii) If the competition coefficient were negative, the implication would be that species two actually benefits the growth rate of species one. The interaction between species one and two would then be **mutualistic**. (iii) Notice that the number of individuals of both species one and two decreases the carrying capacity. N_1 (taken directly from the logistic equation) is unmodified and represents intraspecific competition. N_2 is modified by the competition coefficient and represents interspecific competition. The implied α-value for **intraspecific** competition is therefore 1.0. To express this for the numerator on the right side of the equation, we could write $K_1 - \alpha_{11}N_1 - \alpha_{12}N_2$. Thus, when we say that the usual value of the interspecific competition coefficient is usually less than 1, we are saying that interspecific competition is almost always less intense and diminishes growth less than intraspecific competition, to which we have assigned the coefficient value of 1.0.

If we analyze this assumption in terms of niche overlap, the preceding makes intuitive sense. That is, in intraspecific competition a species is competing with members of its own species, with which is has total niche overlap (Actually, since there should be genetic variation within the species there would not be total niche overlap, and in some species males and females have different niches, particularly when it comes to food sizes or prey taken.) In interspecific competition, by contrast, the two species should have different niches. Thus an individual of species two should compete less intensely with an individual of species one than with an individual of its own species. Growth rate should be diminished less by inter- than by intraspecific competition. Hence the α_{ij}-values are normally less than one and greater than zero.

Solving the Lotka–Volterra equations presents a number of problems. In addition to the large number of parameters that would have to be estimated (r, K and alpha for each species) we would be assuming that changes with density were linear in both sets of differential equations. However, it is possible to do an "equilibrium analysis." That is, we can analyze the equations after the results of competition are complete and both species are no longer growing. At equilibrium, the left sides of the equations, which represent growth, can be set to zero. These equations can then be written as:

$$dN_1/dt = 0 \quad \text{and} \quad dN_2/dt = 0 \tag{7.3}$$

$$0 = r_1 N_1 \frac{K_1 - N_1 - \alpha_{12}N_2}{K_1} \quad \text{and} \quad 0 = r_2 N_2 \frac{K_2 - N_2 - \alpha_{21}N_1}{K_2} \tag{7.4}$$

These equations, again, have no interesting explicit solutions. They can be solved if r_1 or $r_2 = 0$, but that simply means one of the species is not viable in this environment. If N_1 or $N_2 = 0$, then we have competitive exclusion. If K_1 or $K_2 = 0$, we have an undefined situation. What is left is to analyze the following expressions:

$$0 = K_1 - N_1 - \alpha_{12}N_2 \tag{7.5}$$

$$0 = K_2 - N_2 - \alpha_{21}N_1 \tag{7.6}$$

An ensuing graphical analysis, based on the steps outlined below, produces four possible solutions.

First we solve Equation 7.5 for N_1 and for N_2, producing Equations 7.7a and 7.7b:

$$N_1 = K_1 - \alpha_{12}N_2 \tag{7.7a}$$

$$N_2 = \frac{K_1 - N_1}{\alpha_{12}} \tag{7.7b}$$

If species one wins in competition and species two goes extinct, we have Equation 7.7c:

$$N_1 = K_1 \tag{7.7c}$$

If species two wins and species one goes extinct, we have 7.7d.

$$N_2 = \frac{K_1}{\alpha_{12}} \tag{7.7d}$$

From Equation 7.6, we get the following two equations:

$$N_1 = \frac{K_2 - N_2}{\alpha_{21}} \tag{7.8a}$$

$$N_2 = K_2 - \alpha_{21}N_1 \tag{7.8b}$$

If species two wins and species one goes extinct, we get:

$$N_2 = K_2 \tag{7.8c}$$

If species one wins and species two goes extinct, we get:

$$N_1 = \frac{K_2}{\alpha_{21}}$$ (7.8d)

Therefore, if species two is extinct, Equations 7.7c and 7.8d tell us that: $N_1 = K_1 = K_2/\alpha_{21}$ and:

$$\alpha_{21} = \frac{K_2}{K_1}$$ (7.9)

Similarly, if species one is extinct, Equations 7.8c and 7.7d tell us that:

$$N_2 = K_2 = K_1/\alpha_{12}$$

and:

$$\alpha_{12} = \frac{K_1}{K_2}$$ (7.10)

According to these results, the competition coefficients are determined by the ratio of the carrying capacities.

To begin the graphical analysis, we start with Equation 7.5. As we have seen from Equation 7.7c, when species two = 0, then $N_1 = K_1$. Similarly, when species one = 0, $N_2 = K_1/\alpha_{12}$ (7.7d). If we graph N_1 versus N_2 (Fig. 7.2), we can use the two points $(K_1, 0)$ and $(0, K_1/\alpha_{12})$ to produce a line that represents saturation levels for species one. Combinations of populations 1 and 2 below (to the left of) the line are such that population 1 continues to increase. Combinations above the line (to the right) lead to a decrease in the size of population 1 (Fig. 7.2). The line itself is known as the zero isocline for population 1.

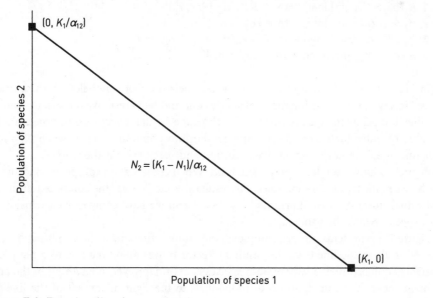

Figure 7.2 Zero isocline for species 1.

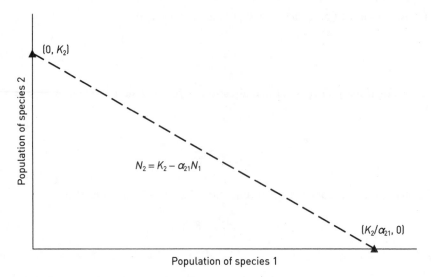

$$N_2 = K_2 - \alpha_{21}N_1$$

Figure 7.3 Zero isocline for species 2.

In a similar fashion, from Equation 7.8c we know that if population $1 = 0$, then $N_2 = K_2$. From Equation 7.8d, we have that if population $2 = 0$, then $N_1 = K_2/\alpha_{21}$. From the resultant points $(0, K_2$ and $K_2/\alpha_{21}, 0)$, we can draw a second line or zero isocline, representing saturation levels for species two. Again, combinations of populations 1 and 2 to the left or below the line result in increases toward the carrying capacity for population 2; combinations above the line lead to a decrease in population 2 (Fig. 7.3).

We now place these zero isoclines or saturation levels on the same graph. Four combinations are possible when drawing these lines:

1 $K_1/\alpha_{12} > K_2$ combined with $K_1 > K_2/\alpha_{21}$ (Fig. 7.4);
2 $K_2 > K_1/\alpha_{12}$ combined with $K_2/\alpha_{21} > K_1$ (Fig. 7.5);
3 $K_1 > K_2/\alpha_{21}$ combined with $K_2 > K_1/\alpha_{12}$ (Fig. 7.6); and
4 $K_1/\alpha_{12} > K_2$ combined with $K_2/\alpha_{21} > K_1$ (Fig. 7.7).

In Fig. 7.4, both populations increase when their numbers are below both saturation lines. However, when combinations of species one and two are above the species-two zero isocline, but below the species-one zero isocline, the resultant vectors move the combined number of individuals toward only one equilibrium point, and that is when $N_1 = K_1$. In case one, then, we have competitive exclusion and species one is the winner.

Figure 7.5 illustrates the opposite situation. When combinations of species one and two are between the two saturation lines, the resultant vector moves the combined number of individuals toward an equilibrium at $N_2 = K_2$. Again we have competitive exclusion, but now species two is the winner.

Figure 7.6 represents a more complex, ambiguous situation. If the combination of N_1 and N_2 is along the line 0–S, the resultant vector moves along the line to a temporary, unstable equilibrium at S, where both species coexist. However, if anything in the environment moves the combinations of individuals to the right or the left of the line 0–S, then the resultant vectors move rapidly toward competitive exclusion. However, in this

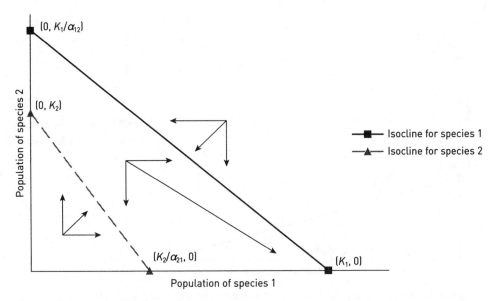

Figure 7.4 Species 1 wins.

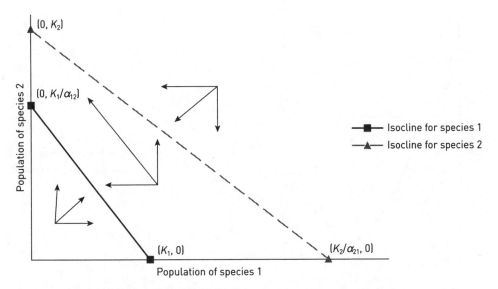

Figure 7.5 Species 2 wins.

case there are two stable equilibrium points. If the combinations of the two species produce points to the left of the line 0–S, the result is competitive exclusion with species two as the winner. If the points are to the right of the line 0–S, species one is the winner. Therefore this set of conditions will produce an unstable equilibrium or competitive exclusion with an indefinite winner (a stochastic result).

Finally, in Fig. 7.7 we have a situation in which each species slows its own growth more than that of its competitor. This allows a stable equilibrium with coexistence of the two species at point E.

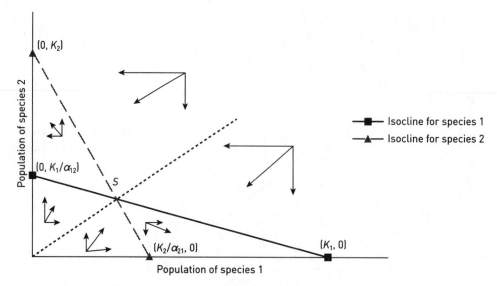

Figure 7.6 Unstable equilibrium (S) or competitive exclusion with an indefinite winner.

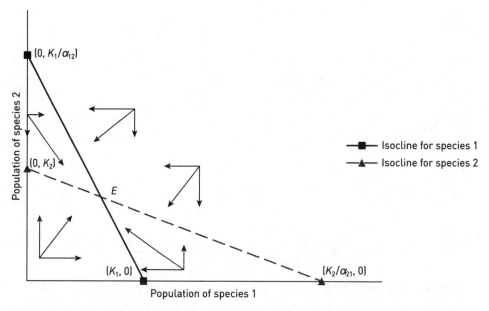

Figure 7.7 Coexistence of both species at the stable equilibrium, E.

7.4 Laboratory experiments and competition

The classic works of Thomas Park and his colleagues (Park 1954, Neyman *et al.* 1956) illustrate cases 1–3, as well as the fact that competitive outcomes are not necessarily deterministic (see Table 7.1). In Park's most often cited study, he raised two species of flour

Table 7.1 Results of competition between two species of flour beetles (*Tribolium confusum* and *T. castaneum*) in six different temperature and humidity combinations. Based on Park (1954).

Temperature (°C)	Percent relative humidity	"Climate"	Percent of replicates in which one species or the other wins	
			T. confusum	*T. castaneum*
34	70	Hot–moist	0	100
34	30	Hot–dry	90	10
29	70	Warm–moist	14	86
29	30	Warm–dry	87	13
24	70	Cold–moist	71	29
24	30	Cold–dry	100	0

beetles, *Tribolium castaneum* and *T. confusum*, in vials of sifted flour under different temperature and humidity regimes (Table 7.1).

In these experiments the initial numbers of individuals were equal. That is, $N_1 = N_2$ at $t = 0$. In the top and bottom rows of the chart we have typical deterministic results, illustrating the first two graphical analyses (Figs 7.4 and 7.5) in which competitive exclusion occurs. The other four results are stochastic in that the result of any single experiment is unpredictable. Neyman *et al.* (1956) conducted further experiments using the cold–moist regime. They found that one species or the other was a deterministic winner when given a large initial numerical edge (Fig. 7.6). But there still existed an "indeterminate" zone (in the region of $N_1 \approx N_2$) where either species could still win.

A natural extension of these two-species competitive interactions was the work of Vandermeer (1969). Building on the work of Gause (1934), Vandermeer raised four species of protozoans in monocultures, thereby determining their r_m and carrying capacities. Next he grew each of the species in pair-wise combinations. From these experiments he estimated the pair-wise competition coefficients. The general results were as follows: *Paramecium aurelia* depressed the growth of *P. caudatum*, drove *P. bursaria* extinct, and depressed the growth of *Blepharisma* sp. *P. caudatum* also drove *P. bursaria* extinct and drove *Blepharisma* sp. to a very low population level. *Blepharisma and P. bursaria* had little effect on each other and coexisted.

Based on the pair-wise competition coefficients and *K*-values, Vandermeer predicted the outcome of placing all four species together in a community, including predicted growth rates and carrying capacities. His predictions and the actual outcomes were surprisingly similar, leading to the conclusion that higher-order interactions (the combined effects of two species on a third) and nonlinear relationships were not significant in this community.

Other multi-species studies, however, did not confirm Vandermeer's findings. For example, Neill (1974) conducted a series of replicated removal experiments in a laboratory microcosm containing four species of microcrustaceans and associated algae and bacteria. Each species of crustacean was grown in pairs and estimates of population density were made under each regime. Computation of competition coefficients showed that the α-values depended on the community composition. The joint effects of two species on a third (in a three-species community) were not as predicted from the separate inter-actions in the two-species systems. Such results helped push ecologists to look for a different theoretical approach to competition.

7.5 Resource-based competition theory

David Tilman (1976, 1987) and others pointed out that the Lotka–Volterra equations were "phenomenological" and not "mechanistic." That is, competition coefficients were merely measures of the effect of one species on the growth rate of another. They are estimated from experiments in which two species are grown together. Therefore they are not an inde-pendently derived value that allows one to predict coexistence or competitive exclusion, or, in the latter case, which of two species should win. Furthermore, a competition coeffi-cient does not help determine the mechanism of competition; we have no information on what resource the species might be competing for. If competition is really concerned with a resource in short supply, we need to understand what the resource is and how each species is using it before we can understand the potential competitive interaction.

Tilman, a particularly strong advocate of a mechanistic approach to competition (1976, 1981, 1982), developed what is now known as **resource-based** competition theory. In so doing he brought together ideas from a variety of disciplines, including microbiology, enzyme kinetics, and agricultural chemistry. For example, the idea that population growth is con-strained by the depletion of critical resources can be traced to the agricultural chemist Liebig (1840) and his law of the minimum. Liebig asserted that a population increases until the supply of a single critical resource becomes limiting. For example, plant growth may con-tinue until the amount of phosphorus, nitrogen, light, or soil moisture becomes limiting. According to Liebig's law, if plant growth is constrained by phosphorus and a farmer adds phosphorus fertilizer, plant growth will continue until another resource, such as nitrogen, becomes limiting. If the farmer adds nitrogen, then soil moisture may become the limit-ing factor. Liebig's law is overly simple in that two or more resources may interact to limit a population, but it puts resource supply into the context of population regulation, and therefore competition.

In the resource-based approach to competition, we need to couple the availability of resources to population growth. We begin by considering a renewable resource. If the supply rate is not affected by the population of potential consumers, the rate of supply (for example, phosphorus arriving at a small lake via a local stream) could be considered a constant (Eqn. 7.11), where R_i is the quantity of the resource i and k_{Ri} is the supply rate:

$$dR_i/dt = k_{Ri} \tag{7.11}$$

If this one resource sets the limit of growth for a population (as in Liebig's law of the minimum), the growth rate depends on both the resource level and the density of the population itself. That is, dN/dt is a function of both the resource supply rate, k_{Ri}, and population size, N. Assume that each individual must consume the resource at rate q to maintain itself. We ignore the possibility of storing the resource for later use. The popu-lation of N individuals will consume the resource at the rate qN. The remaining resources may eventually be lost downstream or in the lake sediments. Alternatively, they may be taken up by the consumers and used for the production of new individuals in the population. The supply rate of the resource available for reproduction is therefore $k_{Ri} - qN$. Suppose each individual converts the resource into new individuals with efficiency b. Population growth can now be written as:

$$dN/dt = bN(k_{Ri} - qN) \qquad (7.12)$$

We can rearrange this equation to:

$$dN/dt = bk_{Ri}N\left(1 - \frac{qN}{k_{Ri}}\right) \qquad (7.13)$$

Equation 7.13 is a form of the logistic equation with $r = bk_{Ri}$ and the carrying capacity equal to k_{Ri}/q. We have now coupled resource supply with population growth rate. The growth rate of the population is proportional to the supply of the critical resource and the carrying capacity is the resource supply divided by the amount needed for maintenance per individual.

Now envision the relationship between resource availability and population growth. As described earlier, in a density-dependent population, the per capita growth rate declines with population size, yielding a negative slope. Logistic-like equations assume that the resources become scarcer as populations grow. However, if we graph per capita growth versus an increasing supply of resources, the predicted slope is positive (Fig. 7.8). As the resource becomes more abundant, the population growth rate increases. Per capita growth, however, finally levels off and declines to zero when some other resource limits the population (as proposed by Liebig).

The next step is to introduce a mortality rate, m. Instead of assuming that resources are simply turned into births, we assume instead that a certain minimum level of the resource is needed to maintain the population. Therefore, as shown in Fig. 7.9, we introduce the concept of R^*, **the level of the resource needed to balance mortality**. If the resource is provided at the rate R^*, then $dN/dt = 0$, growth just balances mortality, and the population

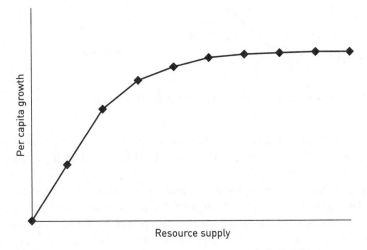

Figure 7.8 Per capita growth as a function of resource availability.

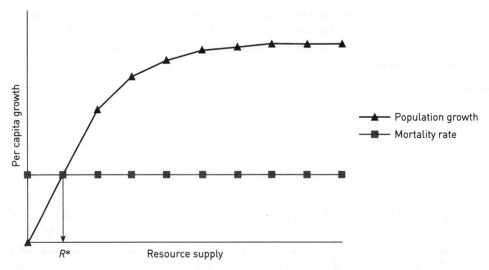

Figure 7.9 Per capita growth as a function of resource availability, but with a constant mortality rate. R^* = the amount of resource producing a per capita growth rate of zero.

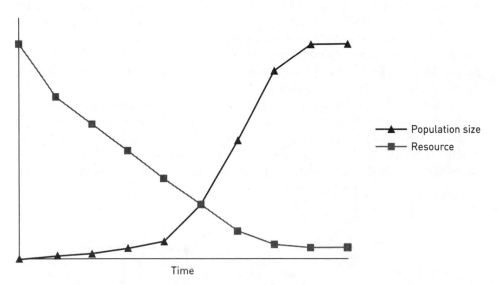

Figure 7.10 Population and resource dynamics.

maintains itself. If the resource is provided at a level less than R^*, growth is less than mortality, and the population declines. Conversely, if the resource is provided at a level greater than R^*, we have growth ($dN/dt > 0$).

Examine Fig. 7.10. As the population grows, the resource is increasingly depleted, and once the resource declines to the level R^*, the population should stop growing. The population thereafter remains steady at a size determined by the resource quantity R^*.

Tilman (1976) realized that an independently derived R^* could be used to predict population dynamics and, ultimately, competitive interactions. He first used the

Michaelis–Menton enzyme kinetics equation, which is normally employed to describe the relationship between cellular metabolism and substrate concentrations. This equation had also been used by microbiologists to describe the growth rate of bacteria on organic substrates (Monod 1950).

In the Michaelis–Menton equation, the growth rate, μ, on a given substrate or re-source, R, is set equal to the maximum growth rate modified by the concentration of the resource and by a value known as K_μ, the **half-saturation constant** for the resource in ques-tion. K_μ **is the concentration of the resource that produces half the maximum growth rate.** If μ_{max} is the maximum growth rate of the population, and R is the resource or substrate, the resultant growth rate (μ) according to the Michaelis–Menton equation is:

$$\mu = \mu_{max}\left(\frac{R}{R + k_\mu}\right) \tag{7.14}$$

In Monod's version of this equation, instead of μ we substitute dN/dt, and we use b instead of μ_{max}. We will also now define the half-saturation constant for any given resource as K_i, rather than K_μ. The result is Equation 7.15:

$$dN/dt = \frac{bNR}{K_i + R} \tag{7.15}$$

Per capita growth is shown in Equation 7.16

$$dN/Ndt = \frac{bR}{K_i + R} \tag{7.16}$$

This equation tells us that at very high resource levels, the expression $R/(K_i + R)$ is very close to one and the per capita growth is simply b. This is the maximum growth rate, ignoring the mortality rate, m, and is equivalent to the unmodified intrinsic rate of increase r_{max}. We can add a constant mortality rate to Monod's equation as follows:

$$dN/Ndt = \frac{bR}{K_i + R} - m \tag{7.17}$$

We can set dN/Ndt (per capita growth) equal to zero and solve for R^*, which is the resource level at which growth stops. The solution, as shown in Equation 7.18, allows us to predict R^* if we know three variables: (i) the half-saturation constant, K_i, (ii) the maximum growth rate, b, and (iii) the mortality rate, m. The advantage of this approach, as opposed to that of the Lotka–Volterra equations, is that a resource is identified and a variable, R^*, can be derived from simple experiments, which can then be used to compare how populations of different species respond to different resource levels.

$$R^* = \frac{mK_i}{b - m} \tag{7.18}$$

For example, Tilman (1976) identified critical resources for which two species of plank-tonic algae were likely to be competing. Two freshwater diatom species were grown under different levels of the important limiting nutrients, phosphate (PO_4^{-3}) and silicate (SiO_4^{-4}). Using a growth vessel known as a chemostat, he determined the growth rates of each species cultured singly and also determined the limiting concentrations of phosphate and silicate. From these experiments he determined the values of the half-saturation constants for each species for both phosphate and silicate. By examining the ratios of silicate to phosphate utilizations he was able to establish boundaries for competition. He found that when both species were limited by phosphate, the diatom *Asterionella* won in competition with *Cyclotella*. When both species were limited by silicate, *Cyclotella* won. However, when both nutrients were simultaneously in short supply, the two species coexisted. Tilman minimized mortality rates and was able to predict which species would win based on growth rates and the half-saturation constants. He found that the species with the lower half-saturation constant (and therefore lower R^* in this case) would win if both species were limited by a single nutrient. However, if each species was limited by a different nutrient, the density of each species was held in check through intraspecific competition and the two species coexisted. The incorporation of the equations originally developed by Michaelis/Menton and Monod, and the use of half-saturation constants, allowed Tilman to predict the results of competition involving two species and two resources.

Subsequent papers by Tilman (1981), Tilman *et al.* (1982), Hansen and Hubbell (1980), and a book by Tilman (1982), have become the foundations for "resource competition" as a distinct theory and an important area of inquiry. Tilman's work expanded from the laboratory work on diatoms and other algae to competition studies of terrestrial plant communities at the Cedar Creek Natural History Area in Minnesota. Hansen and Hubbell (1980) and Tilman (1982) elaborated on the Monod equation, but they still emphasized the critical parameter, R^*. This parameter can be used to predict which one species will survive in a mixed-species culture when there is a single limiting nutrient. According to what is now known as the R^*-**rule**, for any given resource (R), if we determine the R^*-value for each species when grown alone, the species with the lowest R^* should competitively exclude all other species, given enough time and a constant environment.

In deriving their version of the R^*-rule, Hansen and Hubbell (1980) assumed that two competitors are grown in a continuous culture with a continuous input of a nutrient (R) as well as an effluent rate, which is equivalent to a death rate, m. The growth rates for two competing species were defined as:

$$dN_1/dt = \frac{b_1 R N_1}{K_1 + R} - mN_1 \tag{7.19}$$

and

$$dN_2/dt = \frac{b_2 R N_2}{K_2 + R} - mN_2 \tag{7.20}$$

where
b_i = maximum cell division rate (= r_{max});
R = the concentration of the one limiting resource in the culture;
K_i = half saturation constant for the limiting resource;
m = death rate, here due to outflow;
N_i = concentration of cells in the culture (population size).

If we do an equilibrium analysis, and set $dN_i/dt = 0$, the result is:

$$\frac{b_i R}{K_i + R} - m = 0$$

If we set $K_i = R$, then $b_i/2 = m$.

Thus one solution is that growth stops when the concentration R equals the half-saturation constant. At that time the cell division rate is at half of its maximum level ($r_a = 0.5 r_m$) and just equals the death rate. Thus, as explored above in a graphical analysis, the equilibrium resource availability, R^*, occurs when the growth function intersects the line m representing mortality. This is the amount of resource needed to just sustain the population (growth just offsets mortality).

When Hansen and Hubbell (1980) analyzed Equations 7.19 and 7.20, they found that they are globally stable when either: (i) all competitors die out, or (ii) one species survives while the second species dies out – that is, when competitive exclusion occurs. Which species survives depends on the critical parameter, R^*, which we already saw in Equation 7.18 as $R^* = mK_i/(b - m)$.

R^* must be less than R_0, otherwise all species die out because of lack of resources. If all R^*-values for all species are less than R_0, then, according to the R^*-rule, the species with the lowest R^* wins. Again, three parameters, the half-saturation constant, the intrinsic rate of increase, and the death rate, combine to determine which species wins in competition for any given resource. This is not predicted from classical Lotka–Volterra competition theory. A species with a high affinity for the resource can still lose if it has a low growth rate (r-value) and a high death rate.

Hansen and Hubbell (1980) confirmed the expected results with several species of bacteria auxotrophic for tryptophan. In the example in Table 7.2, based on Hansen and Hubbell (1980), we expect population two to outcompete the other three species for this resource. Species three is expected to outcompete both populations one and four, and population one should win in competition with population four.

Tilman and others have tested the R^*-rule on algae, other microorganisms, higher plants, and zooplankton. For example, Tilman (1982) grew two species of diatoms, *Asterionella formosa* and *Synedra ulna*, which require SiO_2 for cell-wall structure, in a laboratory chemostat. The R^*-values for *A. formosa* and *S. ulna* were 1.0 µM and 0.4 µM, respectively. When

Table 7.2 Example of R^* calculations based on Hansen and Hubbell (1980). K, half saturation constant; m, mortality rate; b, maximal growth rate; r_a, actual growth rate $= b - m$. $R^* = mK_i/(b - m) = mK_i/r_a$.

Population, i	K_i (g L^{-1} × 10^{-6})	m per hour	b per hour	r_a ($= b - m$) per hour	R^* (g L^{-1} × 10^{-6})
One	4.0	0.05	0.25	0.20	1.00
Two	4.1	0.05	0.50	0.45	0.46
Three	6.5	0.05	0.50	0.45	0.72
Four	20.0	0.05	0.25	0.20	5.00

grown together, as the silicate levels eventually were reduced to 0.4 µM, *Synedra* hung on at its equilibrium number, whereas *Asterionella* went extinct.

See Grover (1997) for a relatively recent review covering both the status of the theory and relevant field and laboratory studies.

Resource competition: conclusion

Applying the R^*-rule and resource models to higher animals has proved less useful and, as usual, theories developed for simple laboratory settings run into a number of problems and complications when applied to field situations. In addition, we should recognize that other modes of competition, such as interference competition, do not follow the simple rules of resource-based models. Moreover, competition on any given site does not necessarily involve all species in the community. The R^*-rule would only apply to the species which have "shown up" in a given habitat patch. Therefore, as elaborated below (also see Tilman 1994, 1999, Hubbell 2001), a community is not limited to the species that are the superior competitors.

7.6 Spatial competition and the competition–colonization trade-off

The idea that multiple species can coexist in a community without yielding to the superior competitors can traced to the competition–colonization trade-off idea first proposed by Levins and Culver (1971), and elaborated by Hastings (1980), Tilman (1994), Yu and Wilson (2001), and others. Recall that in a metapopulation, two species can coexist if one is a superior competitor and the other is a better colonizer. Remember also that in a metapopulation the increase in the proportion, P, of sites occupied by a species was based on the colonization rate, cP, times the proportion of sites occupied and available $(1 - P)$, minus the local extinction or mortality rate, εP. When Equation 7.21 is set equal to zero and we solve for $\hat{P}$, we have the proportion of habitat sites occupied at equilibrium (Eqn. 7.22). The colonization rate necessary for equilibrium is displayed as Equation 7.23.

$$\frac{dP}{dt} = cP(1 - P) - \varepsilon P \tag{7.21}$$

$$\hat{P} = 1 - \frac{\varepsilon}{c} \tag{7.22}$$

$$c = \frac{\varepsilon}{1 - \hat{P}} \tag{7.23}$$

We can generate equations for two species, one for the superior competitor, P_1 (Eqn. 7.24, based on Eqn. 7.21), and one for the superior colonizer, P_2 (Eqn. 7.25). c_1, c_2, ε_1, and ε_2 represent colonization and mortality rates for species one and two, respectively. Equation 7.24 is the same equation as if the species lived by itself, since it is assumed to be unaffected by the inferior competitor. On the other hand, the inferior competitor can only colonize sites occupied by neither species $(1 - P_1 - P_2)$. Thus its successful colonization of new sites is a product of $c_2 P_2 \times (1 - P_1 - P_2)$. This success rate is decreased by its natural patch extinction rate $(\varepsilon_2 P_2)$ and its displacement through competition with species one $(c_1 P_1 P_2)$.

$$\frac{dP_1}{dt} = c_1 P_1 (1 - P_1) - \varepsilon_1 P_1 \tag{7.24}$$

$$\frac{dP_2}{dt} = c_2 P_2 (1 - P_1 - P_2) - \varepsilon_2 P_2 - c_1 P_1 P_2 \tag{7.25}$$

As noted in Equation 7.23, in order to attain its equilibrium, the colonization rate of the superior competitor must satisfy. $c_1 \geq \varepsilon_1/(1 - \hat{P}_1)$. To find the equivalent value for species two, set Equation 7.25 = 0, and divide all terms by P_2. The result is Equation 7.26a. The equilibrium value of P_2 is found by following steps 7.26b to 7.26d:

$$0 = c_2 - c_2 P_1 - c_2 P_2 - \varepsilon_2 - c_1 P_1 \tag{7.26a}$$

$$c_2 - c_2 P_2 = c_2 P_1 + \varepsilon_2 + c_1 P_1 \tag{7.26b}$$

$$c_2 (1 - P_2) = c_2 P_1 + \varepsilon_2 + c_1 P_1 \tag{7.26c}$$

$$\hat{P}_2 = 1 - \frac{c_2 P_1 + \varepsilon_2 + c_1 P_1}{c_2} \tag{7.26d}$$

For species two to remain viable $\hat{P}_2$ must be greater than zero. Therefore we set the right side of Equation 7.26d > 0, and set $P_1 = \hat{P}_1$, producing the inequalities in Equations 7.27a to 7.27d:

$$1 > \frac{c_2 \hat{P}_1 + \varepsilon_2 + c_1 \hat{P}_1}{c_2} \tag{7.27a}$$

$$c_2 > c_2 \hat{P}_1 + \varepsilon_2 + c_1 \hat{P}_1 \tag{7.27b}$$

$$c_2 (1 - \hat{P}_1) > \varepsilon_2 + c_1 \hat{P}_1 \tag{7.27c}$$

$$c_2 > \frac{c_1 \hat{P}_1}{1 - \hat{P}_1} + \frac{\varepsilon_2}{1 - \hat{P}_1} \tag{7.27d}$$

By using Equation 7.23 and replacing $1 - \hat{P}_1$ with ε_1/c_1 on the far right term, we obtain Equation 7.28:

$$c_2 > c_1 \left(\frac{\hat{P}_1}{1 - \hat{P}_1} + \frac{\varepsilon_2}{\varepsilon_1} \right) \tag{7.28}$$

Based on Equation 7.23, $c_1 = \varepsilon_1/(1 - \hat{P}_1)$, we know that for $\hat{P}_1$ to be positive, $1 > \varepsilon_1/c_1$. Therefore, c_1 must be greater than m_1 (Eqn. 7.29a) for the equilibrium abundance of species one to be stable. Similarly, if we take Equation 7.28 and substitute $1 - (\varepsilon_1/c_1)$ for $\hat{P}_1$, we end up with Equation 7.29b:

$$c_1 > \varepsilon_1 \qquad\qquad (7.29a)$$

$$c_2 > \frac{c_1(c_1 + \varepsilon_2 - \varepsilon_1)}{\varepsilon_1} \qquad\qquad (7.29b)$$

As stated by Tilman (1994), these are the "necessary and sufficient conditions for the stable coexistence of a superior competitor and an inferior competitor in a subdivided habitat."

If the mortality rates were equal for the two species, Equation 7.29b simplifies to $c_2 > c_1^2/\varepsilon$ found by Hastings (1980). Since we know that $c_1 > \varepsilon_1$, it follows that the ratio $c_1/\varepsilon > 1$. Therefore c_2 must be greater than c_1. If the mortality rate of the inferior competitor (m_2) is greater than or equal to that of the superior competitor, an examination of Equation 7.29b should demonstrate to you that species two only persists if $c_2 > c_1$. If the inferior competitor has a lower mortality rate, however, Nee and May (1992) have shown that it may coexist with the superior competitor, but only if at least half of the habitat is left available for colonization. This becomes clear if we assign $\hat{P}_1 = 0.5$ in Equation 7.28, resulting in $c_2 > c_1\left(1 + \dfrac{\varepsilon_2}{\varepsilon_1}\right)$. No matter how small the mortality rate of ε_2, we must still have $c_2 > c_1$.

These basic equations have been generalized to multi-species situations by Tilman (1994) and others. Termed the "**spatial-competition hypothesis**," this theory proposes stable coexistence for inferior competitors in a diverse community. Coexistence by spatial competition does assume a two- or even three-way trade-off among competitive ability, colonization ability, and longevity. Since there is actually little evidence for a trade-off between longevity and competitive ability, the major trade-off is assumed to be between competitive ability and colonizing ability. This is, of course, not inconsistent with the MacArthur and Wilson (1963, 1967) proposal for r- and K-selected species, in which r-selected species were assumed to sacrifice competitiveness for colonizing capacity.

As pointed out recently by Yu and Wilson (2001), the displacement-competition model (equals the "spatial-competition hypothesis" of Tilman 1994), assumes that a propagule of the superior competitor will displace an adult of the inferior competitor. Second, the displacement of the adult occurs rapidly enough to prevent reproduction of the inferior competitor. Yu and Wilson stressed that in many ecosystems neither of these assumptions is true. Rather, the juveniles compete among themselves, while waiting for an adult to die. In a forest, for example, seedlings have little, if any, effect on the death rates of adult trees. Once an opening in the canopy occurs, these juveniles compete with each other for the light gap. Yu and Wilson (2001) designated this **replacement** or "lottery" **competition**, and asserted that the competition–colonization trade-off is not sufficient to produce coexistence of inferior with superior competing species. We will not elaborate on their models here. However, they propose that in a lottery system, all that is necessary to allow coexistence of inferior with superior competitors is some form of environmental heterogeneity in patch density combined with either the competition–colonization trade-off or a trade-off between dispersal and fecundity. Since many highly competitive species produce small numbers of large offspring with low dispersal abilities (plants with a small number of large seeds, for example), the basic idea is that spatial variability is essentially a "niche-axis" (Yu and Wilson 2001) that facilitates the high species richness found in many ecosystems.

7.7 Evidence for competition from nature

Connell's barnacles

The classic experimental demonstration of competition in the field was done by Joseph Connell (1961a, 1961b) on the barnacle species *Chthamalus stellatus* and *Balanus balanoides*. These two species are found growing in the rocky intertidal zone off the coast of Scotland. Intertidal zones frequently show vertical zonation of species based on their abilities to survive periods of exposure to the air during low tides, and wave action followed by submersion during high tides. *Balanus* is consistently found on lower rock surfaces, usually near mean tide level or slightly above. *Chthamalus*, however, is found on the upper rocks, between mean high neap tide and mean high spring tide. While the adults of these two barnacle species have non-overlapping distributions, the larvae of both species settle over a wide variety of rock surfaces, showing a great deal of overlap. The question Connell posed was, is the distribution of adults the result of competition, or is there a difference in the fundamental niches of the two species? Connell performed a variety of experiments in which he moved the barnacles to different levels of the intertidal zone. He also experimentally removed one species or the other where the two were growing together, and observed the results of putting the two species together. He found that whenever he removed *Balanus*, *Chthamalus* was able to survive in the lower regions of the intertidal zone. However, in the presence of *Balanus*, *Chthamalus* was overgrown and eventually displaced. In the upper regions of the intertidal zone, however, *Balanus* was unable to survive the long exposures to air during low tides. Since *Chthamalus* was able to survive this exposure, it survives in the upper intertidal zone. Thus the two species occupy mutually exclusive microhabitats due to a combination of competition and differences in their fundamental niches.

Direct observations of competition in ants

Because both worker and soldier ants are numerous, easy to observe, and usually diurnal, aggressive interactions among ant species, demonstrating interference competition, can be documented throughout the world (Holldobler and Wilson 1990). Placing a food bait of tuna or sugar water will provoke competitive interactions in a matter of minutes to hours. Once bait is put out in the West Indies, where there are few ant species, there is a kind of predictable sequence, reminiscent of ecological succession (a kind of "ant succession"). As described by Holldobler and Wilson (1990), first to arrive are workers of *Paratrechina longicornis*, known locally as "hormigas locas" (crazy ants). These workers are very adept at locating food and often are the first to arrive at newly placed baits. They fill their crops rapidly and hurry to recruit nestmates with odor trails laid from the rectal sac of the hindgut. But they are also very timid in the presence of competitors. As soon as more aggressive species begin to arrive in force, the *Paratrechina* withdraw and search for new, unoccupied baits. *Paratrechina* is an example of an "opportunist" species. They are poor competitors, but excellent dispersers. Next to arrive are species known as "extirpators." These species recruit other workers by odor trails and fight it out with competitor species. Examples include species in the genera *Pheidole* and *Crematogaster*, the fire ant (*Solenopsis geminata*), and the "little fire ant" (*Wasmannia auropunctata*). Some of these species have well-developed soldier castes that play a key role in the aggressive interactions. Injury and death are commonplace, and one species eventually dominates the bait. Pre-emption is usually the deciding factor. The colony whose foragers arrive first typically wins; foragers recruit nestmates, who surround the bait. When worker scouts encounter a large number of workers from another colony, they are easily repulsed (Holldobler and Wilson 1990). Species with a third strategy, called "insinuators," also arrive at the baits. These are small colonies with small-sized worker ants such as *Tetramorium simillimum* and species of *Cardiocondyla*. A scout who discovers the bait will recruit only one nestmate at a time. Small size and stealthy behavior allow these individuals to take some of the bait without provoking a response from the extirpator species, – a situation reminiscent of small animals sneaking in and removing bits of food at a lion kill.

Holldobler and Wilson also emphasize that territorial fighting and "ant wars" are common, especially among species with large colonies. Numerous cases have been documented in which introduced ant species have eliminated other species over a few years' time. For example, on Bermuda *Iridomyrmex humilis* has been replacing *Pheidole megacephala* since the former was introduced in 1953, although the two species may be reaching equilibrium short of extinction of *Pheidole* (Lieberburg *et al.* 1975). As a final example, the red imported fire ant (*Solenopsis invicta*) has virtually eliminated the native fire ant (*S. xyloni*) from most of its range in the United States (Holldobler and Wilson 1990).

Literature reviews of field studies on competition

In the 1980s the importance of interspecific competition in nature was questioned by a number of biologists. Strong *et al.* (1983), among others, challenged much of the evidence usually cited for the prominent role assigned to competitive interactions in structuring natural communities. They asserted that the data were often indistinguishable from random models. Others charged that there were few experimental studies of competition from nature, and that predation was a much more significant ecological interaction. Still others, such as Wiens (1977), asserted that competition is a temporally sporadic, often impotent, interaction. Schoener (1982, 1983) decided to review the literature to determine if competition had been affirmed as an important interaction in nature. He found, to his surprise, over 150 experimental field studies of competition in natural ("field" settings), many of which had been conducted in the previous five years. Schoener carefully defined an interspecific competition experiment as a manipulation of the abundance of one or more hypothetically competing species. All such experiments had to have proper controls. Prior to these experimental studies, there had been a dependence on "natural experiments," which will be discussed below (Diamond 1983).

The "field" was defined as a study in which some major natural factors extrinsic to the organism remain uncontrolled. Schoener did not allow laboratory or greenhouse setups, but did count experiments involving fenced exclosures or caged portions of shorelines to fit the definition of a field study.

Through 1982 Schoener found that 164 published studies fitted the criteria. Of those 164, 90% (148) of the studies and 76% of the species involved did show positive evidence of interspecific competition. In a separate analysis and using different criteria, Connell (1983) found evidence of competition in 40% of the experiments and 50% of the species. There were, however, few studies involving herbivorous insects that demonstrated interspecific competition. Schoener suggested, as had Hairston *et al.* (1960), that herbivores, which occupy an intermediate position in the food web, are controlled by predators and therefore competition is a less important interaction for this trophic position. Schoener's literature review found little evidence to support Wiens' idea that there is a great deal of temporal variability in competition. Competitive variability is especially rare in marine ecosystems. Variability was mostly found in dry, continental habitats. This is interesting in that Wiens developed his ideas after carrying out research on bird communities in North American arid or semi-arid shrub habitats.

A decade after the analyses of Schoener and Connell, Gurevitch *et al.* (1992) analyzed competition studies carried out from 1980 to 1989, using a statistical approach. They found "medium" effects of competition on primary producers, carnivores, and herbivorous marine mollusks. Larger effects of competition were detected on some herbivores and stream arthropods. As found by Schoener, however, studies on herbivorous terrestrial insects usually failed to show significant effects of competition.

One can conclude from these literature reviews that competition is a common event in nature that contributes to the organization of ecological communities. It is, however, not the only important interspecific interaction.

7.8 Indirect evidence for competition and "natural experiments"

Because competition has been difficult to demonstrate directly in the field, and because the historical or evolutionary result of competition may have resulted in coevolution between species to minimize competition (sometimes called the ghost of competition past), Diamond (1978, 1983) proposed that indirect evidence, often the result of "natural experiments," should be given the same credibility as experimental studies in the field or laboratory. As important as field experiments are, a single variable is manipulated; nothing else can be controlled once the field sites have been selected. Diamond lists five different drawbacks or weaknesses of field experiments. (i) The outcome of the experiment often varies from year to year and season to season since weather and predators are uncontrolled. (ii) Most field experiments are not run for enough time. This deficiency is, however, being remedied. For example, the National Science Foundation (NSF) is addressing this problem in its Long Term Ecological Studies (LTER) program. (iii) The importance of large temporal and spatial scales cannot be addressed in contemporary time and space. (iv) A manipulation of two species may incorrectly ignore the importance of a third species. (v) The kinds of experiments that might reveal important information, such as the removal or introduction of a species in an ecosystem, are often "technically impossible, morally reprehensible and politically forbidden" (Diamond 1983).

In order to solve these problems, Diamond (1983) extolled the virtues of "natural experiments" and other kinds of data gathered from field observations as opposed to experiments. According to Diamond, natural experiments have three advantages. First, they permit an ecologist to rapidly gather data. As an example, he described the work of Schoener and Toft (1983). They surveyed spider populations on 92 small Bahamian islands, 48 of which lacked lizards and 26 of which were occupied by at least one species of lizard. They found that spiders were ten times more abundant on the islands without lizards. The explanation was that lizards are both competitors with and predators on spiders. Diamond's point, however, was that this natural experiment (lizards present on some islands, absent on others), would have been very difficult and time-consuming to set up, and we would have waited a very long time (up to several years) before the spider populations reached new equilibrium values. Using the natural experiments, Schoener and Toft completed their fieldwork in 20 days!

Second, natural experiments allow ecologists to examine situations they would not be allowed to set up experimentally. It is likely, for example, that the Bahamian government would have objected to having lizards removed from 48 islands. In another example, Brown (1971a, 1971b) has shown that two species of chipmunk (genus *Eutamias*) divide the forest by altitude when they are sympatric on mountains in the Sierra Nevada range. But on several mountains, probably due to chance colonization or extinction events, only one species is present. When only one species occupies the mountain, without its competitor, it is found at all elevations. A field experiment, in which one species or the other was eliminated from an entire mountain, would never have been approved by the US Fish and Wildlife Service or by any granting agency. Yet this natural experiment is an elegant demonstration of the phenomenon known as ecological release (see below).

Finally, natural experiments allow us to examine the end results of ecological or evolutionary processes operating over a longer period of time than the usual field experiment.

The weakness of natural experiments is, of course, that the investigator does not know what created the observed situation. There are often multiple explanations for the observed differences and it may be impossible to distinguish among them. Furthermore, the use of simple field observations, as opposed to rigorous, controlled experiments, is unacceptable to many scientists. Diamond, however, asserted that a variety of evidence derived from a variety of methodologies provides the most robust results.

Below we will review five different kinds of indirect evidence related to the role of competition, many coming from natural experiments.

Ecological release

In ecological release, a species occupies a broader niche or geographical area in the absence of a closely related competitor. The chipmunk distribution in the Sierra Nevada mountain range fits this description. Another example is the distribution of two species of *Planaria* in streams. When found alone in a stream (allopatric distribution) each species occupies a wide range of stream temperatures. When both species are found in the same stream (sympatric distribution), however, the distribution of both species is restricted. *P. montenegrina* is found from 5 to about 13.5 °C, whereas *P. gonocephala* occupies the warmer portions of the stream from 13.5 to approximately 23 °C (Beauchamp and Ullyott 1932).

Ants also show ecological release. For example, *Paratrechina longicornis* normally nests under objects on the ground in open environments in southern Florida and in the Florida Keys. However, the Dry Tortugas, the outermost islands of the Florida Keys, have been colonized by very few ant species. On the Dry Tortugas, released from competition, *P. longicornis* is extremely abundant and nests in tree boles and open soil, sites normally occupied by other species in the rest of southern Florida. By contrast, *Pseudomyrmex elongatus*, which also has colonized the Dry Tortugas, has expanded neither its population nor its normal nesting habitat of thin twigs near the top of the tree canopy (Holldobler and Wilson 1990).

Contiguous allopatry

In this phenomenon, two species occupy distinctly different geographical areas directly adjacent (contiguous) to one another. Another study on chipmunks on the east slope of the Sierra Nevada mountains in California provides an example of both contiguous allopatry and ecological release. Heller (1971) found four species of chipmunks living at different altitudes. The least chipmunk (*Eutamias minimus*) is found at the lowest elevations within the sagebrush range. When all other chipmunks are absent, the least chipmunk can occupy all altitudes up to the alpine. However, the yellow pine chipmunk (*E. amoenus*) through aggressive behavior restricts the least chipmunk to the hot, dry sagebrush areas. If the least chipmunk is absent, the yellow pine chipmunk does not invade the hot, dry habitats. Evidently this dry area is not part of the fundamental niche of the yellow pine chipmunk. The lodgepole pine chipmunk (*E. speciosus*) is the most aggressive of the four species, but it is most vulnerable to heat stress and is restricted to shady, cool forests. It apparently limits both the upper distribution of the yellow pine chipmunk and the lower distribution of the alpine chipmunk (*E. alpinus*). Therefore, these four species, when all are present, show contiguous allopatry by altitude. When one or another is absent, the least chipmunk shows ecological release.

Diamond (1978) has described other examples of contiguous allopatry and ecological release. In one case he mapped the distribution of two species of warblers in the genus *Crateroscelis* on Mount Karimui in New Guinea. The first species, *C. murina*, became relatively more abundant as Diamond hiked from sea level to 1650 m. From that altitude to the summit of the mountain *C. murina* was abruptly replaced by *C. robusta*. This second species had its greatest relative abundance at just over 1650 m. Diamond found a similar situation when he examined the distribution of three species of nectar-drinking parrots on several New Guinea mountains. When the species *Charmosyna placentis* was present it occupied forests from sea level to about 600 m. The sibling species *C. rubronotata* and *C. rubrigularis* were confined to higher elevations when in competition with *C. placentis*. The distributions were contiguously allopatric. When *C. placentis* was absent from a mountain on a different island, however, the other two species were both able to expand their range to lower elevations. The opposite is also true. In the absence of the two sibling species, *C. placentis* is able to expand its range to higher elevations.

Niche partitioning

In niche partitioning, two or more species coexist while sharing one or more resources in such a way that the niche overlap apparently violates the competitive-exclusion principle. Upon closer investigation, the resources, though shared, are used with different frequencies or are used in different ways so as to allow coexistence. For example, the root systems of coexisting annual plants can be shown to partition the soil by depth, thereby avoiding direct resource competition (Wieland and Bazzaz 1975). In his classic study, MacArthur (1958) showed that five species of *Dendroica* warblers coexisted by foraging in different portions of trees in a coniferous forest. Although there was overlap, each species spent the majority of its foraging time in a unique portion of the trees. Other examples come from Diamond (1978). In one case he describes the coexistence of four species of fruit pigeons on islands of the Bismarck Archipelago. The pigeons range in size from 91 g to 722 g. The fruit sizes they consume range from 3 mm to 50 mm in diameter. Again, while there is a great deal of overlap among the pigeons in terms of fruit sizes consumed, the mean usage rates are clearly different. The smallest pigeon consumes fruits that are, on average, 8 mm in size; the 135 g pigeon consumes fruits that average 11 mm; the 470 g pigeon, 18 mm; and the 722 g pigeon, 25 mm. In a similar analysis, Diamond (1978) showed that eight species of fruit pigeons in New Guinea coexist not only by specializing on different sizes of fruit, but also by the position of the fruit on the branches.

A particularly interesting example of niche partitioning involves the four species of antbirds of the genus *Myrmotherula* that coexist on Barro Colorado Island in Panama. These birds forage by following swarms of army ants. As the army ants move through the forests, insects, lizards, and other small prey flee. Antbirds swoop in and capture unsuspecting prey intent upon escaping from the army ants. This would seem to be a limited niche, yet four bird species of the same genus make their living in this manner. MacArthur (1972), using data supplied by John Terborgh, showed that the mean foraging heights among the four species differed. Not only did these means differ, but they were also separated from each other by at least one standard deviation.

Niche partitioning can also occur within a species. Male red-eyed vireos (*Vireo olivaceus*) forage for insects primarily in the upper canopy of the forest, while females concentrate their foraging in the lower canopy and near the ground (Williamson 1971).

Holldobler (1986) found that two species of ants in Australia display niche differentiation based on foraging at different times of day. *Iridomyrmex purpureus* and *Camponotus consobrinus* use the same food sources and are found nesting near each other. However, *Iridomyrmex* forages mostly during the day and *Camponotus* mostly at night. When each species is found alone its foraging time is increased by 1–2 hours per day. Strangely, these shortened foraging periods are the result of direct interference. In the morning *Iridomyrmex* workers congregate around nest exits of *Camponotus* and close them with pebbles and soil. At dusk the situation is reversed, with *Camponotus* workers harassing *Iridomyrmex* at its nests. In another example of niche partitioning among ants, Levins *et al.* (1973) showed that *Pheidole megacephala* dominates baits in the shade, whereas *Brachymyrmex heeri* takes over the same baits when they are in direct sun. As the shadows move across the forest floor this situation reverses itself regularly, within 30 minutes, accompanied by occasional aggressive encounters among workers.

Character displacement

Character displacement is defined as a situation in which two species, when living in separate geographical ranges (allopatric distributions), have nearly identical physical characteristics (i.e., beak sizes in birds, overall body sizes in lizards and snails, canine sizes in the cat family). When sympatric, however, these physical or morphological characteristics diverge in one or both species. This divergence minimizes competition for food and allows the two species to coexist. Brown and Wilson (1956) appear to have introduced this idea. When examining the overall size and beak lengths of specimens of the eastern (*Sitta tephronota*) and western rock nuthatches (*S. neumayer*), they found that the allopatric populations were almost identical in both average size and in the range of sizes. However, these two species become sympatric in Iran. In sympatry, the eastern rock nuthatch is larger, while the western species has become smaller. In this sympatric zone their beak and body sizes are completely non-overlapping. This allows them to feed on different-sized prey items and therefore coexist.

Another example (Lack 1947, Schluter *et al.* 1985, Grant 1999) comes from studies of Galapagos finches (also known as Darwin's finches). The medium ground finch (*Geospiza fortis*) and the small ground finch (*G. fuliginosa*) are allopatric on the islands of Daphne Major and Los Hermanos. On Santa Cruz they are sympatric. As in the case above, when they are allopatric their beak sizes are very similar (mean size approximately 10 mm in *G. fortis* and 8.5 mm in *G. fuliginosa*). On Santa Cruz, however, the sizes of their beaks do not even overlap. The average for *G. fortis* is between 11 and 12 mm, while that of *G. fuliginosa* is reduced to about 7.5 mm. Beak size correlates with diet in the Galapagos (Grant 1999).

Fenchel (1975) has demonstrated character displacement in shell lengths of two species of mud snails (*Hydrobia ulvae* and *H. ventrosa*). The overall size of the shell is indicative of the particle sizes consumed by the snails. As in the cases above, in allopatry the snail sizes were almost identical, but in sympatry the *H. ventrosa* population is greatly reduced in size. Simultaneously, the *H. ulvae* population includes many very large individuals, not present in the allopatric population. The means and modes of the two species are distinctly different in sympatry. For example, the population mode for *H. ulvae* is 5 mm in sympatry, but only 3 mm in allopatry.

In the above examples only pairs of species were examined. However, size displacements may occur among several coexisting species. Strong *et al.* (1979) called this "community-wide character displacement." A pattern of regular differences in some size-ranked sequence has been accepted as evidence that past competition helped to mold niche differences. In fact, G.E. Hutchinson (1959) proposed that a ratio of between 1.1 and 1.3 among closely related species in a size-ranked sequence was sufficient for coexistence. This is known in the literature as Hutchinson's ratio or rule. There are many who have questioned the meaning or significance of such a rule (Simberloff and Boecklen 1981, Strong and Simberloff 1981). However, an interesting paper by Dayan *et al.* (1990) has shown that among wild felines in Israel, character displacement of canine diameters is remarkably constant at 1.10–1.14. These size ratios include three species of *Felis*, but also include male and female sexual dimorphism in canine sizes (Table 7.3).

Dayan *et al.* (1990) hypothesized that intra- and interspecific competition for food has selected for these canine ratios. The assumption is that canine size is highly correlated with the size of prey taken by these cats. In separating males and females, they were treating each species/gender combination as a distinct "morphospecies" in the community. A great deal more information is needed on: (i) the size distribution of potential prey; (ii) whether prey populations are limiting to predator populations; (iii) whether larger canines correlate with the capture of larger prey; and (iv) habitat use by the cats.

Table 7.3 Mean canine diameters for male and female coexisting species of *Felis* in Israel. From Dayan *et al.* (1990).

Species, sex	Mean canine diameter (mm)	Ratio between adjacent pairs of canine diameters
F. silvestris, female	4.8	–
F. silvestris, male	5.5	1.15
F. chaus, female	6.3	1.15
F. caracal, female	6.9	1.10
F. chaus, male	7.6	1.10
F. caracal, male	8.4	1.11

Nevertheless, the data suggest that competition has played an important role in the evolution of this feline community.

Taper and Case (1992) have cautioned that any study purporting to demonstrate character displacement should meet the following criteria:

1 Morphological differences between a pair of sympatric species must be statistically greater than the differences between allopatric populations;
2 the observed differences between sympatric and allopatric populations must have a genetic basis;
3 differences between sympatric and allopatric populations must have evolved on the site and not be due to different founder populations;
4 variation in the character must have a known, ecologically important function;
5 competition must be known to occur for a resource in short supply and the character must play an important role in that competition;
6 differences in the character cannot be explained by differences in resources available to the sympatric and allopatric populations.

Ecologists have confirmed, however, that the Galapagos finch study (Grant 1999) satisfies all of these criteria.

Historical replacement

Finally, over human history there are documented cases of one species invading and eliminating another species from its original range. For example, Diamond (1978) has described both historical replacement and niche segregation in two species of tits (*Parus*). The blue tit (*P. caeruleus*) was found in Europe west of the Ural Mountains where it occupied a wide variety of habitats from forest to riparian thickets. The azure tit (*P. cyanus*) was formerly found in Asia east of the Urals. But in the late nineteenth century it spread 1600 km west across Russia to the Baltic Sea. Over the first 10 years of the 20th century, however, the azure tit retreated several hundred kilometers eastward from the Baltic. At present, the two species overlap on the eastern edge of the geographical range of the blue tit. However, in the overlap zone the two species segregate by habitat. The azure tit is found in riparian thickets, while the blue tit is in upland forest.

In recent years the black duck (*Anas rubripes*) has been declining and is being replaced by the mallard (*Anas platyrhynchos*) in many areas of Eastern North America. It is unclear whether this is being caused by hunting, environmental and habitat changes, captive breeding and release programs, or competition between the two species. A study by Merendino *et al.* (1993) tentatively concluded that mallards were competitively excluding black ducks from the most productive wetlands.

There are many known examples where introduced species have outcompeted the native flora or fauna. Starlings (*Sturnus vulgaris*) and house sparrows (*Passer domesticus*) are infamous for evicting native cavity-nesting birds such as eastern bluebirds (*Sialia sialis*) and northern flickers (*Colaptes auratus*) from nest sites in North America. The purple loosestrife (*Lythrum salicaria*), a native of Europe, has invaded wetlands in temperate North America and is crowding out native wetland plants. Finally, Australia and Hawaii are living laboratories demonstrating the effects of introduced competitors (as well as predators and herbivores) on native flora and fauna.

7.9 Conclusions

How important are competitive interactions to the functioning of communities and ecosystems? There has been a long tradition of theory, laboratory studies, and field studies that have emphasized the potential importance of competition in population regulation and in the shaping of population and community relationships among species. Darwin, Liebig, Tansley, Lotka, Volterra, Gause, Park, Connell, MacArthur, Tilman, and many others have contributed to the rich literature on competitive interactions. The competitive-exclusion principle established an expectation that competitive interactions not only shape present-day ecological interactions, but also have directed past evolutionary events (the so-called "ghost of evolution past"). Resource-based competition theory has also laid out certain ground rules (the R^*-rule) for coexistence or exclusion in resource-limited environments. On the other hand, the competition–colonization trade-off concept sets up conditions whereby inferior competitors may persist in communities. For example, ant species known as "opportunists" and "insinuators" are able to coexist with the aggressive dominant competitor species, known as "extirpators," based on their ability to find food sources first, or by their stealthy foraging style. And even among the dominant competitors, the species that finds a food resource first and dominates it through recruitment of nestmates is the winner at that resource. Therefore, as suggested by Tilman (1994) and Yu and Wilson (2001), coexistence of inferior competitors should be expected in a spatially diverse environment.

Some ecologists have complained that the fascination with competition has resulted in an underestimation of the importance of parasitic, predatory, or even mutualistic relationships in shaping ecological communities. American politicians have even turned economic competition into a kind of religion. Let us now, therefore, turn to other kinds of interactions.

8

Mutualism

- Mutualism or parasitism?
- Modeling mutualism
- The costs of mutualism

8.1 Introduction

Mutualism or parasitism?

As outlined in the introduction to Part II, mutualism is an interaction in which both species benefit. In facultative mutualism individuals in a population are able to survive and reproduce without their presumptive mutualist, although their fitness is enhanced when they participate in the mutualism. By contrast, in obligatory mutualism, individuals in a species are unable to survive without their mutualistic partner. As pointed out by Vandermeer and Goldberg (2003), mutualistic relationships are complex, and do not necessarily fit into the two categories of facultative and obligatory. For example, a mutualism may be obligatory for one partner, but not for the other. The mutualism may be very weak (provide few benefits) and therefore may only be found in very specific environments. In fact, many types of mutualisms are much more common in tropical environments. For example, Neotropical ants and African termites both raise mutualistic fungi in "gardens" within their nests. Though fungus gardening does occur outside of the tropics, it is much more common and conspicuous in the humid tropics. In ant–plant mutualisms, ants defend plants from herbivores or perform other services in exchange for nest sites and nutrition provided by the plants. Although these mutualisms exist in the temperate zone, almost all of the obligatory ant–plant mutualisms are found within the tropics. The number of plant species providing extra-floral nectar, a low-cost method of attracting ants to plants, declines with increasing latitude and altitude, and is rare in the North Temperate zone.

Animal pollination and fruit dispersal are also much more common in tropical latitudes. For example, bats that provide pollination and fruit dispersal for higher plants are only found south of 33° N latitude. No bees that are obligatory pollinators of orchids are found north of 24° latitude. At Monteverde, a cloud forest in Costa Rica, animals as opposed to wind pollinate more than 90% of the dicots and 88% of the monocots. A study by Murray et al. (2000) of fruit dispersal at Monteverde showed that more than 81% of tree species are adapted for seed dispersal by vertebrates. This compares well with 89% at Alto Yunda, Colombia, 92% at La Selva, Costa Rica, and 92% at Rio Palenque, Ecuador. At Monteverde about 80% of animal-dispersed trees are specifically adapted for bird dispersal. Adaptations for bat and ant dispersal are less common. Not all life forms, however, are adapted for animal dispersal of their fruit at Monteverde. A majority of epiphytes (66%) and herbs (73%) are adapted for wind or other abiotic means of dispersal. Most of the wind-dispersed seeds, however, are orchids (among epiphytes) and weedy species of Asteraceae (among herbs). Lianas and shrubs are intermediate: the fruits of the majority of species are bird-dispersed while 35–45% is abiotically dispersed.

Mutualistic relationships are wonders of natural history and prime examples of co-evolution. Yet, as Bronstein (2000) put it, "Mutualism is the most poorly understood form of interspecific interactions." Others such as Law (1988) and Watkinson (1997) have made the same point. Furthermore, Watkinson (1997) lamented that "There is not even a sound theoretical framework for the treatment of mutualistic interactions." Recent research, however, has begun to emphasize the fact that in most mutualisms, the relationship is simultaneously beneficial and harmful to one or both participants. Rather than thinking of the mutualistic species as happily entering a partnership, Bronstein (2002) has asserted that mutualism is more likely a "reciprocal parasitism" in which each partner obtains what it can at the lowest possible cost to itself.

Consider, for example, the fig-wasp pollination system. Figs (species of *Ficus*) have an obligatory pollination system with wasps (Hymenoptera: Chalcidoidea: Agaoninae). There are five species of figs at Monteverde in Costa Rica, but the reproductive biology of only one of them has been worked out (Bronstein 2000). The female pollinators (*Pegoscapus silvestrii*) are drawn to volatile odors released by the female florets of figs. These females will already have mated with males and will also have ripped open the anthers of male flowers in their natal fig. They pack pollen into pockets in their abdomens. Thus, inseminated and pollen laden, they arrive at a new tree where they squeeze into the 1 cm fig flower. Once within a fig, the wasp deposits pollen on the stigma of a floret. She then lays a single egg in each of the ovaries she can reach with her ovipositor. The female is fatally trapped inside the fig flower, but her offspring will develop in the fig ovaries, feeding on the seeds as they develop. Some seeds escape wasp predation because the flower's style was too long for the female's ovipositor. The seeds and seed-eating fig-wasp larvae develop over a two-month period. Mature males eventually emerge and search out females, still developing inside the growing figs, for mating. The male then chews an exit hole through the wall of the fig, through which the females can also depart their natal fig. Thus, in payment for pollination the figs lose a large number of potential seeds to wasp larvae. The female wasps would lay eggs on all of the ovules if they could. If they succeeded they would indeed be parasites.

Or consider the relationship between East African whistling thorn acacias (*Acacia drepanolobium*) and ants. These acacias are some of the most common plants of the savannas of central Kenya. Almost all of these trees host thousands of ants. The ants provide defense against herbivores in exchange for food and shelter. A pair of thorns lies at the base of every leaf cluster, and each branch is lined with two types of thorns. The slender, white, needle-sharp thorns, which may be 75 mm in length, are the most abundant. Intermingled with these thorns are pairs of thorns with bulbous, hollow bases. These "swollen thorns" house the ant colonies; each thorn can harbor hundreds of ants. As in Central American acacias (Janzen 1966), when an herbivore disturbs the tree, the ants stream out of the thorns and bite the intruder, and when Stanton and Young (1999) experimentally removed ants from trees, they found that herbivore damage by both browsing mammals and insects increased significantly.

The acacia provides the ants with nectar from glands along the leaves. These extra-floral nectaries are particularly abundant on new leaves. However, *Acacia drepanolobium* does not provide lipid or protein food sources for these ant colonies. The ants must forage for insects and other protein-rich foods. Therefore the mutualism is weaker than the *Acacia–Pseudomyrmex* relationship described by Janzen (1966) in Central America, where the acacias provide lipid- and protein-rich structures, known as Beltian bodies, on the tips of the leaflets.

There are four different ant species that colonize *A. drepanolobium*. The red-and-black cocktail ant (*Crematogaster mimosae*) and the black-and-white cocktail ant (*C. nigriceps*) are most effective against herbivores. Intruders are immediately attacked by a horde of biting ants. The ants race around emitting alarm pheromones with their abdomens held high (hence the term "cocktail ant"), and this recruits more workers to the scene. Herbivores such as goats, which are bitten while attempting to browse on a defended tree, refuse to approach those trees again. Trees defended by these two ant species are rarely damaged by herbivores.

A third species, the slender black acacia ant (*Tetraponera penzigi*) has a nasty sting, but is more passive and only attacks if the swollen thorns themselves (the home of larvae, pupae, adults, and winged reproductive ants) are attacked by monkeys or other animals. On the other hand, these ants patrol leaves day and night, removing pollen and probably fungal spores. Therefore these workers may provide the acacias protection from disease.

The fourth species, the black cocktail ant (*C. sjostedti*), provides no services to the trees whatsoever. Stanton and Young (1999) found that long-horned beetles could girdle stems and kill entire sections of the tree while this ant was present. The reason for this may be that this ant species does not even live in the swollen thorns. Instead it nests in hollow spaces within dead and drying branches. The beetles actually provide this ant with nest space, and the mutualistic relationship has shifted to one in which these two insect species "cooperate" in exploiting the acacias.

Although all four species of ants may occupy trees on a given hectare of land, an individual tree is almost never occupied by more than one species of ant. The species are intolerant of each other and engage in aggressive, mortal combat. Experiments have shown that the fights continue until one species has wiped out the second on a given tree. Unfortunately for the acacias, the black cocktail ant, which is functionally a parasite, is the dominant competitor, winning most battles with the other three species.

The red-and-black cocktail ant, as well as the black cocktail ant, tend scale insects that feed on the phloem of the acacias. Thus, both species have found an additional method of draining energy from the trees. Worse still is the habit of the black-and-white cocktail ant workers of removing the tips of most growing shoots. These workers also remove stem tissue containing leaf and flower buds. New branches are only allowed to grow in prox-imity to swollen thorns. These black-and-white cocktail ants have therefore changed the architecture of these trees from one with large open canopies to one of compact masses of branching stems. In addition, by chewing off its flower buds, it prevents these trees from reproducing.

In summary, the ant–acacia symbiosis in Africa, although potentially mutually beneficial, appears to have tipped in favor of the ants, which function more like parasites than mutualists. As we examine mutualisms, we should ask ourselves how these mutualisms differ from host–parasite relationships.

8.2 Modeling mutualism

In his pessimistic remarks on our understanding of mutualistic interactions, Watkinson (1997) points out that when modifications of the Lotka–Volterra competition equations (Eqns. 8.1 and 8.2) are employed to model mutualism, the result is "unbounded exponential growth" of both populations. Robert May (1981b) called this result an "orgy of mutual benefaction." In Equations 8.1 and 8.2, as in the Lotka–Volterra equations, the growth rates of the two mutualistic species are determined by their present population sizes (N_1 and N_2) and their intrinsic rates of growth (r_1 and r_2), and diminished by intraspecific competition. These carrying capacities, however, are set for each population when living without its mutualist. Obviously, these equations only apply to facultative mutualism, since in obligatory mutualism there can be no positive carrying capacity in the absence of the partner species. The main difference between these equations and the competition equations is that the terms c_1N_2 and c_2N_1 have a positive sign. The terms c_1 and c_2 are mutualism coefficients and replace the competition coefficients in the Lotka–Volterra competition equations. The term c_1 measures the rate at which an individual of N_2 benefits the growth rate of population N_1. Similarly, c_2 measures the rate at which an individual of N_1 benefits the growth rate of population N_2.

$$dN_1/dt = r_1N_1\frac{K_1 + c_1N_2 - N_1}{K_1} \tag{8.1}$$

$$dN_2/dt = r_2N_2\frac{K_2 + c_2N_1 - N_2}{K_2} \tag{8.2}$$

An equilibrium analysis can be done, similar to what we did with the competition equations, by setting dN_1 and dN_2 equal to zero. Ignoring situations where r, N, or K are equal to zero leaves $K_1 + c_1N_2 - N_1 = 0$ and $K_2 + c_2N_1 - N_2 = 0$. The basic results are as indicated in Equations 8.3 and 8.4. As you can see, the population of each species is increased beyond its carrying capacity with additional individuals of its partner species. The more intense the mutualism (the greater the benefits provided to the partner species) the larger the equilibrium population becomes.

$$N_1 = K_1 + c_1N_2 \tag{8.3}$$

$$N_2 = K_2 + c_2N_1 \tag{8.4}$$

Furthermore, the "orgy of mutual benefaction" referred to above should be evident. Greater numbers of N_2 produce greater numbers of N_1, which, in turn, produce larger numbers of N_2, and so on. Strangely enough, the interaction is stable when only one partner benefits from the interaction. If $c_2 = 0$, for example, N_2 cannot exceed K_2 and N_1 stabilizes at $K_1 + c_1 K_2$.

Another approach is to substitute the carrying capacity after mutualism into Equations 8.1 and 8.2. The new carrying capacities, K_1^* and K_2^*, are shown below (Eqns. 8.5 and 8.6). The resultant equations (8.7 and 8.8), however, are only stable if we specify either a numerical limit to K_1^* and K_2^* or that the mutualism coefficients decline toward zero as the populations approach the new K^* carrying capacity. In an obligatory mutualism we would have to specify that both N_1 and $N_2 > 0$. Otherwise, we have mutual extinction. In fact, it is reasonable to assume, as in the Allee effect, that each population has a minimum viable population size, below which the mutualism falls apart and both species go extinct.

$$K_1^* = K_1 + c_1 N_2 \tag{8.5}$$

$$K_2^* = K_2 + c_2 N_1 \tag{8.6}$$

$$dN_1/dt = r_1 N_1 \frac{K_1^* - N_1}{K_1^*} = r_1 N_1 \frac{K_1 + c_1 N_2 - N_1}{K_1 + c_1 N_2} \tag{8.7}$$

$$dN_2/dt = r_2 N_2 \frac{K_2^* - N_2}{K_2^*} = r_2 N_2 \frac{K_2 + c_2 N_1 - N_2}{K_2 + c_2 N_1} \tag{8.8}$$

When trying to model mutualism by starting with the Lotka–Volterra equations, we encounter the same problems as those described in Chapter 7, on competition. The Lotka–Volterra approach is a phenomenological one. What we need is a mechanistic approach based on the rates of exchange of relevant resources between the two mutualists. However, since each mutualistic interaction is both unique and complex, a model based on one mutualism would lack generality. While specific models describing specific mutualisms have been written, none has entered the literature as a standard approach.

8.3 Conclusions: the costs of mutualism

The study of mutualism has passed through a number of stages, but has yet to move very far from the descriptive phase (Bronstein 2001). Many early naturalists were skeptical about mutualistic relationships. For example, before Janzen (1966) established the obligatory mutualism between *Pseudomyrmex* ants and *Acacia* trees, the ants were said to be of no more use to the plants than "fleas on a dog." Once mutualisms were described they were often pronounced to be mostly confined to the tropics. However, Janzen (1985) has pointed out that every organism is involved in at least one mutualism in its life. Given that the eukaryotic cell evolved as the result of mutualistic relationships, the phenomenon is ubiquitous. And extra-tropical locations do not lack mutualisms. Forest trees with mycorrhizal fungi dominate boreal habitats, deserts are populated with legumes and their nitrogen-fixing bacteria, and tundras are dominated by lichens.

While mutualistic interactions have become accepted as comparable in importance to ecosystems as competition and predator–prey relationships, the view that mutualism represents "cooperation" between species has been challenged. Bronstein (2001) has stressed that mutualisms involve costs for each species, as well as benefits. Costs of mutualism are only now being tabulated, and there is little consistency in how data are gathered. Bronstein (2001) cites the following examples: (i) 20% of the total carbon budget of forest trees may be consumed supporting mycorrhizae (Johnson et al. 1997); (ii) 3% of the energy budget of many plants is devoted to providing floral nectar for pollinators (Harder and Barrett 1992); (iii) extrafloral nectar costs about 1% of the energy budget of the plants involved (O'Dowd 1979, 1980). In the obligatory interaction between figs and their wasp pollinators, Bronstein (2001) estimated that 53% of ovaries of Ficus aurea are lost to the wasps, while yuccas (Yucca spp.) evidently lose 5–20% of their seeds to their moth pollinators. Finally, Wolfe (2001a, 2001b) estimates that nitrogen-fixing bacteria consume 20% of the carbohydrates produced by legumes.

Rather than assuming that species have somehow entered into permanent, mutually agreeable contracts, we need to analyze mutualism with the following points in mind:

1 Mutualism always involves costs as well as benefits.
2 Costs set limits on the evolution of mutualisms.
3 There is a conflict of interest between the mutualistic species.
4 Organisms that "cheat" on the mutualism by reaping the benefits without reciprocation will often enjoy an advantage.
5 Related organisms (such as non-mutualistic ant species) may take advantage of a mutualism and act as parasites on the relationship.
6 Mutualisms may evolve toward a host–parasite relationship from a mutualistic one.
7 Continuous coevolution is necessary to maintain a mutualism.

9

Host–parasite interactions

- Factors affecting microparasite population biology
- Modeling host–microparasite interactions
- Dynamics of the disease
- Endangered metapopulations and disease
- Social parasites

9.1 Introduction

Given the prominent role of medicine in today's world, it is amazing to realize that the theory that microbes (germs) were the cause of many diseases was not really established until the 1870s. Attempts to create models of epidemiology did not begin until the twentieth century. Hamer (1906) formulated a discrete time model in an attempt to understand epidemics of measles. By the late 1920s Kermack and McKendrick (1927) had published models showing that the density of individuals susceptible to a disease must exceed a critical number before an epidemic was possible. A rich literature of mathematical epidemiology developed during the middle of the twentieth century. But most ecologists were only dimly aware of this literature until the late 1970s, when Anderson and May published a series of articles with titles such as "Population biology of infectious diseases" (Anderson 1982, Anderson and May 1979, 1981, 1982). A recent review of Hethcote (2000) is recommended for students interested in a review of the history of infectious-disease models.

In this chapter we will examine models for interactions between hosts and microparasites such as viruses and bacteria. Parasitoid–host interactions (the Nicholson-Bailey models) will be covered in Chapter 10, on predator–prey relationships. Here we are primarily concerned with models for microparasitic diseases such as measles, in which the parasite reproduces quickly and reaches tremendous populations within the host. The duration of the acute infection is limited by either host defenses or host mortality, and is short relative to the host life span. Recovered hosts may have lifelong immunity. The dynamics of this relationship are driven by transmission between hosts and the essentials can be modeled by classifying host individuals as **susceptible**, **infected**, or **recovered** (**immune**). This is known as the SIR model, although there are many modifications and elaborations on the basic SIR model (Hethcote 2000). For example, in the MSEIR model, individuals are born with passive immunity (M); over time they lose this immunity and transfer to the susceptible class (S); individuals exposed (E) to the disease then become part of the infected (I) class; those who survive become part of the recovered (R) class. The actual abundance of the parasite within the host is ignored. Recall that many metapopulation models also ignore population size and dynamics within habitat patches.

By contrast, macroparasites (such as intestinal worms) typically cause chronic, persistent infections. (This is also a feature of certain microparasites such as herpesviruses and the malaria parasite.) Disease severity depends on the number of parasites present. Not only do infected vertebrate and invertebrate hosts accumulate parasites throughout their adult lives, but both the number and diversity of parasitic species also increase with host age (Dobson *et al.* 1992). For example, in brown pelicans (*Pelecanus occidentalis*), the number of helminth parasite species increases by age class, reaching 13 for birds aged over three years (Humphrey *et al.* 1978). The same study showed that the number of individual parasites peaked at 8000 for one-year-old pelicans, and then declined to about 4000 in birds older than three years. A survey by Dobson *et al.* (1992) of North American mammals showed that the average individual carried 369 macroparasites of three different species. The survey examined four orders: carnivores, lagomorphs, rodents, and artiodactyls. Carnivores carried the most diverse parasite fauna, lagomorphs the least.

Models of macroparasite dynamics must account for parasite abundance within hosts as well as host-to-host variation in parasite abundance (May and Anderson 1979). Another complication is that many macroparasites such as *Schistosoma mansoni*, which causes the disease known as schistosomiasis, have an asexually reproducing stage in an intermediate host. Schistosomiasis is one of the most important human diseases in tropical regions. It is estimated that 100 million people carry at least one worm. The adult schistosomes live in pockets of the intestinal blood vessels or veins of the bladder in the vertebrate host, where males and females carry out sexual reproduction. The eggs are passed out of the host through the feces and urine into a body of water, where the eggs hatch into free-swimming larvae known as miracidia. A successful miracidium penetrates a snail. Once inside the snail it undergoes asexual reproduction. After 4–7 weeks new free-swimming stages, known as cercariae, are shed from the snail into the water. The cercariae must find the vertebrate host within 48 hours and each cercaria is capable of penetrating the skin of a vertebrate host. Once inside the host they travel through the circulatory system in a journey that can take 6–12 weeks. Once they locate the proper tissue they mature into adult worms, completing the life cycle. An important aspect of this kind of life history is the asexual-reproduction stage in the snail. The large numbers of cercariae shed from the snail make it much more likely that the vertebrate host will be located and the life cycle completed. Finally, perhaps because of the complexity of the life cycle of the parasite, vertebrate hosts generally have ineffective immune responses. In fact, the pathology of schistosomiasis is a consequence of the immune response of the host, in which shed eggs are attacked and calcareous deposits laid down around them. The accumulation of these deposits blocks the spleen and excretory organs of the infected hosts. For models of macroparasites with complex life cycles, such as the parasitic helminths, see Dobson *et al.* (1992).

9.2 Factors affecting microparasite population biology

As outlined by Nokes (1992), three main factors govern the interaction between a host and a microparasite: (i) the course of the infection in the host; (ii) the mode of transmission between host individuals; and (iii) the behavior and demography of the host population.

The course of an infection includes a latent period after exposure to the source of the infection. During this period the virus, for example, will increase exponentially. The next stage includes the infectious period, during which time the host develops the symptoms of the disease. Meanwhile, as the parasite population is building its numbers, the host immune system begins developing specific antibodies. As the antibody numbers increase, the parasite population plummets and the symptoms of the illness subside. The host ceases to be infectious at some point during the illness and the previously susceptible individual passes from S (susceptible) to I (infected) to R (recovered and immune). Of course some infected individuals may die during the course of the disease, and some recovered individuals may eventually lose their immunity. Other important factors affecting the natural history of the infection include the length of the infectious period, the time-lag derived from the latent period, the development of immunity by the host (thereby removing susceptible individuals from the host population), and the ability of some parasites to remain latent and undetected in host individuals (for example, herpes), only to reappear at some later date.

The two basic modes of transmission are: (i) vertical, in which the disease is passed from mother to offspring (cytomegalovirus and hepatitis B virus); and (ii) horizontal, in which diseases are passed from one individual to another in the environment. Most infections disease organisms are passed through the horizontal method, although some disease organisms can be passed both vertically and horizontally (hepatitis B). These include direct and indirect transmission. Direct transmission includes: (i) close-contact diseases (common cold, influenza, measles); (ii) sexual-contact diseases (hepatitis B virus, HIV, syphilis); and (iii) contaminative-contact diseases (cholera, tetanus, typhoid). Indirect-contact diseases include those that involve transmission from one animal host to another (malaria, rabies, Lyme disease, or the plague) or via needles (HIV or hepatitis B). Human and animal borne or transmitted diseases have been the hardest to control, just as macroparasitic diseases with intermediate hosts, such as schistosomiasis, have yet to be successfully controlled in many tropical countries.

Finally, the frequency and severity of disease outbreaks are also highly dependent upon the behavior of the host population. As discussed below, many diseases such as measles and influenza are dependent upon the size and density of the host population. Large and dense host populations lead to a high disease transmission rate. Public health measures such as isolation of infected individuals, the treatment of water and sewage wastes, and pasteurization of milk have been instrumental in curbing the spread of many diseases. Behavioral changes related to sexual activity are also essential in limiting the spread of other classes of disease.

9.3 Modeling host–microparasite interactions

In the SIR model, we assume the following:

N = total host population density
S = susceptible host density
I = infected host density
R = recovered (immune) host density
b = host birth rate
m = natural host mortality rate unrelated to disease mortality
α = disease-induced mortality rate
β = transmission rate of disease from one host to another
v = recovery rate, or the per capita rate of passage from the infected (I) to the recovered (R) classes. This is usually the inverse of the average infectious period
y = rate at which recovered individuals lose their immunity. That is, the rates at which individuals return to the susceptible class (S) from the R class.

We assume that the rate of transmission of the disease, which is the rate by which susceptible individuals become infected, equals the product, βSI. This means that the rate of infection depends on the rate of population mixing, which is a simple function of the density of both types of individuals. β is the parasite-specific transmission rate, which will differ depending on the life history and type of the infectious agent and the host. The expression βSI is often referred to as the "law of mass action" because of its similarity to the laws governing the mixing of gases (Nokes 1992).

The total population consists of susceptible, infected, and recovered individuals:

$$N = S + I + R \tag{9.1}$$

The growth rate of each segment of the population is written as a differential equation. The increase in the number of susceptible individuals (Eqn. 9.2) is based on the birth rate (bN) and the rate at which recovered individuals lose immunity (γR). Losses are due to the host mortality rate unrelated to the disease (mS), and to the conversion of individuals from the susceptible to infected classes (βSI):

$$dS/dt = bN - mS - \beta SI + \gamma R \tag{9.2}$$

The growth rate of infected individuals (Eqn. 9.3) equals the product, βSI, minus losses due to the combined effects of natural and disease-caused mortality ($m + \alpha$), as well as recovery rate (from I to R), v:

$$dI/dt = \beta SI - (m + \alpha + v)I \tag{9.3}$$

The growth rate of the recovered class (Eqn. 9.4) equals vI minus the death rate unrelated to disease and the loss of immunity, γR:

$$dR/dt = vI - mR - \gamma R \tag{9.4}$$

If we add up all of these equations, we have:

$$dN/dt = dS/dt + dI/dt + dR/dt$$

$$dN/dt = b(N) - mS - \beta SI + \gamma R + \beta SI - (m + \alpha + v)I + vI - mR - \gamma R$$

Let $N = S + I + R$. Expanding, we have:

$$bS + bI + bR - mS - \beta SI + \gamma R + \beta SI - mI - \alpha I - vI + vI - mR - \gamma R$$

After canceling terms, we have:

$$dN/dt = bS + bI + bR - mS - mI - \alpha I - mR$$

Rearranging, letting $N = S + I + R$, and substituting the intrinsic rate of increase, r, for $b - m$, we have:

$$dN/dt = bS - mS + bI - mI + bR - mR - \alpha I = (b - m)(S + I + R) - \alpha I$$

The result is Equation 9.5:

$$dN/dt = bN - mN - \alpha I = rN - \alpha I \tag{9.5}$$

Therefore we see that the growth of the population is diminished by the "natural" death rate, m, and by the mortality rate due to infections, αI.

9.4 Dynamics of the disease

Spreading of a disease, that is, an epidemic, requires that the number of infected individuals remain steady or increases. This means that $dI/dt \geq 0$. Since $dI/dt = \beta SI - (m + \alpha + v)I$ (Eqn. 9.3), we have:

$$\beta SI - (m + \alpha + v)I \geq 0 \tag{9.6}$$

This simplifies to:

$$\beta S - (m + \alpha + v) \geq 0 \tag{9.7}$$

The rates m, α, and v are all time-dependent. They represent rates at which a susceptible individual either dies or moves from the infected to the recovered class. The inverse of $(m + \alpha + v)$ can be thought of as the length of the infectious period, D (Nokes 1992). If we substitute D for $m + \alpha + v$, we have:

$$\beta S - 1/D \geq 0$$

Through rearrangement, we then have:

$$\beta S \geq \frac{1}{D}$$

and then:

$$\beta SD \geq 1 \tag{9.8}$$

We define the product, βSD, as equal to R_0, the **basic reproductive number** (BRN) or parameter for this disease. The result is Equation 9.9:

$$R_0 = \beta SD > 1 \tag{9.9}$$

R_0, the **mean number of new infections caused by a single infective individual**, is an important parameter. If this value is > 1, then $dI > 1$ and the disease incidence will increase. If $R_0 < 1$, the epidemic fails. R_0 is directly proportional to the rate at which an infection spreads, βS. It also depends on the mean amount of time an infection is active, D. For an infection to spread, it must have the right combination of susceptible hosts (S), a reasonably high transmission rate (β), and a sufficiently long period of transmission. One conclusion we may draw from this analysis is that for a disease to succeed, it needs a dense population of hosts. For mosquito-borne diseases like malaria, BRN depends on: (i) vector (mosquito) abundance; (ii) focused feeding (the tendency to bite specific hosts and nothing else; and (iii) vector longevity (the equivalent of D) (Spielman and D'Antonio 2001).

Since the maximum number of susceptible individuals is N, to have an epidemic we must have:

$$\beta ND > 1 \tag{9.10}$$

By rearranging, we find that the **minimum size for an epidemic** is such that:

$$N > \frac{1}{\beta D} \tag{9.11}$$

This allows us to conclude that highly infectious and long-duration diseases can have a low minimum number of infected individuals, but the value of N must be fairly high for directly transmitted microparasitic diseases such as common viral infections, which have short durations of infection and produce long-lasting immunity.

Another way of expressing R_0 is to allow $N = 1/\beta D$, and to **redefine the minimum size for an epidemic as** S_T. If $S_T = 1/\beta D$, then $1/S_T = \beta D$. Substituting $1/S_T$ for βD in Equation 9.9, we have:

$$R_0 > S/S_T \tag{9.12}$$

This means that when the population density of susceptible hosts is above S_T, then $R_0 > 1$ and the infection spreads. There exists a fair amount of data on human epidemics that confirm the theoretical prediction that epidemics occur mainly in areas with populations greater than a particular threshold. For example, measles, mumps, influenza, and polio can be totally absent from small, isolated communities (Nokes 1992). In Table 9.1, for example, are data reported by Anderson (1982) on the number of cases of measles in North American cities of various sizes from 1921 to 1940. Observe that with one exception, Cleveland, no months without reported cases of measles were reported until the city size was below 300,000.

Island populations show the same trend. During the same period, 1921–40, when the population of an island was less than 100,000 there were no reported cases of measles

Table 9.1 Reported cases of measles by month in North American cities, sorted by size, in the period 1921–40. Based on Anderson (1982).

City	Population Size × 10^5	Number of years with at least one month in which no cases of measles were reported
New York	75	0
Chicago	34	0
Philadelphia	19	0
Detroit	16	0
Los Angeles	15	0
Montreal	10	0
Cleveland	9	1
Baltimore	9	0
Boston	8	0
Toronto	7	0
Washington, DC	7	0
Pittsburgh	7	0
Milwaukee	6	0
Buffalo	6	0
Minneapolis	5	0
Vancouver	3	20
Rochester	3	3
Dallas	3	18
Akron	2	8
Winnipeg	2	7

in about 50% of the months. The proportions increase until, in Hawaii, with a population of 550,000 at that time, cases of measles were reported in 100% of the months (Cliff *et al.* 1986).

Host population size also has an effect on the average age of infected individuals. In larger populations the rate of transmission (βSI) and the basic reproductive number, R_0, will be larger since S is larger. Therefore the mean age at which individuals are infected will be lower. This idea is illustrated in Table 9.2. The mean age of the infection decreases

Table 9.2 The effect of population size on the mean age A (in years) at the time of infection of various childhood diseases in New York State communities in the period 1918–19. Based on data from Smith (1983), published in Nokes (1992).

Population size	Measles	Whooping cough	Scarlet fever	Diphtheria
Less than 2500	12.9	8.2	12.3	14.2
2500–10,000	10.7	6.9	11.2	12.5
10,000–50,000	9.0	5.7	10.2	11.5
50,000–200,000	9.0	6.3	10.5	10.6

Table 9.3 Mean age of infection and R_0 for measles in England and Wales, 1950–55. Based on Smith (1983) and Nokes (1992). Reproductive rate based on an average life span of 69.5 years.

Population size	Mean age at infection in years, A	Basic reproductive number, R_0
"Rural"	5.0	15.0
<50,000	4.5	16.5
50,000–100,000	4.2	17.5
>100,000	3.9	18.7

as communities get larger, although the last two population categories do not show the trend consistently.

Table 9.3 illustrates the drop in mean age of infection for measles in England and Wales during the period 1950–55. Notice the increase in the value of R_0 with population density. Note also the contrast between the New York and British data. Children in Britain contracted measles between the ages of 4 and 5 on average, as compared to ages 9 to 13 in New York State. The infection rate depends on "mass action" or mixing. The New York populations in the early twentieth century were much less mobile than the British populations in the 1950s. In addition, a larger percentage of children were likely attending school regularly in the 1950s. Increased rates of infection accompany school terms when children are aggregated together.

Hethcote (2000), however, asserts that human contact rates, and therefore the potential for the spread of diseases, are now only weakly dependent on population size. He believes that the patterns of daily encounters are largely independent of the size of the community within a given area. This may be especially true now for school children who are routinely bused long distances to population centers, at least in the United States.

Finally, prior to massive immunizations, microparasitic diseases were well known for their regular oscillatory patterns. Data from New York, Britain, and elsewhere have shown that in the twentieth century measles had a regular two-year epidemic cycle, mumps a three-year cycle, and rubella a four- to seven-year cycle (Nokes 1992, Hethcote 2000). Infections, such as certain influenzas, are seasonal, and are therefore on a yearly cycle. These yearly cycles are related to seasonal climatic variations and/or to seasonal patterns of population mixing. The greatest factor in the longer-term oscillations is the reduction in the susceptible population during epidemics as they either die or become immune. As new births increase the density of susceptible individuals, the epidemic returns when the value of S exceeds S_T. The tendency for host–parasite as well as predator–prey interactions to undergo oscillations is a confirmation of Turchin's (2001) third principle of population ecology (Chapter 1).

9.5 Immunization

One goal of immunization is to reduce the number of susceptible individuals in a host population, thereby lowering the net reproductive number of the disease to less than one.

Table 9.4 The estimated fraction needed for successful immunization against several diseases, in a dense (>50,000) New York State population in 1918–19. A is average age of infection (from Table 9.2), R_0 is the reproductive parameter for the disease, and p is the proportion that would need to be immunized for eradication of the disease. Average life span estimated at 55 years.

Disease	A	R_0	p
Measles	9.0	7.1	0.86
Whooping cough	6.3	9.7	0.90
Scarlet fever	10.5	6.3	0.84
Diphtheria	10.6	6.2	0.84

Recall that S_T is inversely proportional to D, the average infectious period, and to the rate of infection (β). Therefore, a high rate of infection and/or a long course of the disease make effective immunization more difficult.

Suppose we immunize a fraction, p, of the population. The new reproductive parameter, R_0^*, becomes:

$$R_0^* = (1-p)R_0 \tag{9.13}$$

We need to drive S to a level below S_T to eradicate the disease. If we set:

$$R_0^* = S/S_T = 1 = (1-p)R_0 \tag{9.14}$$

and we solve for p, we have:

$$p = 1 - 1/R_0 \tag{9.15}$$

Therefore, for the disease to be extinguished, p (the proportion immunized) must be greater than $1 - 1/R_0$.

For example, let us re-examine Table 9.2 for New York State in 1918–19. As you can see from Table 9.4, the proportion of the population that would need to be immunized for successful eradication is between 84% and 90%. By comparison, May (1983, as summarized in Alstad 2001) estimated the proportion immunized needed for successful eradication of measles and whooping cough in England and Wales as 92% and 94%, respectively. Successful eradication of rubella requires 86% immunization (Hethcote 2000). These percentages are not atypical for highly contagious diseases. Because about 5% of those vaccinated do not become immune, Hethcote (2000) estimated that eradication of measles requires 99% vaccination, and that 91% is needed for rubella. By 1998 measles was no longer an indigenous disease in the United States (Hethcote 2000) and in March of 2005 the Centers for Disease Control and Prevention pronounced rubella eradicated as an indigenous disease in the United States.

More complex models would include equations for a host that has a very long latent period during which time it is not yet infectious, or for hosts that never reach a state of total immunity and continue to be infectious throughout the remainder of their lives (various venereal diseases or typhoid, for example).

9.6 Endangered metapopulations and disease

As discussed previously, metapopulation ecology has come to the fore as a theoretical framework for conservation planning. High dispersal rates (high movement between patches) have been predicted to increase the proportion of patches occupied at equilibrium, the time of metapopulation extinction, and the effective population size (Hanski and Gilpin 1997). Conservationists have favored measures, such as habitat corridors, which increase connectivity among patches. Hess (1996), however, has suggested that easy migration can have the negative effect of spreading disease among patches, causing extensive local extinctions. Using a metapopulation analysis, Hess found that high migration rates, by facilitating the movement of disease organisms, could reduce patch occupancy and increase the probability of metapopulation extinction.

Gog *et al.* (2002), however, disagreed. They believe that most of the infections that threaten wildlife are not caused by migration of diseased organisms, but by "spillover" from other, more abundant hosts already present in the habitat patches. The reservoir for these "spillover" diseases is often domestic animals. For example, domestic dogs are the probable source of diseases that have threatened African wild dogs (*Lycaon pictus*), African lions (*Panthera leo*), Baikal seals (*Phoca sibirica*), grey wolves (*Canis lupus*), and arctic foxes (*Alopex lagopus semenovi*) (Gog *et al.* 2002).

In the deterministic model of Gog *et al.* (2002), S is the proportion of susceptible host patches (host population present, no disease), and I is the proportion of infected patches (host population and disease present). The extinction rates of susceptible and infected populations are x_S and x_I, respectively. The migration rate between susceptible and infected populations is ψ. When an infected disperser arrives at a susceptible patch, it infects the resident population with the probability of δ. Infection spreads at the rate of $\psi\delta IS$. The preceding is identical to the Hess (1996) model. What Gog *et al.* (2002) added is an extension of the Hess model in which they simulated various parameters of an infection rate from an "outside source," g. Starting with the Hess model, Gog *et al.* set g at zero, then ran a number of simulations showing the important effects when g is a non-zero parameter.

The equations for mean proportion of patches occupied in the S and I states are:

$$\hat{S} = \psi S(1 - I - S) - x_s S - \psi\delta IS - gS \tag{9.16}$$

$$\hat{I} = \psi I(1 - I - S) - x_I I + \psi\delta IS + gS \tag{9.17}$$

Equations for stable equilibrium values of S and I for different values of g and m are found in appendix A of Gog *et al.* (2002). Figure 9.1 has been produced based on their equations for $g = 0$ (representing no outside sources of disease) and $g = 0.4$ (representing a moderately large background infection rate). Other parameters are the same as in Hess (1996): $x_s = 1.4$, $x_I = 2.4$, $\delta = 0.5$.

As we see from Fig. 9.1, Gog *et al.* (2002) found that when g is zero or very small (as in Hess), occupancy rates first increase but then decrease with increased migration as more and more patches experience extinction due to disease. Eventually, patch occupancy increases again with more migration, as all patches become infected. This result (Fig. 9.1a) led Hess (1996) and others to suggest that increased migration between patches can have a negative effect on patch occupancy and can increase the probability of metapopulation

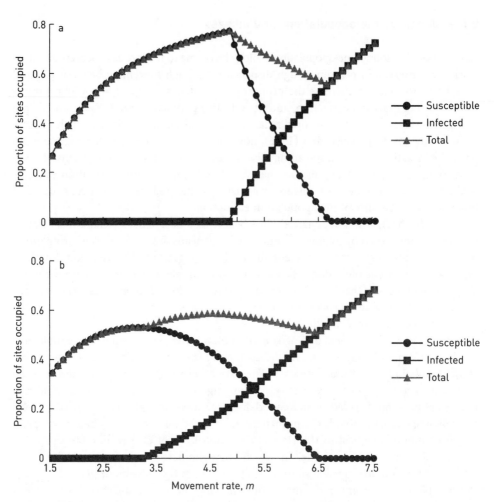

Figure 9.1 Proportion of suitable patches occupied as a function of movement rate: (a) with the parameter $g = 0$; (b) with $g = 0.4$. When $g = 0.4$ there is a reasonably large chance of infection from an "outside source." Adapted from Gog *et al.* (2002) and Hess (1996).

extinction. However, for larger values of g, increasing the migration rate results in little if any depression in the rate of patch occupancy. In other words, the decrease in patch occupancy at intermediate levels of migration is minimized (Fig. 9.1b). Gog *et al.* (2002) concluded that the net effect of migration is almost always positive, and that at high rates of infection from external sources the benefits of migration will always outweigh the costs. The major application of these models is that in wild populations suffering from a high rate of infection from alternative host species, patch occupancy should increase, rather the decrease, with migration rate.

Gog *et al.* (2002) concluded that corridors between suitable habitats are likely to benefit metapopulation persistence, a conclusion also reached by Laurance and Laurance (2003; see Chapter 5). However, they pointed out that the Hess model might apply well to captive populations, where transfer of animals from one facility to another is often a cause of disease epidemics. They stressed the importance of veterinary screening and quarantining procedures before transferring animals from one captive population to another.

9.7 Social parasites

By definition, parasites reduce the fitness of the host on which they live. This is usually the result of the consumption of host tissues. There are, however, parasites that reduce fitness by means of behavior. **Social parasites** include brood or nest parasites. Birds such as cowbirds and cuckoos lay their eggs in the nests of other species (Davies 2000). When the eggs hatch, the host parents feed the parasitic chicks, even in preference to their own offspring. The chicks are large and aggressive, and since the parents often make no distinction among the chicks, the cowbird or cuckoo chicks get a majority of the food provided by the parents. In most cases the host raises few young of their own species when a parasitic chick is present in the nest. Moreover, cuckoo and cowbird females often remove host eggs prior to the laying of their own eggs, and in a nest still containing host eggs a cuckoo chick will eject host eggs from the nest (Davies 2000). Social parasitism among birds has evolved on every continent except Antarctica and there are about 100 species of obligate brood parasites worldwide (Davies 2000). Brood parasitism has probably evolved six times since it is found in six different bird families. Brood parasites include Old World cuckoos (Europe, Asia, Africa, Australia, and New Zealand), New World cuckoos (North and South America), cowbirds (North and South America), a duck (South America), honeyguides (Africa and Asia), and finches (Africa) (Davies 2000).

In some instances, the host makes an attempt to distinguish among the eggs and dumps the parasite eggs from the nest. Social parasites, including several species of cuckoos, cowbirds, African honeyguides, and finches have responded by laying "mimetic" eggs. That is, eggs which look like those of the host species. Furthermore, according to the "Mafia hypothesis," (Zahavi 1979, Zahavi and Zahavi 1997), some nest parasites retaliate against "dumpers" by destroying all of the eggs in the nest and perhaps even the nest itself. This parasite makes the host birds an offer they "can't refuse." That is, "raise one of my chicks or lose all of your children and your house!" Evidence, however, suggests that common cuckoos (*Cuculus canorus*) do not adopt Mafia-like tactics (Davies 2000), since female cuckoos do not usually revisit nests they have parasitized. However, Davies suggests this hypothesis could work in parasitic species that leave one or more host eggs in the nest.

Soler *et al.* (1995a, 1995b) experimentally tested the Mafia hypothesis between great spotted cuckoos (*Clamator glandarius*) and their magpie (*Pica pica*) hosts in Spain. They found that when a magpie accepted a cuckoo egg, the nest was successful, but when the host rejected the cuckoo egg the nest was often destroyed. Soler and his team experimentally removed cuckoo eggs while simultaneously visiting nests without destroying the cuckoo egg (controls). They found that more than 50% of the experimental nests were preyed upon, either at the egg or young chick stage, compared with only 10% for the control nests. The predation was most likely by cuckoos. Furthermore, in a follow-up experiment Soler *et al.* (1999) showed that the magpies learned from the experience. If their eggs were destroyed after they removed a cuckoo egg, when they re-nested they were 50% less likely to remove the cuckoo egg the second time around. Curiously, however, the cuckoo lays eggs that mimic those of the magpie. Davies (2000) suggests that the Mafia tactics do not work that well, and on average magpies still do better to reject cuckoo eggs. However, the behavior of individual hosts varies and some are more susceptible to Mafia tactics than others. Therefore efforts to intimidate the host are successful often enough to have become part of the relationship between cuckoos and magpies in Spain.

Social parasitism also occurs among social insects. In the so-called slave-making ants, a queen of the species *Lasius reginae* enters the nest of another species (*L. alienus*), kills the resident queen, and forces the workers to care for her own offspring. In such a case the workers are unable to distinguish the foreign ant queen from their own (Faber 1967). This parasitism is temporary, however, during colony foundation. *L. reginae* workers eventually take over foraging and management of the nest. In many other cases, however, the parasitic species produces no workers, a condition termed inquilinism. The socially parasitic species spends its entire life in the nest of its host species. Workers are lacking or are degenerate in normal foraging behavior. Holldobler and Wilson (1990) have described the "ultimate social parasite," *Teleutomyrmex schneideri*, which is a social parasite of *Tetramorium caespitum* and *T. impurum*. This parasite is only found in the nests of its hosts. It lacks a worker class and the queens contribute nothing to the host colonies. Finally, the parasitic queen spends its life as an ectoparasite on the back of the host queen!

In more dramatic cases of interspecific slave making, species such as *Formica sanguinea* raid other species of the same genus (*F. fusca*). As described by Wheeler in Holldobler and Wilson (1990), *F. sanguinea* heads to an *F. fusca* colony 50 to 100 m away, following an amazingly direct route. Once they have arrived, they surround the nest and wait for reinforcements to arrive. The raiders snatch larvae and pupae from the nest and from the jaws of the host workers attempting to flee with their young. They kill adults of *F. fusca* only if they offer resistance. Otherwise the host colony is left to rebuild itself as long as the queen has not been killed. The brood of *F. fusca* is brought back to the *F. sanguinea* colony and enslaved. There are also numerous examples of intraspecific slave making, in which large colonies raid small ones, kill or drive off the queen, and carry or drag larvae, pupae, and young workers to the home nest, where they are put to work (Holldobler and Wilson 1990).

9.8 Conclusions

The complexities of parasite–host interactions rival those of mutualisms. Parasites range from endoparasitic viruses to ectoparasitic ticks, and we lack an understanding of the life cycle of most parasites in organisms other than humans. What we do know is often based on models of human diseases and their modes of transmission. Parasitism can also involve complex behavioral interactions such as brood parasites or slave-making ants. As we will discuss in the next chapter, the activities of parasites can have major impacts on populations, communities, and ecosystems. Some ecologists consider parasite–host interactions the newest and least explored ecological frontier. Parasite–host interactions are much harder to study, and less obvious, than predator–prey relationships. Therefore we know little about them, and ecologists may well have been guilty of underestimating their importance.

10

Predator–prey interactions

- The Lotka–Volterra equations
- Functional responses
- Functional responses and the Lotka–Volterra equations
- Graphical analyses
- The half-saturation constant in predator–prey interactions
- Nicholson–Bailey models
- Field studies of predator–prey interactions
- Trophic cascades
- Types of escape from predation

10.1 Introduction

The relationship between predators and their prey has provided a lively topic of discussion for groups of humans ever since they began gathering around the fire or, now, the seminar table. Historically, people have seen large predators not only as dangerous to themselves and their families, but also as competitors for the prey they were seeking. The large number of fables and fairy tales involving wolves, lions, and tigers attests to their prominent role in human culture. In the twentieth century the perception of large predators for many people, particularly in developed, affluent countries, shifted from "vermin" to "charismatic megafauna."

In population ecology the basic question remains, what determines distribution and abundance? For trophic levels above that of producer, what is the role of predation in controlling herbivore populations? Are prey populations limited primarily by available habitat and food supply, or by their predators? That is, are prey limited by what they eat, or by what eats them? In terms of species diversity, do predators allow more species to exist in a community, or, by limiting prey populations to low levels, do they often drive them locally extinct and thereby limit diversity? Do predators primarily kill very young, very old, and/or sick individuals, such that their effect on population growth is insignificant? Do they actually benefit the prey population by eliminating the spread of disease and eliminating genetically inferior individuals? Do predators take prey individuals when the prey population has exceeded the carrying capacity, thereby helping stabilize the prey population? Or, alternatively, are predators the cause of the periodical population cycles seen in many prey species?

There are no simple answers to these questions, and the answers probably change from one ecosystem to the next. Throughout human history, although we have domesticated members of the cat and dog family, we have also tried to extirpate canine or feline predators perceived to be dangerous and/or which interfere with the management of prey species. The Alaska Department of Fish and Game engages in wolf (*Canis lupus*) and bear control projects in order to ensure an abundance of moose (*Alces alces*) and caribou (*Rangifer tarandus*) for human hunters (National Research Council 1997). In Europe, brown bears (*Ursus arctos*), leopards (*Panthera pardus*), gray wolves, and other large predatory species have been eliminated from most of their prehistoric ranges. On the other hand, predators such as gray wolves and grizzly bears (*Ursus arctos horribilis*) are major tourist attractions in National Parks in the western United States and Canada, and the US Fish and Wildlife Service reintroduced wolves in Yellowstone National Park in 1995, after they had been deliberately extirpated in the 1920s. One of the rationales for reintroducing wolves was the assertion that, without wolves, the elk (*Cervus elaphus*) population had grown too large. Over-browsing by the elk supposedly has led to a decline in willow (*Salix* sp.) and quaking aspen (*Populus tremuloides*), an increase in stream bank erosion and a decline in cutthroat trout (*Salmo clarki*) (Huff and Varley 1999). According to this scenario, the reintroduction of wolves will have a salubrious effect on the Yellowstone ecosystem. Local cattle and sheep ranchers have begged to differ, and Pyne (1997) might argue that fire suppression is much more likely the cause of a decline in willow and quaking aspen rather than the elimination of wolves. In addition, Meagher and Houston (1998) have noted that willow and aspen are minor components (1–2%) of the vegetation in the northern range of the Greater Yellowstone ecosystem. Ecosystems are complex and the specific roles played by predators are difficult to disentangle from the complexities of the food web.

In spite of the much more sophisticated methods of gathering and analyzing data that we have today, when trying to understand predator–prey relationships we are still bedeviled by the fables, simplistic theory, and inadequate analysis of data published in the first half of the twentieth century. For example, the fable of the Kaibab deer herd, first published in a *Wisconsin Wildlife Bulletin* by Aldo Leopold (1943), influenced the opinion of at least one generation of ecologists on the role of predators in ecosystems. The story goes like this. Prior to 1906, a population of mule deer (*Odocoileus hemionus*) shared the Kaibab plateau in northern Arizona with cattle (*Bos taurus*), sheep (*Ovis aries*), coyotes (*Canis latrans*), wolves, mountain lions (*Felis concolor*), bear, and bobcats (*Lynx rufus*). When this area was declared a game refuge as part of the new Grand Canyon National Park by President T. Roosevelt, federal agents not only removed the cattle and sheep, but also did their best to eliminate all the predatory species. In contrast to our views today, in which we advocate the maintenance of an ecosystem, including the restoration of predators, Leopold and other wildlife biologists had a very negative view of predators. Although Leopold changed his view later in life (Leopold 1949, Botkin 1990), he and other government biologists initially agreed with a policy of predator removal. Hunters, in the period 1906–31, killed approximately 781 mountain lions, 30 wolves, 4338 coyotes, and 554 bobcats on the Kaibab. These are impressive numbers, but no one actually knows what effect this had on the predator populations. The deer population, approximately 4000 strong in

1906, was released from both competition and predation, and rapidly increased. In the fable, as published in many ecology and general biology textbooks as late as 1972, the deer population increased to 100,000 by 1924, crashed due to overgrazing and ended up at a population lower than that of 1906. This was thought to be due to permanent damage to the soil and plant life in the ecosystem. The problem is that no one actually counted the deer herd in a systematic manner. The estimate of 100,000 was by one individual who was visiting the area. Other individuals estimated peak abundance at 50, 60, and 70 thousand. Forest supervisors on site thought the peak population was 30,000. In any event, the data are totally unreliable (Caughley 1970). Unfortunately, Leopold based his graphical analysis on that of Rasmussen (1941), who used 100,000 as the deer population estimate for 1924. Commenting on this story, Botkin (1990, pp. 78–80) writes, "an examination of the facts about counts (of mule deer) leaves us up in the air. The famous 'irruption' of mule deer . . . may or may not have occurred, and if it did occur the cause may have been completely unrelated to the presence or absence of predators." The Kaibab story was reprinted repeatedly because it fitted into the "new" view that predators were part of the "balance of nature" (Botkin 1990).

This dichotomous view of predation lives on. To some people predators are pests, undesirable and unworthy of existing except in someone else's ecosystem. To others, predators are part of the balance of nature and we eliminate them at our peril. Since human modification of ecosystems is so pervasive, it is almost impossible for large predators to survive throughout much of their previous range. Yet prey populations, such as white-tailed deer (*Odocoileus virginianus*), have prospered. Hunters who view themselves as filling the necessary role of the wild predator often employ the balance-of-nature argument. Alaskan hunters see themselves in competition for big game with wolves.

The prevailing views of predation in the first half of the twentieth century were colored by the fact that theory and most early studies isolated two species, one prey and one predator, from their community and ecosystem. As we will see, simple, and even many complex models, predict that predator–prey relationships are "inherently oscillatory." An intensively studied phenomenon in population ecology is the regular population cycles displayed by rodents, hares, and other small animals living in boreal and arctic communities. After 80 years, there is still disagreement as to whether lemming (*Dicrostonyx groenlandicus* and *Lemmus lemmus*) cycles, for example, are caused by the interaction of lemmings with their food plant (Turchin *et al.* 2000) or are the result of an interaction with their predators: stoats (*Mustela erminea*), arctic fox (*Alopex lagopus*), snowy owl (*Nyctea scandiaca*), and long-tailed skua (*Stercorarius longicaudus*) (Gilg *et al.* 2003). What separates these modern studies from the simple models that are discussed below are the assumptions that: (i) prey populations are at least partially determined by their food supplies, not just by their predators; (ii) prey populations respond to the entire community of predators, not just a single species; and (iii) predator populations are affected by factors other than just the prey population density. Although most models and theoretical treatments we will discuss have been based on a single predator and a single prey, you should keep in mind that more realistic treatments should be community- and ecosystem-based.

The overall predation rate, that is the number of prey killed per unit area during a specified time period, is dependent on both the **numerical response** of the predator population and the **functional response** of individual predators in the population. As prey density increases, a numerical response is an increase in the numbers of predators per unit

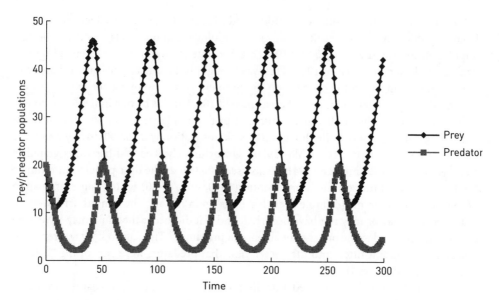

Figure 10.1 Stable limit cycle. Prey and predator populations versus time.

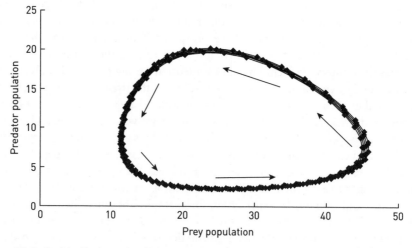

Figure 10.2 Stable limit cycle. Prey population versus predator population.

area, based on both immigration and reproduction. A functional response is an increase in the number of prey consumed per unit time by each individual predator as a function of prey density. The **total response**, which is the combined effect of both the numerical and functional responses, is density-dependent at low prey densities, but can become inversely density-dependent at high prey densities (Holling 1959, Messier 1994), as explained in more detail below.

When Robert May (1976c) reviewed existing predator–prey models and theories, he found that they all lead to one of the outcomes enumerated in the list below. In a stable limit cycle, both the prey and the predator go though regular, predictable cycles (Figs 10.1 and 10.2). In a stable point, both the prey and the predator populations settle at a fixed number (Figs 10.3 and 10.4). In parentheses are early studies that have illustrated these outcomes:

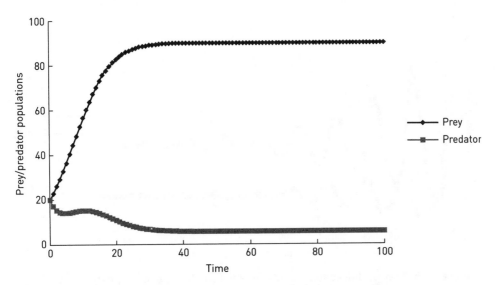

Figure 10.3 Stable point. Prey and predator populations versus time.

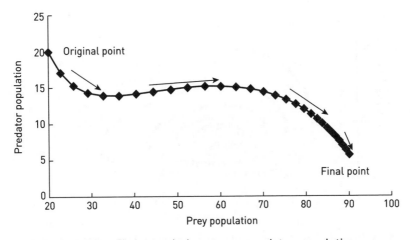

Figure 10.4 Stable point. Prey population versus predator population.

1 Extinction of the predator, survival of the prey (Gause 1934);
2 extinction of the prey followed by extinction of predator (Gause 1934, Huffaker 1958);
3 the prey and the predator populations go through oscillations, which dampen to a stable limit cycle or a stable point (Figs 10.5 and 10.6) (Dodd 1940);
4 the prey and the predator go through increasing oscillations, leading to extinction of the prey and/or the predator (Figs 10.7 and 10.8) (Hassell 1978);
5 immediate stable limit cycle (Hudson *et al.* 1998);
6 immediate stable point.

Metapopulation theory also predicts that a predator–prey interaction may consist of coexistence in a complex heterogeneous environment (Huffaker 1958, Dodd 1940)

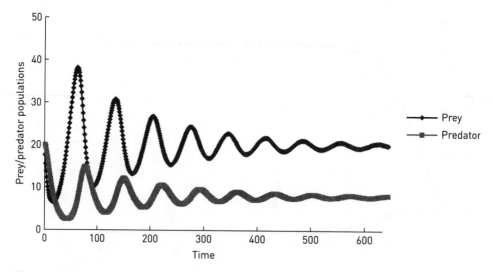

Figure 10.5 Prey and predator populations versus time, showing dampened oscillations leading to a stable point.

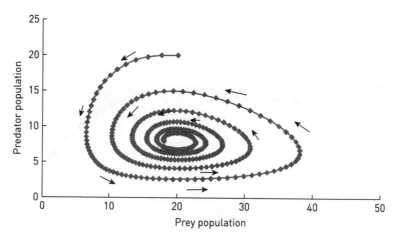

Figure 10.6 Prey and predator populations, showing dampened oscillations leading to a stable point.

combined with extinctions of prey and/or predators at any given site. Prey populations have also been shown to fluctuate while the predator population remains stable. For example, a stable population of predators may switch from one prey species to another, depending upon prey availability (Southern 1970).

As we review various attempts to model predator–prey relationships, it will be obvious that a number of simplifying assumptions have been made. Below is a short list of factors that should be considered when evaluating the reality of a predator–prey model.

 1 Errington (1946), based on his field work with muskrats (*Ondatra zibethicus*), asserted that prey usually have a refuge; only when prey numbers are suffi-ciently large that individuals must leave the refuge are they subject to predation.

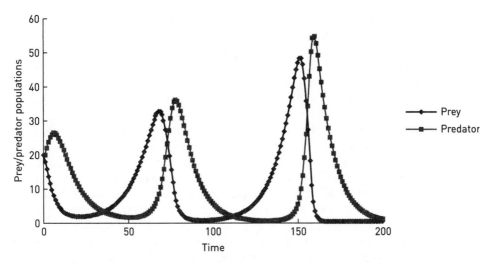

Figure 10.7 Prey and predator populations versus time, showing increasing oscillations leading to extinction of both the prey and the predator population.

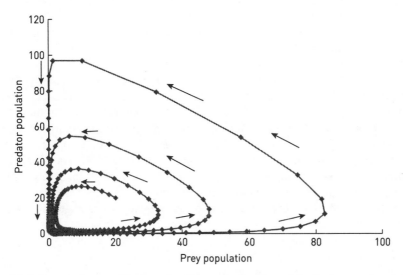

Figure 10.8 Increasing oscillations leading to extinction of both the prey and the predator populations.

2 Some models assume that predation is random. In fact, predation is almost always nonrandom. In addition, predators often specialize on certain age classes of the prey, or on weakened or diseased individuals.

3 The generation times of the predator and the prey populations are often very different. If the growth rate of the prey is much higher than that of the predator, and/or the generation time is much shorter in the prey species, the predator may rarely have an effect on the prey population. Alternatively, there may be a time lag between growth of the prey population and the numerical increase in the predator population. As we saw in Chapter 2, time lags tend to produce population cycles.

4 Predators may be generalists and switch from one prey species to another, depending on prey population size. Such behavior tends to stabilize prey populations. In fact, recent studies discuss the idea of a "predator pit" for the prey (Hudson and Bjørnstad 2003) when there is a rich community of generalist predators. When there are many predator species present, prey populations are controlled at a stable number and do not go through cycles typical of the same prey species in less predator-rich communities.

5 Alternatively, the predator population may remain relatively constant in spite of large-scale fluctuations in the prey population. For example, when Southern (1970) studied the relationship between the tawny owl (*Strix aluco*) population and its rodent prey near Oxford, UK, he found a relatively constant breeding population of owls, in spite of the fact that the rodent prey base oscillated. Over a 13-year period the number of rodents per 12 acres (4.9 ha) varied from fewer than 10 to more than 250. During the same time the number of breeding pairs of the owl changed from a low of 17 to a high of 30. In years of low rodent density, however, no owls attempted to breed.

6 The predator may have a carrying capacity unrelated to the number of prey individuals. Thus the predator may never increase beyond a certain number and the prey population "escapes" from predation once it reaches a certain population threshold.

7 Density-independent mortality of the prey and/or the predator population must be evaluated.

8 Predator–prey interactions may have multiple equilibriums. As pointed out by Messier (1994), a predator–prey interaction may have a low-density equilibrium and high-density equilibrium. At low prey densities the predation rate is density-dependent and the prey population is maintained at the level K_1. At densities just above K_1 is the "predator pit" and the prey population is pushed back down to K_1 by the functional and numerical responses of its predator(s). However, if the prey population escapes to density K_2 it has escaped the pit. At densities above K_2 the overall predation rate is inversely density-dependent due to limitations in the functional and/or numerical responses of the predators. The prey population is able to increase to level K_3, the high-density equilibrium, where it is limited by its food supply or some other environmental variable.

The introduction of predators into "naïve" communities has provided us with evidence that predators can have a powerful destabilizing effect on ecosystems. One well-known example is the inadvertent introduction of the sea lamprey (*Petromyzon marinus*) into the Great Lakes (Baldwin 1964, Smith and Tibbles 1980). Marine sea lampreys are found off the Atlantic coast of North America, but migrate into fresh water to spawn. Adult lampreys feed by rasping a hole in the host and sucking out fluids. Lampreys were able to move up the St Lawrence River but were not found in the Great Lakes due to the Niagara Falls. But, beginning in 1829, various canals were built to allow ships to pass from the St Lawrence into the Great Lakes. The Welland Canal was finished in 1829, allowing ships and lamprey passage to Lake Erie. However, the systematic movement of lampreys into the Great Lakes did not begin until 1921. Once established in the Great Lakes, the lamprey began preying upon lake trout (*Salvelinus namaycush*), an economically important fishery. Lampreys were found in Lakes Huron and Michigan in 1936–37 and in Lake Superior in 1938. Catches of lake trout declined from a high of 3000 tons per year in Lake Huron in 1936 to virtual extinction by the late 1950s. The story was repeated in Lakes Michigan and Superior, but took more time. The peak of 3500 tons of lake trout was harvested in the middle 1940s in Lake Michigan and dropped to virtually zero by 1950–51. In Lake Superior the peak harvest was around 2500 tons in 1950 and dropped to less than 500 tons by 1960. Massive efforts to eliminate the sea lamprey failed, but did reduce its population. The result was only a partial recovery by the lake trout to its former numbers.

To take another example, Like most Pacific Islands, Guam lacked large predatory snakes. The only native snake was a blind, wormlike snake, which fed on termites and ants (Fritts 1988). The brown tree snake (*Boiga irregularis*), a member of the family Colubridae and native to Australia, Papua New Guinea, and other islands in northwestern Melanesia, arrived in Guam shortly after World War II, probably as a stowaway in military cargo from Papua New Guinea (Fritts 1988). These snakes did not become conspicuous until the 1960s, but by 1968 had colonized the entire island. Brown tree snakes feed on a wide variety of bird, mammal, and lizard species. The population of snakes increased for over 35 years, reaching a density at one point of 100 per hectare in some areas (Rodda *et al.* 1992). By 1963, several formerly abundant native bird species had disappeared from areas where the snakes were most populous. By 1986, nine native forest bird species were extinct due to snake predation and several other bird species were endangered. Small mammals are also very rare in most areas inhabited by the snake. Finally, having depleted birds and mammals, the brown tree snake now feeds to a large extent on lizards. Between three and six lizard species have been or are being extirpated on Guam (Savidge 1988, Rodda and Frits 1992).

Over evolutionary time, of course, predators and their prey reach an accommodation through coevolution or else they cannot coexist. Extinction and re-colonization in metapopulations occurs continuously over shorter periods of time, as do extinction and speciation over evolutionary time. Therefore we should not necessarily expect, based on a "quasi-religious idea of the balance of nature" (McCullough 1997), that predator and prey populations should be "stable." Nevertheless, we can all agree that the rapid extinction of entire communities of native organisms by introduced predators is highly undesirable.

One last word, before we proceed. The term "stable" as applied to a population has a variety of definitions. Here we will use the word to simply mean a population, or a pair of populations when applied to predator–prey relationships, which have a "return tendency" to a particular density. Recall that a density-dependent population is one in which there is some mean level of density around which the regulated population fluctuates. In addition, over time the population does not wander increasingly away from this level. If we apply this to several populations simultaneously, each will fluctuate in a cloud of points around some point to which it tends to return. We will distinguish this from a cyclic population in which there is a regular, identifiable pattern of population sizes over time (Fig. 10.1). We will call this, as described above, a "stable cycle" or "stable limit cycle." These two patterns are to be distinguished from those of increasing oscillations leading to extinction, or other patterns of extinction of the predator or mutual extinction of the predator and the prey. These interactions are usually called "destabilizing" or "unstable." The cases described above, of the introduction of the sea lamprey or the brown tree snake, qualify as destabilizing. Some authors characterize certain predators as destabilizing and other as stabilizing. For example, Gilg *et al.* 2003 describe the stoat, a specialist predator, as destabilizing, while the generalist predators stabilize the interaction with lemmings.

10.2 The Lotka–Volterra equations

The first well-known models of predator–prey interactions can be traced to Lotka (1925) and Volterra (1926). Although the Lotka–Volterra model has been critiqued on many grounds (May 1975a, 1976c), it is still well respected, and forms the basis for models still in use today. As Hudson and Bjørnstad (2003) put it, "the fundamental theory of predator–prey interactions encapsulated in the worthy Lotka–Volterra model predicts cycles in prey and predator abundance."

The Lotka–Volterra model, like most predator–prey models, consists of two parts. The prey population grows according to a simple exponential or logistic model. Subtracted from this are losses due to predation. These losses are due to the **overall predation rate**, which itself consists of two parts. The **numerical response** of the predator is a function of an increased rate of reproduction, an increase in immigration, or both. The second factor, which increases the overall predation rate, is the rate of consumption of prey per individual predator, the **functional response**. In the prey equation, the rate of growth is decreased by the overall predation rate, which is a function of both the numerical and functional responses of the predator.

The predator equation also consists of two parts. The growth of the predator population is a function of the overall predation rate, and is similar to the negative part of the prey equation. The growth rate of the predator is then decreased by a mortality factor, which can be either density-independent or density-dependent.

The relatively simple Lotka–Volterra model was based on the following assumptions:

1 In the absence of predators, the prey population grows either exponentially or logistically.
2 The population growth of the predator is limited only by the availability of the prey.
3 Both predator and prey reproduce continuously, have no age structure, and all individuals are identical.
4 The predation rate is proportional to the rate of encounter between predators and prey. Encounter rate is a random function of population density. That is, both prey and predator individuals move at random.
5 The predator has a density-independent, constant mortality rate.

Most of the eight points raised in Section 10.1 are ignored in the Lotka–Volterra model. Predation is random; there is no refuge for the prey; and there is no carrying capacity for the predator independent of that set by the prey population.

In the version of the Lotka–Volterra model in which the prey population grows exponentially in the absence of predators, the prey growth rate is:

$$dN/dt = r_n N \tag{10.1a}$$

Where r_n is the prey intrinsic growth rate and N is the prey population size.

If the prey population grows according to the logistic, the prey equation is:

$$dN/dt = r_n N \frac{K_n - N}{K_n} \tag{10.1b}$$

where K_n is a carrying capacity for the prey population.

Without prey, the predator population (P) dies off based on the instantaneous density-independent mortality rate, m_p, and the population declines according to Equation 10.2:

$$dP/dt = -m_p P \qquad (10.2)$$

The chance of an encounter between the predator and prey is the product:

$$ENP \qquad (10.3)$$

E is a number less than one, which measures the searching (and capturing) efficiency of the predator. Equation 10.3 assumes that the number of prey taken varies linearly with prey abundance. The coefficient E is a functional-response term based on the rate of predation per individual predator per unit time. The increase in the predator population is the encounter term (10.3) times a constant χ_p, which measures the efficiency by which the food (prey) is turned into new predator individuals. In essence this is the **assimilation efficiency** of the predator. The population growth term for the predator is: $(\chi_p)(E)(N)(P)$. This product, therefore, includes the numerical response to an increase in the prey population, and a linear functional response. The Lotka–Volterra model assumes that an encounter leads to the death of a prey individual. The prey population is thus decreased by the term ENP. Starting with Equation 10.1a, the equations for prey and predator are as shown below.

$$dN/dt = r_n N - ENP \qquad (10.4a)$$

$$dP/dt = \chi_p ENP - m_p P \qquad (10.5)$$

The behavior of this model at equilibrium can be analyzed by setting both dN/dt and $dP/dt = 0$, leading to Equations 10.6 and 10.7:

$$P^* = \frac{r_n}{E} \qquad (10.6)$$

$$N^* = \frac{m_p}{\chi_p E} \qquad (10.7)$$

Equation 10.6 expresses the predator equilibrium in terms of the growth rate of the prey and searching efficiency of the predator. Equation 10.7 describes the prey equilibrium in terms of the mortality rate, the searching efficiency, and the assimilation efficiency of the predator. These equations seem to indicate that the equilibrium values are independent of the numbers of the other population. However, we can also see that the per capita growth rate of the prey becomes zero or negative when the predator population exceeds a fixed number. Similarly, the per capita rate of predator increase becomes zero or negative when the prey population drops below a specific density. Finally, the functional response for the predator is unrealistic in that it assumes that at high prey densities the predator has an unlimited appetite.

As pointed out by May (1975a, 1976c) the Lotka–Volterra model has a peculiar "neutral stability" that can be compared to that of a frictionless pendulum. Populations are predicted to oscillate forever based on the initial conditions, with no mechanism for increasing or decreasing the amplitude of the oscillations. Furthermore, both the functional and the numerical responses of the predator are fixed. This is a significant flaw in either a predictive or a descriptive set of models.

Long ago, Volterra (1931) recognized that adding a density-dependent component to the prey equation would add realism to the model. Therefore the prey equation (10.4a) was modified to 10.4b. The Volterra model is characterized by a stable equilibrium point (Figs 10.5 and 10.6).

$$dN/dt = r_n N \frac{K_n - N}{K_n} - ENP \qquad (10.4b)$$

10.3 Early tests of the Lotka–Volterra models

According to the Lotka–Volterra equations, the response of a predator population to an increase in a prey population is to increase its own numbers. This increase may be through an increase in the birth rate of the predator (perhaps combined with a decrease in death rate) or through immigration. Again, this is termed a **numerical response**. According to the Lotka–Volterra equations, the result of this increase in predation is a coupled numerical response in the prey population, which declines. Once the prey population has declined sufficiently, the negative consequences for the predator population results in its decline. Once predator numbers have decreased sufficiently, the prey population begins to recover, leading eventually to an increase in the predator population, and so on. A graph of the Lotka–Volterra results versus time look like a stable limit cycle (Fig. 10.1), though it is not. In a true limit cycle, if the populations are pushed out of the cycle by density-independent factors, the populations return to the original limit cycle. Due to the neutral stability of the Lotka–Volterra equations, they would have no return tendency.

The Lotka–Volterra equations led some ecologists to adopt the view that predator–prey interactions were "inherently oscillatory," and research was directed to test this proposition. As discussed previously, some parasite–host and various small mammal and bird populations in boreal and arctic regions display regular oscillations in number. Do predators cause these cycles? Already in 1924 Charles Elton had published "Periodic fluctuations in the numbers of animals: their causes and effects." In this and later publications (Elton and Nicholson 1942) he presented the now famous and infamous data on the populations of snowshoe hare (*Lepus americanus*) and the lynx (*Lynx canadensis*). As in the case of the Kaibab deer, these data were not based on systematic population surveys, but rather on the numbers of pelts brought into the Hudson's Bay Company by trappers. The data do show regular fluctuations of great magnitude, but we must ask, how reliable are these data? Are the fluctuations due to predator–prey interactions as envisioned by Lotka and Volterra? We will not try to answer these questions now, but note here that after 70 years of field experiments and time-series analysis, Krebs *et al.* (2001) concluded that the hare cycle can only be understood as an interaction involving the hare population, its food supply, and a community of predators (not just the lynx).

As discussed earlier, the competition equations of Lotka and Volterra were tested in the laboratory by Gause (1934). Gause also tried to test the predictions of the Lotka–Volterra predator–prey equations using microorganisms. He attempted to produce the predicted oscillations using as his prey populations of *Paramecium caudatum* grown in test tubes. To these tubes he introduced another ciliated protozoan, *Didinium nasutum*. *Didinium* is a voracious predator on *Paramecium* and it reproduces by binary fission, just as does its prey. In the simple test-tube environment *Didinium* was able to hunt down all of the *Paramecium*. Once its food supply was gone, *Didinium* starved. Thus, in any one tube, mutual extinction was assured. Gause next tried adding sediment to the bottom of the tubes as a refuge for *Paramecium*. This ensured the survival of the *Paramecium*, but the *Didinium* population eventually went extinct. With its predator eliminated, the *Paramecium* population rapidly grew to the expected carrying capacity. But Gause had more tricks. He now added one *Paramecium* and one *Didinium* every third day to each test tube. This finally resulted in coexistence of the prey and its predator for more than two weeks. Both the prey and the population went through two oscillations during this period.

Did Gause see his work as confirming the equations developed by Lotka and Volterra? Just the opposite. Gause stated that predator–prey interactions are not inherently oscillatory, and that coexistence was possible only through adding heterogeneity to the simple test-tube environment, or through constant interference of the system through the addition of immigrants.

Another early laboratory experiment illustrates the weaknesses in the simple Lotka–Volterra model. In Chapter 5 on metapopulations we described the work of Huffaker (1958), who was a California entomologist interested in biological control of pests in orange orchards. The prey species was the six-spotted mite (*Eotetranychus sexmaculatus*), which feeds on oranges. The predator was a carnivorous mite (*Typhlodromus occidentalis*) which preys on the six-spotted mite. Both species reproduce rapidly through parthenogenesis. In each experiment, Huffaker began with 20 prey females and introduced two predator females 11 days later.

In one experiment, Huffaker concentrated the food (oranges) in one area. The results mirrored those of Gause. The prey population rapidly increased, then was located by the predators, which also rapidly increased. Within a short time (25–30 days) both populations were extinct. Huffaker then began creating a heterogeneous environment. He set up a complex laboratory environment consisting of three 40-cell trays with a total of 120 feeding positions. Although each position contained one orange, he controlled the feeding surfaces by dipping the oranges in wax, leaving only 5% of the orange available for feeding. This forced the herbivorous mite to constantly seek out new feeding surfaces. He added small wooden pegs as launching pads for the six-spotted mites to speed their dispersal from one orange to another. And he added a maze of Vaseline™ barriers across the trays to slow the dispersal of the predatory mites, which could travel only by foot. Once a predator arrived on an orange already colonized by the prey species, it quickly killed and consumed all of the herbivorous mites on that particular orange. But the rapid immigration and emigration of the herbivorous mite, along with the complex, heterogeneous environment created by Huffaker, allowed the two species to coexist in this laboratory environment for over 200 days.

In both of these laboratory studies, the coexistence of the prey with the predator depended upon environmental heterogeneity. Secondly, both systems required regular immigration of the prey and/or the predator to avoid extinction. Neither of these requirements was anticipated by the Lotka–Volterra approach.

Another instructive laboratory experiment was that of Utida (1957). In this case stable oscillations between the azuki bean weevil (*Callosobruchus chinensis*) and a parasitic wasp

(*Heterospilus prosopidis*) were maintained for more than 25 generations in a laboratory Petri dish (1.8 cm high by 8.5 cm diameter). The wasp is actually a specialized predator known as a parasitoid, which paralyzes its prey without killing it. The wasp lays an egg on the paralyzed host, usually an insect larva. The egg hatches and the wasp larva slowly consumes the host, leading to its eventual death. In this case, the wasp only lays eggs on certain instar larvae. The wasp does not parasitize adult or pupal beetles. Therefore, although these two populations undergo the regular oscillations of a limit cycle, both populations persist due to the nonrandom predation by the wasp. Again, the violation of the Lotka–Volterra assumption of random hunting is what allows the coexistence of prey and predator.

10.4 Functional responses

The Canadian ecologist C.S. Holling concluded that our understanding of predation needed a more realistic, empirical approach. He believed that to understand predator–prey relationships it was first necessary to understand the act of predation. That is, he asserted that the first response of an individual predator to an increase in a prey population is not to increase its growth rate, but rather to increase its per capita predation rate, often by selective behavior. Using terminology from Solomon (1949), Holling (1959, 1961, 1966, 1973) termed this per capita increase in predation the **functional response** of a predator. He analyzed the act of predation and broke it down into behavioral units termed the **components of predation**. In his papers Holling described many other potential factors that might influence predation, but we will limit our discussion to the time and/or energy devoted to these four major components: (i) search, (ii) capture, (iii) handling, and (iv) digestion. Theory and laboratory testing suggest that the functional response of an animal may take one of three forms called the type I, type II, and type III functional responses. In all of these functional-response graphs, the x-axis consists of number of prey per unit area (prey density), while the y-axis consists of the number of prey eaten per individual predator per unit time. Thus the functional response is the result of changes in consumption rate per individual predator. In all three types of functional responses the predation rate will rise with prey density, but will eventually level off. At low prey densities, predation rate is influenced mostly by the amount of time and/or energy devoted to search and capture, while at very high prey densities predation rate is bounded by the last two components of predation, time or energy devoted to handling and digestion of prey.

   The simplest functional response (**type I**) is one in which prey consumption rises linearly with prey density. This is sometimes called the filter feeder's functional response, and is often pictured as a simple straight line. However, this picture is misleading. It omits the fact that increases in prey density beyond a threshold do **not** result in an increased per capita predation rate. This is due to the handling and digestion components of predation. For example, Porter *et al.* (1982, 1983) maintained concentrations of the green alga *Chlamydomonas* in the laboratory at concentrations up to 10^6 cells per cubic cm. They then introduced the arthropod *Daphnia magna* (the "water flea"), a filter feeder. As illustrated by Fig. 10.9, *Daphnia* needed almost 1000 (10^3) cells per cm³ to maintain itself. As the concentration of *Chlamydomonas* increased, so did the consumption rate of *Daphnia*. Once about 20,000 cells per cm³ were reached, however, the increase in consumption decreased rapidly and the water flea processed a maximum of about 7 cells per second or

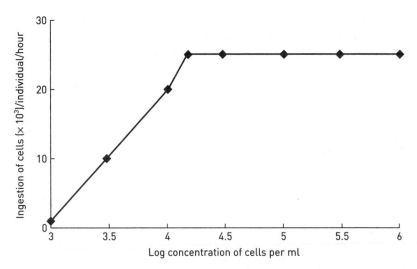

Figure 10.9 Type I functional response. Ingestion of the alga *Chlamydomonas* by *Daphnia magna.* From Porter *et al.* (1982).

25,000 per hour. There is evidently a limit to what *Daphnia* can process, and the type I functional response has a plateau, as do all functional responses, due to the handling and digestive components of predation. Picture yourself eating popcorn. If individual pieces of popcorn were spread about the house and you had to search for them, your consumption rate would be limited by the search component of predation (capture time for popcorn is presumably zero). But if you were surrounded by an unlimited number of tubs of popcorn, your consumption rate would be limited by the time it would take to handle and chew. And eventually, your stomach would tell you, no more! Until you had time to digest what you had stuffed into your mouth, your consumption rate would plummet. Next time you go to a movie theater, take notes on the rates of consumption of your fellow predators!

Porter *et al.* (1982, 1983) also found that when the algal concentration went up, the amount of effort by *Daphnia*, as measured by filtering rate, went down. As food concentration went up the movement of filtering appendages declined from 6 to 3.5 per second, resulting in a decrease in filtering rate from 4 to 1 cm^3 per hour. This means that the food resource was so concentrated that maximum ingestion was possible with much less effort. Therefore, since less energy was expended, the *Daphnia* population was able to turn more assimilated energy into reproduction. As we can see from Table 10.1, an increase in food consumption led directly to a numerical response in *Daphnia magna*, at least up to a point. As the density of *Chlamydomonas* increased, several reproductive parameters also increased. However, at the highest prey density (10^6 cells) reproductive parameters decreased and the intrinsic rate of increase declined from 0.28 to 0.20. These data illustrate the relationship between the functional and the numerical response of a predator, as well as the fact that at very high prey densities other factors may limit the numbers of a predator.

In the **type II** functional response, as prey density increases, prey consumption increases more rapidly than in a type I response, and is nonlinear, but also reaches a plateau.

Table 10.1 Reproductive parameters of *Daphnia magna* as a function of the concentration of its prey, the green alga *Chlamydomonas*. Experiment conducted at 20 °C. Based on Porter *et al.* (1983), and adapted from Ricklefs (1990).

	Concentration of *Chlamydomonas* (cells per cm³)			
	10^3	10^4	10^5	10^6
Percent reproducing	50	87	97	50
Eggs per brood	2.8	2.6	15.5	21.1
Broods per female	1.7	7.5	8.2	3.4
Days between broods	5.4	3.6	3.1	3.3
Age of first brood (days)	23.4	16.9	9.8	9.1
Net reproductive rate (R_0)	2.25	16.23	99.33	34.80
Intrinsic rate of increase (r)	0.03	0.10	0.28	0.20

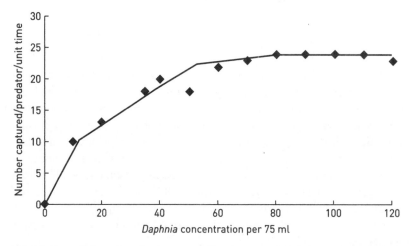

Figure 10.10 Type II functional response. The larvae of the damsel fly *Ischnura elegans* feeding on *Daphnia magna*. From Hassell (1976).

This is called the "general invertebrate curve," since Holling found this curve applied to a number of invertebrate predators that he tested (Fig. 10.10).

In the **type III response**, prey consumption remains low until a threshold density is reached. The predation rate then increases exponentially until it levels out. The shape of this curve is described as "sigmoid" and it looks something like the logistic population growth curve. Sometimes called the "vertebrate" curve or the "learning predator's" curve, the predator ignores the prey when densities are very low in order to make hunting energetically profitable. Another interpretation of the low predation rate at low prey densities is that the predator does not see the potential prey often enough to recognize it as prey. Once the prey population reaches the threshold density, the predator develops a "search image" for the prey and begins actively seeking it. At this point, the predation rate increases exponentially until reaching the maximum rate, which occurs at the plateau due to handling and digestion time.

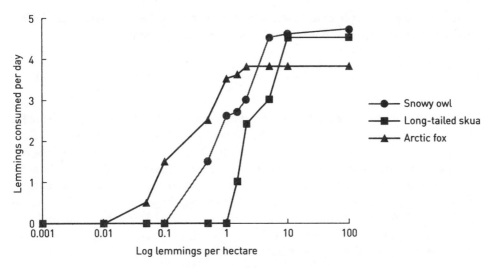

Figure 10.11 Type III functional response curves for three predators of lemmings in Greenland. Adapted from Gilg *et al.* (2003).

Working in the high-Arctic tundra of Greenland, Gilg *et al.* (2003) have shown that the snowy owl, the arctic fox, and the long-tailed skua all have type III functional responses to increasing densities of collared lemmings (Fig. 10.11).

The next step is to integrate the components of predation into the Lotka–Volterra equations. The functional response, *f*, is set equal to *EN* for a type I response, as was actually incorporated into the original Lotka–Volterra equations (Eqns. 10.4a 10.4b, 10.5).

The type II and III functional responses involve variables for **search and capture** rates (E_2N and E_3N^2) and **handling and digestion** rates (h_2 and h_3). As described above, the plateau or asymptote found in each functional-response curve is the result of the handling and digestion rates, *h*. When prey densities get very large, little time or energy must be spent searching or capturing, and the predation rate depends on handling time and digestive pause. Thus we have the Holling equation for the type II response (10.8a) (See Case 2000 for a derivation):

$$f_n = \frac{E_2N}{1 + h_2E_2N}$$

(10.8a)

If we take its inverse, we have:

$$\frac{1}{f_n} = \frac{1 + h_2E_2N}{E_2N}$$

(10.9)

We can rewrite this as:

$$\frac{1}{f_n} = \frac{1}{E_2N} + \frac{h_2E_2N}{E_2N} = \frac{1}{E_2N} + h_2$$

(10.10)

The second term simplifies to h_2; when the prey population becomes very large, the first term becomes very small and we can set it to 0. Thus $1/f = h_2$ and:

$$f_n = \frac{1}{h_2} \tag{10.11}$$

Equation 10.11 simply states that the functional response is the inverse of the handling time component. A similar analysis applies to the type III functional response. A useful interpretation here is that **the inverse of handling time equals the maximum killing rate** (c), when the search and capture components are essentially zero. An alternative form for equation 10.8a sets the parameter $c = 1/h_2$ and substitutes **the half-saturation constant,** d (the prey density at which killing rate is half of the maximum) for $(E_2 h_2)^{-1}$. Substituting $1/c$ for h_2 and c/d for E_2 in Equation 10.8a, we have 10.8b, the alternative equation for the functional response. The advantage of this version of the functional response is that we can relatively easily estimate these parameters from field data:

$$f_n = \frac{cN}{d + N} \tag{10.8b}$$

When prey densities are very low, the search and capture rates become very large compared to the handling time. If we set h equal to 0 in equation 10.8a, the denominator becomes equal to one and the functional-response equations become:

$$f_n = E_2 N \text{ (type II functional response)} \tag{10.12a}$$

$$f_n = E_3 N^2 \text{ (type III functional response)} \tag{10.12b}$$

Therefore, at a low prey density, the functional response is completely determined by the search and capture components, and it is essentially linear in the type II response.

10.5 Adding prey density dependence and the type II and III functional responses to the Lotka–Volterra equations

Now let's return to the Lotka–Volterra equations as modified by a type II functional response. For the prey equation, we will not only add the functional response, but we will also add the carrying-capacity term as suggested by Volterra (1931). The prey equation again consists of two parts: (i) recruitment and (ii) death due to predators. The recruitment section is the traditional logistic, and the losses due to predation are based on the numerical response term (ENP) modified by the type II functional response (Eqn. 10.8a). The resultant equation is:

$$dN/dt = r_n N\left(1 - \frac{N}{K_n}\right) - \left(\frac{E_2 NP}{1 + E_2 h_2 N}\right) \tag{10.13a}$$

In the predator equation the left term is again modified by the type II functional response, but is otherwise unchanged:

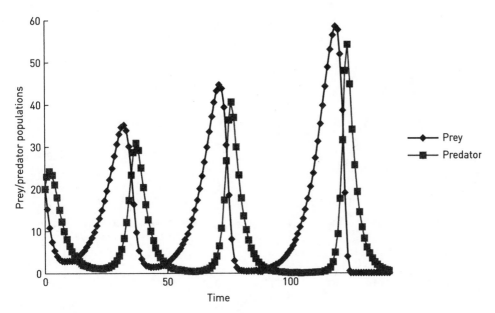

Figure 10.12 Predator–prey interaction following the Lotka–Volterra equations with a type I functional response.

$$dP/dt = \frac{\chi_p E_2 NP}{1 + E_2 h_2 N} - m_p P \qquad (10.14a)$$

Recall that χ_p measures the efficiency at which the food (prey) is turned into new predator individuals (= **assimilation efficiency** of the predator), and m_p is the mortality rate of the predator.

In the type III functional response, the term N is replaced by N^2 in the expression describing death due to predation in Equation 10.13a. That is:

$$\frac{E_3 N^2 P}{1 + E_3 h_3 N^2}$$

Similarly, in Equation 10.14a N is replaced by N^2 on the right side of the equation.

How does the functional response affect the behavior of the predator–prey interaction? Figures 10.12, 10.13, and 10.14 are based on Equations 10.13a and 10.14a. In each case $r_n = 0.2$, $K_n = 100$, $N_0 = P_0 = 20$. E, the coefficient measuring the efficiency of search and capture = 0.02. χ_p (assimilation efficiency) is set at 1.0. The mortality rate $m_p = 0.25$.

In Fig. 10.12 we have removed the functional-response terms from both equations. This means the populations are growing according to the type I functional response, which is equivalent to the basic Lotka–Volterra equations. What we see is mutual extinction such as Gause and Huffaker found. In Figs 10.13 and 10.14, by contrast, we have introduced the type II and III functional response terms with $h = 3.0$ in Fig. 10.13 and $h = 4.0$ in Fig. 10.14. The populations move toward a stable point in each case. Handling time puts a cap on the number of prey that each predator can consume, thereby limiting both its growth rate and the losses to the prey population. The effect is to help stabilize the interaction.

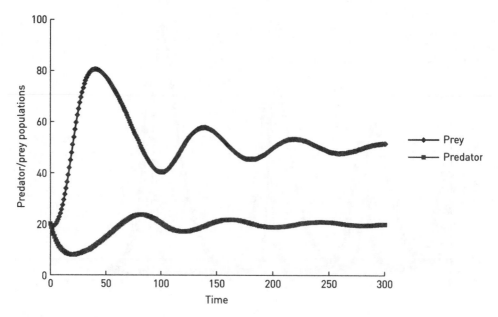

Figure 10.13 Predator–prey interaction with a type II functional response.

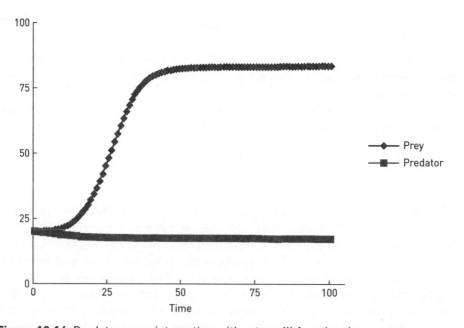

Figure 10.14 Predator–prey interaction with a type III functional response.

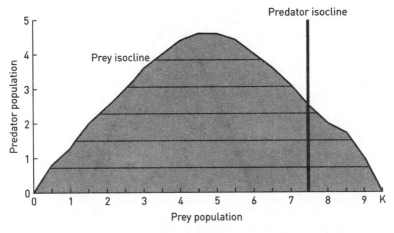

Figure 10.15 Prey and predator isoclines. Adapted from Rosenzweig and MacArthur (1963). The prey population has positive growth in the shaded area. The predator population has positive growth to the right of the vertical line. K = carrying capacity.

10.6 The graphical analyses of Rosenzweig and MacArthur

Rosenzweig and MacArthur (1963) founded a general approach for studying predator–prey interactions known as the isocline or graphical analysis. The graphical analysis consists of a plot whose x-axis is the prey population and whose y-axis is the predator population. Two isoclines are drawn on the graph. The first line, known as the predator isocline, may be a vertical line, or may be modified to any linear representation with a positive slope. This line represents the prey density that results in zero population growth of the predator. Areas to the right of the line allow the predator population to grow; to the left predator populations decline. The second line, the prey isocline, represents the predator and prey combinations that result in zero population growth of the prey. The prey isocline (Fig. 10.15) is based on the per capita growth rate of the prey. At the far right is the point (K, 0). There are zero predators, but the prey growth rate is limited to zero by its carrying capacity. The growth rate has a positive slope at low prey densities and reaches its maximum value at $K/2$. Regions above the prey isocline represent combinations of predator and prey resulting in a negative prey growth rate due to predation. The prey isocline is often shown with the Allee effect included (Fig. 10.16). As discussed in Chapter 2, very low prey populations may result in negative growth. Thus the prey isocline is shifted away from the point (0, 0) to a point representing a minimum viable population size.

This graphical approach allows us to envision a variety of situations in which extinctions, stable cycles, and stable points are produced. In Fig. 10.17 the predator isocline is far to the right, intersecting the prey isocline in an area where it has a negative slope. This represents an "inefficient" predator, which needs large numbers of prey. The result is a series of decreasing oscillations resulting in a stable point at the intersection of the two isoclines. By pushing the predator isocline to the left, however, such that the predator

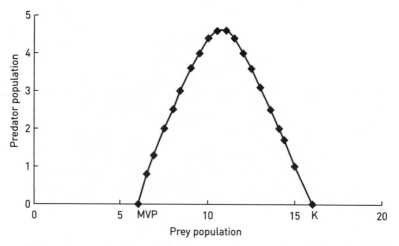

Figure 10.16 The prey isocline with the Allee effect. MVP, minimum viable population; *K*, carrying capacity.

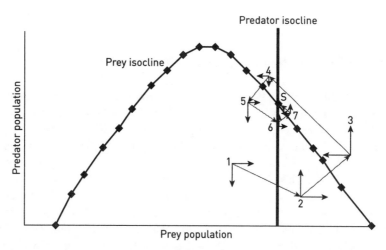

Figure 10.17 Graphical analysis of predator–prey interactions of Rosenzweig and MacArthur (1963). An inefficient predator isocline produces decreasing oscillations to a stable point *S*, where the two isoclines meet.

isocline intersections the prey isocline at a right angle at *K*/2, the result is a stable cycle (Fig. 10.18), rotating around the intersection point, *S*. Finally, if we push the predator isocline further left, so that it intersects the prey isocline where it has a positive slope, simulating a "highly efficient predator," the result is mutual extinction of the predator and its prey (Fig. 10.19). If the predator is limited by a resource other than the prey population, the result is a stable point as illustrated in Fig. 10.20.

This analysis also allows us to examine what is known as the "paradox of enrichment". Laboratory experiments and theoretical simulations show us that predator–prey interactions are often stabilized by lower prey growth rates. In his experiments with oranges and

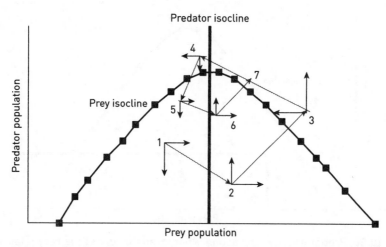

Figure 10.18 Predator–prey isoclines for a moderately efficient predator, resulting in a stable limit cycle.

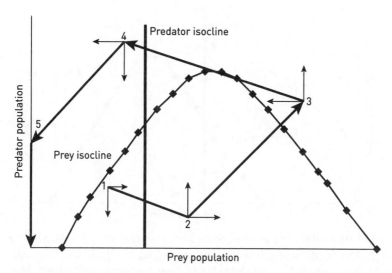

Figure 10.19 A highly efficient predator, resulting in increasing oscillations leading to extinction of both the predator and the prey (point *E*).

orange mites, described earlier, Huffaker (1958) found that to stabilize the predator–prey interaction he had to diminish the prey growth rate. He reduced the amount of feeding surface available for the prey on the oranges by dipping them in wax. Similarly, when Luckinbill (1973) repeated Gause's experiments (substituting *P. aurelia* for *P. caudatum*, but still using *Didinium nasutum* as the predator) both prey and predator went extinct within a few hours. However, by adding methyl cellulose to the medium, which slowed down the movement of both predator and prey, the two species coexisted for several days. He then cut the food supply for the prey in half, and the two species coexisted for 33 days before the experiment was terminated. Rosenzweig (1969) had predicted this result on

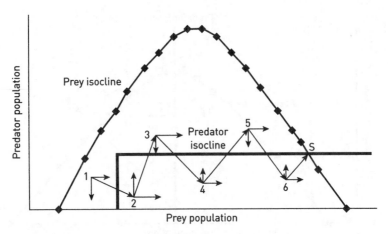

Figure 10.20 Predator–prey interaction when predator growth is restricted by a factor other than the prey population. The result is a stable point, *S*.

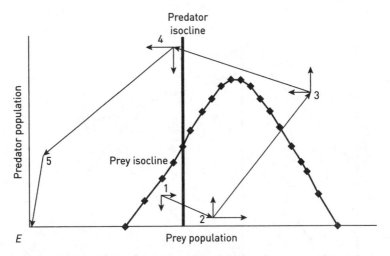

Figure 10.21 The effect of the "paradox of enrichment" on a predator–prey interaction. The predator isocline is identical to that of Fig. 10.17. The prey isocline has been shifted to the right, simulating resource enrichment. The final point is *E* (0, 0).

theoretical grounds. In Fig. 10.21 we see that, although the predator isocline is the same as in Fig. 10.18, where the result was a stable cycle, by moving the prey isocline to the right, simulating a higher growth rate for the prey due to an increase in resources, the result is mutual extinction.

Finally, we can simulate an invulnerable prey refuge (Fig. 10.22). In this case the result, depending on the efficiency of the predator, is a stable point or a cycle (limit cycle), although the cycle may be rather complex. We can easily add the idea of prey refuge in Equations 10.13a and 10.14a by adding a constant M^*. This number represents a minimum prey population which is never exposed to predation. The number of prey available for predation will then be only $N - M^*$, and we replace N by $N - M^*$ in the appropriate places:

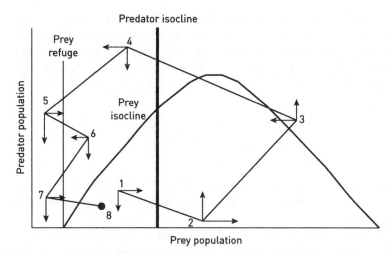

Figure 10.22 Predator–prey interaction when the prey has a refuge. The result is a stable cycle.

$$dN/dt = r_nN\left(1 - \frac{N}{K_n}\right) - \left[\frac{PE(N - M^*)}{1 + Eh(N - M^*)}\right] \qquad (10.13b)$$

$$dP/dt = \frac{\chi_pE(N - M^*)P}{1 + Eh(N - M^*)} - m_pP \qquad (10.14b)$$

As discussed by Case (2000) and Crawley (1992), these types of analyses underscore the point that in a predator–prey relationship there can be more than one stable point, and predator–prey relationships that produce cycles in one ecosystem may result in a stable point in another.

10.7 Use of a half-saturation constant in predator–prey interactions

Turchin and Ellner (2000) have asserted that the predator–prey equations used by Rosenzweig and MacArthur (1963) should be adopted as "the standard" for predator–prey interactions, since they eliminate the assumption of the linear functional response and are "perhaps the simplest model that can actually be applied to real life systems" (Turchin 2003, p. 95). The only differences between these equations and Equations 10.13a and 10.14a are that the handling term in the functional response has been replaced by $d + N$, where d is the half saturation (half maximum killing rate) parameter, and E has been replaced by c, the maximum killing rate when the search and capture components have been minimized. In other words, we are using the alternative functional response shown as Equation 10.8b:

$$dN/dt = r_nN\left(1 - \frac{N}{K_n}\right) - \frac{cNP}{d + N} \qquad (10.15)$$

$$dP/dt = \frac{\chi_pcNP}{d + N} - m_pP \qquad (10.16a)$$

An alternative form of the predator equation (10.16a) is to add what is known as the ZPG component consumption rate (μ_p). This parameter represents the minimum rate of prey consumption needed for a predator to survive and just replace itself. This is usually easier to estimate than the predator death rate in the absence of prey, m_p. The result is Equation 10.16b, a version of which we will use in Chapter 11 on herbivore–plant interactions:

$$dP/dt = \chi_p P\left(\frac{cN}{d + N} - \mu_p\right) \qquad (10.16b)$$

Now compare the predator equation (10.16a) to Equations 7.19 and 7.20, which describe the effects of resource depletion on the growth rate of a consumer in the context of competition.

In a revised version of Equation 7.20, instead of a competing species, N_2, we are substituting P, the predator population. The death rate, m, is now m_p. We have replaced $\chi_p c$, which measures the maximum rate of conversion of the resource (prey) into predators, by b, which was the maximum growth rate of the competing species.

$$dP/dt = \frac{bRP}{K_R + R} - m_p P \qquad (7.20\ \text{revised})$$

The prey population (N in equation 10.16a) is considered a resource, so is replaced by R, the concentration of the resource. The half-saturation constant d is replaced by K_R. The two equations are now identical. Equation 7.20 measures the growth rate of a consumer in terms of its maximum growth rate, the concentration of the resource, the half-saturation constant for that resource, and the death rate of the consumer. The predator equation (10.16a) measures the growth rate of the predator in terms of maximum efficiency in turning prey into predator individuals ($\chi_p c$), the concentration of the resource (N), the half-saturation constant, and the death rate of the predator. We now have a **general mechanistic equation**, useful in both **competitive interactions** and **predator–prey interactions** for the **growth rate of a consuming population**.

Because the half-saturation parameter is more easily estimated in the field than is a specific handling time value, this approach (Eqns. 10.15 and 10.16) was employed by Gilg *et al.* (2003) in their study of lemming–predator interactions in Greenland, as well as by Turchin and Ellner (2000).

10.8 Parasitoid–host interactions and the Nicholson–Bailey models

As pointed out earlier, a parasitoid is a special type of predator. Rather than killing and consuming the prey immediately, a female parasitoid lays its eggs on a particular stage of a host species (usually a larval stage of an insect). The egg hatches into a larva which consumes the host over a period of time, eventually killing it. The parasitoid emerges from the host ready to hunt for more prey. The parasitoids are usually very host-specific and must time their life cycles to those of their hosts. In addition, since there is a time delay between egg laying and emergence of the next generation, discrete time equations are used to model these interactions.

Nicholson and Bailey (1935) developed models for parasitoids and their hosts. These were the starting point for a series of more complex models developed by and expanded to book length by Hassell (1976, 1978).

The assumptions of the basic Nicholson–Bailey model include:

1 The number of encounters, N_e, between P_t parasitoids and a host or prey species is proportional to the host density, N_t.
2 The encounters are randomly distributed among the hosts. This means that some host individuals will be encountered more than once, so that N_e can be larger than N_t.

Given the assumption of random distribution of encounters, the proportion of hosts **not** parasitized is given the zero term of the Poisson distribution:

$$p_0 = e^{(-N_e/N_t)} \tag{10.17}$$

The number of hosts actually parasitized is N_a:

$$N_a = N_t(1 - e^{-N_e/N_t}) \tag{10.18}$$

The number of encounters between P_t parasitoids and their hosts, N_t, can be restated as:

$$N_e = aN_tP_t \tag{10.19}$$

where a is constant called the parasitoid's rate of discovery. It is a measure of searching efficiency.

It follows that $N_e/N_t = aP_t$, and the number of hosts that are parasitized can be rewritten as:

$$N_a = N_t(1 - e^{(-aP_t)}) \tag{10.20}$$

This expression implies that the rate of parasitism will reach a saturation level as parasitoids find fewer and fewer hosts not previously attacked. Nicholson–Bailey called this relationship a competition curve.

As mentioned above, because of the discrete seasonality of most arthropods and because there is no age structure in most populations, the equations defining host and parasitoid dynamics are written as difference equations. The number of progeny a female parasitoid leaves behind is based on the number of hosts attacked.

The prey equation is therefore:

$$N_{t+1} = \lambda N_t \, e^{-aP_t} \tag{10.21}$$

where λ = the finite rate of increase per generation and e^{-aP} represents the proportion of hosts escaping attack by the parasitoid.

If we assume that only one parasitoid emerges from each host, the parasitoid equation is:

$$P_{t+1} = N_t(1 - e^{-aP_t}) \tag{10.22a}$$

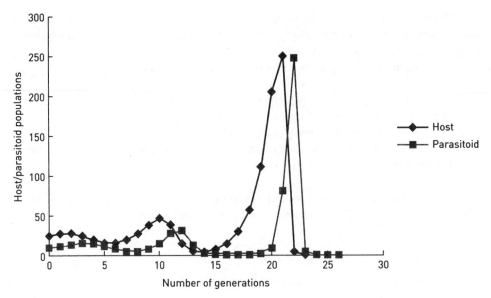

Figure 10.23 Density-independent host–parasitoid interaction. $\lambda = 2.0$, $a = 0.06$ and $n = 1.0$.

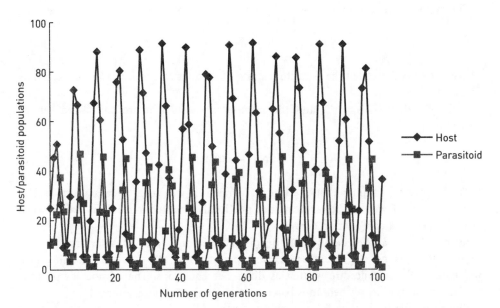

Figure 10.24 Density-dependent host–parasitoid interaction. Parameters as in Fig. 10.23, but $K = 50$.

However, we can allow more than one parasitoid to emerge from each host by adding the parameter n to this equation. In Equation 10.22b, the parameter n equals the number of parasitoids emerging from each host:

$$P_{t+1} = nN_t[1 - e^{-aP_t}]$$

(10.22b)

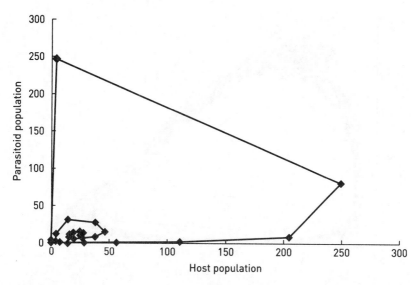

Figure 10.25 Density-independent host–parasitoid relationship showing mutual extinction. Based on data from Fig. 10.23.

The number of parasitoids in the next generation, then, is the product of n times the number of host individuals times the proportion of host individuals successfully attacked.

Unfortunately, this model does not improve on the Lotka–Volterra model; the main difference is the built-in time lag associated with the use of difference equations. While the Lotka–Volterra model produces a neutrally stable cycle, the Nicholson–Bailey model predicts unstable, increasing oscillations (Fig. 10.23).

However, the introduction of a density-dependent host growth factor will stabilize the Nicholson–Bailey model (Hassell 1976) as in Fig. 10.24, in which a carrying-capacity term has been added (Eqn. 10.23). The result is a stable cycle, although the oscillations are very large.

$$N_{t+1} = \lambda N_t \, e^{[(1-N/K_n)-aP_t]} \qquad (10.23)$$

If we look at these two interactions by graphing the host population on the x-axis and the parasitoid population on the y-axis, we get Figs 10.25 and 10.26. Here we clearly see the mutual extinction of the two populations in Fig. 10.25 and the ellipse in Fig. 10.26, which illustrates the stable limit cycle more clearly than does Fig. 10.24.

The introduction of the carrying-capacity term for the host produces a number of interesting results, depending on value of the other parameters. Under a large number of conditions the parasitoid–host interaction produces a stable cycle. For example, as we have shown in Fig. 10.26, if $\lambda = 2.0$, $K_n = 50$, $n = 1.0$, $a = 0.060$, $N_0 = 25$ and $P_0 = 10$, the result is a stable cycle, whereas without the density-dependent term the result is mutual extinction (Fig. 10.25). However, if we run simulations using a computer program like Populus, we find a wide range of results from mutual extinction to a stable point. For example, if the initial host population is too large or too small relative to its carrying capacity, the result can be extinction of the parasitoid population. If the host population

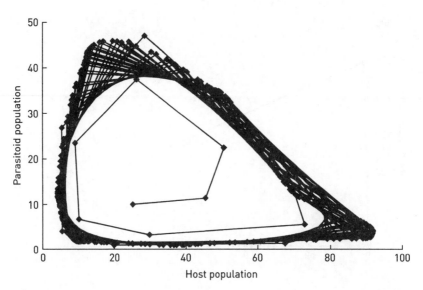

Figure 10.26 Host–parasitoid relationship showing stable limit cycle. Based on data from Fig. 10.24.

has a finite rate of increase that is too small, it can also lead to extinction of the parasitoid. But an extremely high host rate of increase does not destabilize the interaction. A λ-value of over 12 (= r of 2.5) simply leads to a stable limit cycle. On the other hand, given the values shown above, lowering the carrying capacity of the host population to 25 leads to a stable point instead of a stable cycle. But either lowering K_n to 10 or raising it to 75 leads to extinction of the parasitoid. The stability of this interaction is also affected by the searching efficiency parameter, a. A highly efficient parasitoid produces extinction of both the parasitoid and the host. An a-value of about 0.09 results in extinction of only the parasitoid, whereas values between 0.03 and 0.06 result in a stable point rather than a limit cycle (Figs. 10.27 and 10.28). If the parasitoid becomes even less efficient (values much less than 0.03), it goes extinct.

Other factors that have been found to stabilize the parasitoid–host interaction include (Hassell 1978):

1 Interference competition among the parasitoids. If female parasitoids avoid each other or avoid laying eggs where others have laid theirs, the rate of parasitism rises more slowly.
2 A refuge for the host. As in other predator–prey models, if the prey has hiding spots or habitats that cannot be found by the parasitoid, the interaction is more stable.

10.9 Predator–prey interactions in practice: field studies

As we indicated at the beginning of this chapter, simple predator–prey models inevitably predict an oscillatory interaction between predators and their prey. However, by adding

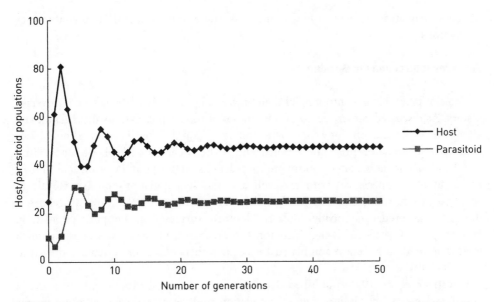

Figure 10.27 Host–parasitoid interaction with searching efficiency $a = 0.03$. Result is a stable point.

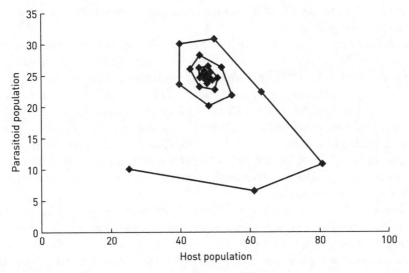

Figure 10.28 Host–parasitoid interaction with a searching efficiency $a = 0.03$. Result is a stable point.

a density-dependent factor for the prey population, and some version of a functional response by the predator, depending upon the values of the parameters, these models predict a range of responses, from extinction to cycles to stable points. Adding a carrying-capacity term for the predator, independent of the prey population, may also help stabilize a predator–prey interaction. More realistic models would include the ability of the predator to choose from among several prey species and/or among age classes within a species. With these thoughts in mind we will now examine several examples of real-life predator–prey interactions.

Fox–prey interactions in Sweden

The role of predators in controlling bird and mammal populations has had a controversial history. As discussed earlier, based on his studies of rodent populations in England, and after reviewing data on snowshoe hare and lynx populations, Charles Elton (1924) believed that predator–prey interactions tend to be oscillatory (display a stable limit cycle) and that predators often control prey populations. At one point Peterson *et al.* (1984), based on their experiences with the wolf and moose populations on Isle Royale in Michigan, seem to have agreed. On the other hand, Errington (1946) and his followers thought that predators mostly took a "doomed surplus" of the prey population. However, for 40 years following Errington's work, there had been few studies in which vertebrate predators were manipulated to discover their effect on vertebrate prey (but see NRC 1997).

Lindstrom *et al.* (1994) published data based on a "natural experiment" in which a Scandinavian red fox (*Vulpes vulpes*) population underwent a severe sarcoptic mange epidemic. The mange mite (*Sarcoptes scabiei*) appeared in Sweden in 1975 and had a dramatic effect on fox populations throughout the 1980s. The mite caused hair loss, skin deterioration, and death, and the fox populations declined by 70%. This allowed Lindstrom *et al.* to survey the effects on the small-game community of a dramatic drop in a top-carnivore population.

The "alternate prey" hypothesis states that certain prey species are cyclic (3–4 years in voles and lemmings, 10 years in snowshoe hares, for example) and are mostly uncontrolled by predation. Such species are little affected by predation and the oscillations are likely due to interactions with the vegetation (Turchin and Batzli 2001), while other species are controlled by predation. In years when these cyclic prey numbers are in the low phase of their cycle, predators switch to "alternate prey." These alternate prey species **are usually limited** by predation, especially during the years when the cyclic prey species are in the low phase. In Scandinavia, Lindstrom *et al.* (1994) proposed that while voles are largely uncontrolled, hare, grouse, and other small mammals and birds are alternate prey species subject to control by predation.

Lindstrom *et al.* (1994) therefore examined the effect of the reduced fox populations on voles, hares, grouse, and deer. Methods were as follows: (i) rodent populations were surveyed by snap-trapping; (ii) the investigators checked red fox reproduction by entering dens and determining litter sizes; (iii) populations of hare, grouse, and deer were checked by the pellet-count method; (iv) bird populations were estimated by using imitation birdcalls and recording responses; (v) hunters were given questionnaires concerning populations of all of the above and hunting records were consulted. Fawn/doe ratios were estimated from "incidental observations."

In the 1970s, previous to the mange introduction, red fox litter sizes followed the vole (Cricetidae) densities, increasing during the vole peaks. The fox population declined during the mange epidemic and then recovered after 1990 to the level before the mange appeared. Vole populations were not affected by fox numbers. They continued their normal cycles before, during, and after the fox mange epidemic. Lindstrom *et al.* hypothesized that these cycles were in response to variations in plant productivity and snow depth. Mountain hare

(*Lepus timidus*) and grouse populations went up 40–100% during the fox mange epidemic and their populations dropped as the fox populations recovered.

Previous to the mange epidemic, black grouse (*Tetrao tetrix*) populations peaked one year after the peak of vole density. As foxes concentrated on consuming voles, predation on the hares dropped, and hare numbers increased accordingly. European hare (*L. europaeus*) and black grouse populations followed the same patterns, consistent with the alternate-prey hypothesis.

For roe deer (*Capreolus capreolus*) the fox population affected the mortality of fawns. After the mange epidemic the number of fawns per doe increased 30% and the average deer density went up 64%. However, the production of fawns was also correlated with the previous year's winter weather. With a delay of 1.5 years, snow depth explains 48% of the variation in fawn/doe ratios. Roe deer usually go into estrous in the second summer of life and, with delayed implantation, give birth at age two. A fawn undergoing a hard winter, however, would delay puberty by one year. Thus, there are fewer births 1.5 years after a hard winter, introducing a time lag into the deer population response. Still, fox density did have a significant effect on fawn production, even accounting for winter weather.

Summarizing: (i) All prey populations except voles increased in density as the mange struck the fox population. All but roe deer returned to previous population levels as the fox population returned to normal. (ii) Consistent with the alternate-prey hypothesis, whereas voles are not limited by fox predation, hare and grouse are. (iii) Red fox is a "keystone" species in structuring the small-game community in Scandinavia. It conveys the 3–4 year vole cycle to the hare and grouse populations. It shows both a functional response and a numerical response, since vole populations affect fox litter size. (iv) The mange mite, since it affects red fox populations so drastically, could also be called a keystone species. (v) The role of humans cannot be ignored. First, by exterminating wolves from this area, the role of the fox has been enhanced. Second, by opening up the forest through clear-cutting practices, man has also created excellent fox habitat. In a closed forest the role of the fox is likely to be of much less significance.

Are population cycles caused by predation?

Other work, however, seems to indicate that vole cycles **are** caused by predation. In the paper by Lindstrom *et al.* (1994), although red fox predation was temporarily eliminated, weasel-type (Mustelidae) and avian predators were not controlled. Recent research suggests that population cycling by prey species is only eliminated when all potential predators or parasites are eliminated (May 1999).

For example, Hudson *et al.* (1998) studied population cycles in the red grouse (*Lagopus lagopus*). Red grouse have cycles of population abundance with an average period of seven years. Moreover, these cycles are synchronized over large portions of northern England (Cattadori *et al.* 2005). These cycles are of special interest because grouse are a traditional English/Scottish shooting bird. About 75% of the British populations of red grouse undergo these cycles. Hypotheses for causation included vertebrate predators, food supplies, territory size, or parasite infections. According to Hudson *et al.* (1998), the answer is that the cycles are caused by parasites and the synchronization is set by climatic conditions affecting the transmission of the parasite (Cattadori *et al.* 2005).

The parasite is a nematode (*Trichostrongylus tenuis*). Hudson *et al.* (1998) began a study of six grouse populations, in which they predicted cyclic troughs in 1989 and 1993. In two of these populations birds were caught and treated with an anthelminthic medicine that eliminated the parasite. These treatments were in advance of the 1989 and 1993 predicted crashes. Two of the populations were not treated, and two of the populations were treated prior to 1993, but not prior to 1989.

The results were remarkable (May 1999). In the two untreated populations the predicted crashes occurred, on schedule, in 1989 and 1993. The magnitude of these crashes was a drop of more than three orders of magnitude. In the two populations treated twice, one population remained steady throughout the period 1988–96, while the second showed a small drop (threefold) in 1993. The populations treated only in 1989 had a major crash in 1993, but remained relatively steady in 1989. For this vaccination program to work, however, at least 20% of the birds had to be injected. This is based on models of epidemiology, discussed in Chapter 9.

In a second example, Korpimaki and Norrdahl (1998) studied two species of field voles (*Microtus agrestis* and *M. rossiaemeridionalis*) and a species of bank vole (*Clethrionomys glareolus*) that show persistent three-year cycles in Finland. Korpimaki and Norrdahl set up six large study areas of 2–3 km². They reduced the numbers of weasels (*Mustela nivalis*) and stoats, the major mustelid predators. They also removed kestrels (*Falco tinnunculus*) and owls, the major avian predators. In 1992, prior to an anticipated population crash, they were successful in removing most of the weasels but not the stoats, and they did not attempt to remove the birds of prey. In 1995, however, they were able to remove all of the predators. None of the predators were removed in the control areas. In 1992, with only the weasels removed, the vole populations crashed, as did the control populations in both 1992 and 1995. However, in 1995, when all the vertebrate predators were removed, the vole populations remained steady without a crash. In earlier work, Norrdahl and Korpimaki (1995) had removed only avian predators. Here again, the voles crashed in both manipulated and control plots. We can conclude that vole cycles are indeed caused by predators. To eliminate cycles, all of the predators must be removed; removal of only a subset of the predator community leaves other predators to drive the cycle.

Turchin *et al.* (2000) drew a large distinction between vole and lemming populations. They agreed that voles, such as *Microtus agrestis*, are limited by their predators. They concluded, however, that lemming populations in Norway and other regions of Scandinavia are controlled by their food supply. They make the case that because voles feed on grasses and other herbs with the ability to quickly recover after defoliation by mobilizing underground food reserves, the vole–vegetation interaction is highly stable. According to Turchin *et al.*, lemmings are primarily moss-eaters. Since mosses grow very slowly, the inherent time lag is highly destabilizing. Lemmings tend to deplete their forage in the arctic and alpine habitats where they live before predator populations begin to affect their dynamics. Lemmings are highly mobile during these population peaks, leading to the myths about lemming suicidal behavior (Chitty 1996). Thus, although both voles and lemmings undergo periodic oscillations, the cycles are driven by different ecological mechanisms, a predator–prey interaction in one case (voles) and a food–herbivore interaction in the other case (lemmings). This is a classic top-down (voles) versus bottom-up (lemmings) population control paradigm.

All of these works stress that cyclic prey species are actually preyed upon by a community of predators, again reminding us of the inadequacy of a single predator/single prey model. True cycles appear to be confined to northern temperate and subarctic ecosystems and are unknown in the tropical latitudes or southern hemisphere (Sinclair and Gosline

1997). Furthermore, it is widely believed that the amplitude of these cycles decreases at lower latitudes. However, a review of 700 long-term (25+ years) animal population studies found that this expected latitudinal gradient in population cycles can only be confirmed for North American carnivores, British lagomorphs, and Fennoscandian rodent populations (Kendall *et al*. 1998). The database used by Kendall *et al*. did not include studies from the tropics or the southern hemisphere. Nevertheless, they found periodicity in 33% of the mammal and 43% of the fish populations reviewed. Periodicity was less common among bird (13%) and insect (16%) populations, but was particularly common among grouse. Since almost 30% of all populations in their review showed regular cycles, this phenomenon cannot be dismissed (Kendall *et al*. 1998).

One reason that cycles may be more common at high latitudes among mammals is the relative simplicity of the predator–prey community. For example, when Gilg *et al*. (2003) studied lemming cycles in the high arctic tundra of Greenland, they found one major prey species, the collared lemming, and four predator species. Three species, as shown in Fig. 10.11, were generalist predators (snowy owl, arctic fox, and long-tailed skua) and one was a specialist predator, the stoat, a member of the weasel (Mustelidae) family. Mustelids are prolific predators and prolific breeders, capable of a very strong numerical response. In their study, Gilg *et al*. (2003) concluded that the lemming cycle in Greenland is driven by predation and there is no evidence of food or space (nest site) limitation. The specialist predator, the stoat, produces the cycle. There is a one-year delay in the numerical response of the stoat, resulting in a cyclical dynamic between lemmings and stoats. But the authors emphasize that generalist predators are necessary to stabilize the lemming–stoat interaction. The three generalist predators focus on lemmings only when the lemming population is very dense. As seen in Fig. 10.11, all three have strong functional responses to lemming density. Snowy owls and arctic foxes, in addition to stoats, also display a strong numerical response. The generalist predators are thought to be necessary to slow down the growth rate of the lemming at its highest densities, thereby allowing the specialist predator to catch up and begin driving the lemming population downward.

Some Canadian lemming populations, however, do not cycle (Reid *et al*. 1997). The collared lemmings at Pearce Point in the Northwest Territories are preyed upon by a specialist predator (rough-legged hawk *Buteo lagopus*), a semi-specialist predator (red fox), and by several generalist predators (golden eagle *Aquila chrysaetos*, grizzly bear, arctic ground squirrel *Spermophilus parryii*, peregrine falcon *Falco peregrinus*, and gyrfalcon *Falco rusticolus*). The interactions are governed by differences between summer and winter predation rates. The major stabilizing factor, however, is the presence of ground squirrels, which become the primary prey or principal alternative prey for the predators when lemming densities are low. Not only do ground squirrels provide a prey base to keep a diverse array of predators in the ecosystem, but they also prey on lemmings themselves. Without ground squirrels, Reid *et al*. (1997) hypothesize the loss of three of the generalist predator species. The complex guild of generalist predators plus the ground squirrels found in this ecosystem never allow the lemmings to go through an irruption.

Snowshoe hare cycles

As discussed earlier, snowshoe hare populations undergo 10-year cycles in North America, and, based on pelts returned to the Hudson's Bay Company, Charles Elton proposed that lynx controlled these hare populations. Recently Krebs *et al*. (1995, 2001), set up 1 km^2

plots in the boreal forests of the Yukon in Canada. Fertilizer was added to promote plant growth in some plots, and in some plots predators were excluded. To summarize this very complex study, they found that both food and predation help drive the snowshoe hare cycle, and that the two effects are likely linked. Although Krebs *et al.* (2001) emphasize that "the hare cycle is not driven primarily by plant–herbivore interactions," food limitation increased predation by forcing hares to search more extensively for food, by reducing their health, and by making them less likely to escape predation. This result reinforces the predator-sensitive food hypothesis described below for wildebeest (Sinclair and Arcese 1995).

When radio-collars were placed on hares, Krebs *et al.* (2001) found that 95% of the hares died as a result of predation. In the Yukon, predators included lynx, coyotes, goshawks (*Accipter gentilis*) and great horned owls (*Bubo virginianus*). In this study, Krebs *et al.* (2001) could not identify a predominant role for any one predator species. As phrased by Krebs *et al.* (2001) "the hare cycle is not strictly a lynx–hare cycle, as many textbooks claim." The entire community of predators drives the hare cycles, not just lynx. In fact, in areas such as the Anticosti Island in eastern Canada, where no lynx are found, the hare cycles continue.

Another interesting twist is that when snowshoe hares are in a down portion of their cycle, the predatory species turn on each other (O'Donoghue *et al.* 1995). These data also come from the Yukon, where hare populations cycle every 8–11 years. Prey species included red squirrels (*Tamiasciurus hudsonicus*) and voles. The primary predators included lynx, coyotes, great horned owls, and goshawks, as well as red fox, wolverines (*Gulo gulo*), and wolves. Since the primary predators appear to go through population cycles related to those of the hare, O'Donoghue *et al.* put radio-collars on lynxes and coyotes and radio-tagged owls, hawks, and other birds of prey in order to determine their fate.

Snowshoe hare populations peaked at 450 per square mile (174 per km^2) in 1989–90 and declined to 7 per square mile (3 per km^2) by the spring of 1992. The lynx population declined from 60 to 15 per square mile (23 to 6 per km^2). Did the lynx leave the area? If they died what were the causes of death?

Of 15 radio-collared lynx in 1991–92, by the spring of 1993 only seven remained. Six individuals emigrated and two deaths were recorded. Collars were found as far away as 800 km in Alaska, British Columbia, and the Northwest Territories.

In 1992–93, of nine radio-collared lynx, all had died by April. One lynx was found healthy and well-fed, but had been killed by a wolf. A female lynx was killed by another lynx. One young male starved. A male and female pair was migrating into the mountains when a wolverine killed the female. The male was killed by either a wolf or a wolverine, based on evidence from blood and tracks in the snow. The investigators actually witnessed a lynx being killed by a coyote. Other scientists have also reported witnessing lynx being killed by other lynx or by wolverines. All of this does not happen when hare are abundant. Furthermore, when hare populations are low, lynx themselves become very aggressive, routinely killing red fox.

Birds of prey followed the same pattern. In 1989 and 1990, when hare were abundant, 10 of 11 goshawk nests succeeded in fledging their young. In 1991, 50% nest failure was reported. All failed because of **predation**, evidently by great horned owls. The owls ate the adult female birds as well as their young.

In conclusion, snowshoe hare cycles are not simply driven by the lynx population. The vegetation and the entire community of predators are involved in these cycles.

Furthermore, these trophic interactions involve the community of predators itself. When prey populations are scarce, predatory animals turn on each other, and mortality of the predators due to starvation was, in fact, relatively rare.

Moose—wolf interactions

Isle Royale is a 574 km² island in Lake Superior that is a protected National Park. Moose populations have been present on the island for about 100 years (Pastor *et al.* 1988). Shortly after 1900 moose arrived on the island from Minnesota or Canada. They found abundant food and no major predators. The Isle Royale ecosystem is, in fact, relatively simple in terms of large herbivores and carnivores. As described by Smith *et al.* (2003), when moose arrived, the only mammalian carnivores were coyote and red fox, and the large herbivores included red squirrel, snowshoe hare and beaver (*Castor canadensis*). Ravens (*Corvus corax*) were also present. The moose population rapidly increased to some 3000 individuals by the early 1930s (Murie 1934). As they over-browsed their food supply the population declined. Several forest fires in the 1930s resulted in the regeneration of aspen and paper birch (*Betula papyrifera*), which allowed the moose population to increase again since these are preferred foods. In the late 1940s, gray wolves arrived and evidently quickly extinguished the coyote population, maintaining a relatively simple food web. According to Mech (1966), predation limited the moose population below its food supply. More recent research (Peterson 1999), however, indicates that both vegetation and wolves play a role in moose population dynamics.

Between 1948 and 1950 four exclosures (each 100 m² in size) were set up to evaluate the effect of moose on the vegetation and other ecosystem properties. Pastor *et al.* (1988) concluded that intensive moose browsing led to fewer deciduous trees and a forest dominated by spruce (*Picea glauca*) and balsam fir (*Abies balsamea*), which are less palatable to moose. In addition, lower levels of nitrogen remained in the soil. The net result was an ecosystem less able to support moose populations, since moose do less well on this diet.

Although there seems to be some agreement that moose populations are affected by wolves, the question remains, how much of an effect? Peterson *et al.* (1984), based on about 20 years of data on the wolf population at Isle Royale, concluded that moose and wolves would go through population cycles typical of predator—prey interactions involving rodent and hare populations. However, since the body mass (*M*) of moose is larger, they concluded that moose—wolf interactions would cycle with considerably longer period lengths. In fact, after noting that the intrinsic rate of increase of mammals scales as $M^{-0.26}$, they proposed that vertebrates cycle as the fourth root of the body mass, $M^{0.25}$. From this they predicted that moose populations should cycle with a period length of 38 ± 13 years. Given the paucity of data, however, this prediction cannot be confirmed. Moreover, their prediction that "oscillating elephant populations" should cycle with a period of 71 ± 21 years seems without merit.

Making predictions about prey and predator populations based on allometric relationships or on prey-induced vegetation changes also ignores the role of climate. Mech *et al.* (1987) showed that Isle Royale moose and northeastern Minnesota deer populations were both significantly affected by the snow accumulation of the previous winter. In severe winters moose populations were not able to find browse, and their physical condition deteriorated. Many of the moose and deer that were killed by wolves would have starved to death anyway. Therefore, although wolf predation was a direct mortality agent, Mech *et al.* believe it was secondary to winter weather in influencing deer and moose populations. In addition, snow accumulation during previous winters affects maternal nutrition to such a degree that fecundity and/or calf survivorship during the next growing seasons is seriously influenced.

The wolf population on Isle Royale decreased from 1981 through the mid-1990s, because of a canine parvovirus, probably introduced from the mainland on the hiking boots of visitors. During this period the moose population increased until it reached a record level in 1995 of 2400. Before the wolf population could recover (there were approximately 50 in 1980 and 24 in 1995), the moose population crashed. Almost 80% died, mostly from starvation, but a severe winter tick infestation contributed to the crash (Peterson *et al.* 1998). Peterson (1999) concluded that although the moose population density is influenced by wolves, the population level is ultimately set by available food.

Smith *et al.* (2003) have pointed out how much more complex the food web is at Yellowstone National Park in Wyoming, where wolves have recently been introduced, than is the case at Isle Royale. Whereas the only mammalian predators at Isle Royale are wolf and red fox, at Yellowstone coyote, mountain lion, grizzly bear, and black bear (*Ursus americanus*) are found along with wolves. Human hunting on American elk (*Cervus elaphus*, called red deer in Europe) must also be factored in. Avian predators at Yellowstone include bald eagles (*Haliaeetus leucocephalus*), golden eagles, magpies (*Pica pica*) and ravens, whereas only ravens are found on Isle Royale. Additional large prey species, beside moose, found at Yellowstone are pronghorn antelopes (*Antilocapra americana*), bighorn sheep (*Ovis canadensis*), mule deer, American elk, and bison (*Bison bison*). At Yellowstone only 26 instances of wolves killing moose have been recorded since the wolves were reintroduced. Wolves mainly prey on elk, and are predicted to reduce elk populations. However, the elk herds have only declined by approximately 18% so far (Smith *et al.* 2003, Ripple and Beschta 2003).

After reviewing 27 studies on moose–wolf interactions, Messier (1994) was able to generate both functional and numerical for wolves as a function of moose density. He concluded that when wolves are the single predator, a moose population will stabilize at 1.3 moose per km², whereas the equilibrium density with no predators is 2.0 moose per km². His analysis is consistent with a model proposing that under these circumstances moose populations are regulated at low densities due to density-dependent wolf predation. However, Messier comments that Isle Royale is an exception: it has a high-density moose population limited by food competition, and wolves are present but do not regulate moose density (Messier 1991, 1994).

Predator–prey relationships in Africa

Sinclair and Arcese (1995) studied the interaction between food supply and predation in the regulation of large herbivore populations on the Serengeti Plain in East Africa. They distinguished among three hypotheses:

1 The **predation-sensitive food** (PSF) hypothesis states that both food and predation limit prey populations. As food becomes limiting, animals take greater risks to obtain food and become victims of predation.
2 The **predator regulation** hypothesis states that predators hold prey populations well below starvation levels.

3 The **surplus predation** hypothesis states that predators kill only those prey individuals that are excluded from optimal habitat and are already dying of starvation. This is similar to Errington's (1946) ideas about the "doomed surplus" of muskrat populations.

These three hypotheses were tested by examining the body condition of Serengeti wildebeest (*Connochaetes taurinus*) over a 24-year period. Two phases of population growth were examined: (i) 1968–73, when food was abundant and prey populations were increasing; and (ii) 1977–91, when the wildebeest population was stationary and partially regulated by competition for food. Sinclair and Arcese examined live animals, predation kills, and non-predation deaths. Body condition was measured by an examination of bone marrow, the last reservoir of fat in these animals.

The predator-regulation hypothesis predicts that bone-marrow condition will be similar in predator-killed and live samples, while the surplus-predation hypothesis predicts that bone marrow will be similar in predator-killed and non-predator deaths. The PSF hypothesis predicts that bone marrow condition of animals killed by predators should be: (i) poorer than that of live animals; (ii) better than that of the animals who die of causes, such as disease, unrelated to predation; (iii) better when food is limiting than when it is abundant (because when food is abundant, predators only kill sick or young animals).

Analyses of the bone marrow from animals dying due to predation or from non-predation-related causes showed that these animals were in poorer health than the live population. In both the increasing and stationary phases, the animals dying from predation were in better condition than the animals dying from non-predation causes. These results are consistent with the PSF hypothesis, and inconsistent with the other two hypotheses. Lions (*Panthera leo*) and hyenas (*Crocuta crocuta*) killed animals in similar condition, but lions took younger animals.

The results suggest that: (i) body condition affects vulnerability of individual wildebeest to predation, and (ii) predation jointly limits the population along with intraspecific competition for food resources.

Sinclair *et al.* (2003) have also analyzed the community-wide patterns of predation in the Serengeti ecosystems of Tanzania and Kenya. Twenty-eight species of ungulates and 10 species of carnivores inhabit these areas, consisting mostly of open grassland (savanna). In any one habitat as many as seven species of canid and felid carnivores are present. The predators range from 8 kg (Golden jackal, *Canis aureus*) to 150 kg (lions), while prey sizes range from small gazelles (*Gazella* sp.), which weigh 18–20 kg, to elephants (*Loxodonta africana*), rhinoceros (*Diceros bicornis*), and hippopotamus (*Hippopotamus amphibius*), which come in at over 3000 kg. Long-term studies of the causes of ungulate mortality show that the proportion of adult mortality due to predation is above 80% in the smallest species such as oribi (*Ourebia ourebia*), impala (*Aepyceros melampus*), topi (*Damaliscus lunatus*), and wildebeest. By contrast, in a heavier species such as zebra (*Equus quagga*), adult mortality due to predation is 70%. There is a "threshold" in body size of about 150 kg, after which deaths due to predation decline substantially. Only about 23% of adult buffalo and 5% of adult giraffe (*Giraffa camelopardalis*) mortality is caused by predation; rhinos, hippos, and elephants (the "big three") suffer virtually no adult predation. While the smallest ungulates are preyed upon by as many as seven different predators, the number of potential predators falls off linearly as a function of the log of the herbivore weight, to zero for the big three. Thus the smallest ungulates (less than 150 kg)

are limited by the diverse array of predatory species, while the large ungulates are basically food-limited. A "natural experiment" confirmed this generalization. In the northern
Serengeti poaching and poisoning eliminated most of the carnivores, including lions,
hyenas, and jackals, whereas in the nearby Mara Reserve the predator community remained
intact. During the years when predator populations were eliminated five of the species below
150 kg in weight increased their populations conspicuously as compared to the Mara Reserve.
Once the predators returned, their populations declined. By comparison, the 800 kg
giraffes did not increase in the predator-removal area. Sinclair *et al.* (2003) concluded that
the mammalian herbivore populations in the Serengeti ecosystem are subject to top-down
(predation) or bottom-up (food limitation) processes depending on their size.

10.10 Trophic cascades

A trophic cascade is a "progression of indirect effects by predators across successively lower
trophic levels" (Estes *et al.* 2001). Predators may have non-lethal effects on a prey population when prey changes its behavior due to the presence of predators. This has been referred
to as the "ecology of fear" concept (Brown *et al.* 1999). In the predation-sensitive-food
hypothesis discussed above, both predation and food availability limit prey populations,
and prey occupy risky sites only when suitable vegetation is limited. Ripple and Beschta
(2003, 2004) have proposed that the reintroduction of wolves in 1995 into Yellowstone
National Park has already had a profound effect on the ecosystem. Elk will only forage at
sites that allow early detection and escape from wolves. The result has been re-growth of
cottonwood (*Populus* sp.) and willow in riparian areas along Soda Butte Creek and the
Lamar River. Both areas are frequented by wolves and are avoided by elk. Ripple and Beschta
(2003, 2004) have proposed that the return of willow and cottonwoods has provided shade
and cover, which will benefit both trout and migratory birds. Furthermore, several new
beaver colonies have colonized these riparian zones. The number of beaver colonies in
Yellowstone's northern range has increased from one in 1996 to seven in 2003. Through
their dam building, these beaver populations will produce a multitude of effects within
these watersheds. Thus the reintroduction of wolves into Yellowstone has potentially
produced a cascade of effects in both terrestrial and wetland environments, the full scope
of which will be unveiled in the next few decades.

10.11 The dangers of a predatory lifestyle

Ross (1994) has described the fate of solitary predators, such as cougars. Cougars, also
known as pumas or mountain lions (*Felis concolor*), are almost unique among solitary predators in that they consistently seek prey larger than themselves. African lions, hyenas, wild
dogs (*Lycaon pictus*), and wolves all practice cooperative hunting when attacking large prey.
Most solitary hunters, such as weasels or foxes, generally prey on creatures smaller than
themselves. Among the cats, leopards, cheetahs (*Acinonyx jubatus*), jaguars (*Panthera onca*),
and lynxes also usually follow this pattern of attacking smaller animals.
    Attacking large prey has the drawback that it can sometimes prove fatal for the hunter.
Among Alberta cougars, there are some dramatic examples of what can go wrong in a
violent struggle with a large prey animal. Aside from human hunters, these struggles are
the main cause of death in this cougar population.

One young adult female suffered a broken back when the mule deer she was riding down a steep slope slammed into a pine tree. Another female was speared by a sharp branch when the elk that she eventually killed tried to shake her loose from its throat. An adult male cougar that attacked a bighorn sheep lost his footing in the struggle, and both fell to their death over a 27-meter cliff.

10.12 Escape from predation

Let us now change our focus to an evaluation of predator–prey relationships from the perspective of the lower trophic level. What mechanisms have evolved allowing organisms to escape predation? These means of escape can be classified into the following four general categories: (i) escape in time; (ii) escape in space; (iii) escape through behavior; (iv) escape through physical/chemical mechanisms.

Escape in time or predator satiation

Life cycles have evolved so as to minimize predation or (for plants) herbivory. Usually one part of the life cycle is highly vulnerable to predation, such as newly produced offspring or adult stages involved in reproductive behavior. Examples include mast-fruiting in oaks, periodical cicadas (Williams *et al.* 1993, Karban 1997), bamboos that reproduce once a century (Janzen 1976), century plants, mayflies, and the mating flights of ants and termites. In all of these cases reproduction is synchronized in a local population or in many populations over a large geographical area. So many potential prey organisms are produced in such a short time period, or in such an unpredictable manner (e.g. periodical cicadas reproduce once every 13 or 17 years), that potential predators can eat their fill (**satiation**) without seriously reducing the prey population. Since predator populations cannot build up due to the brevity of the vulnerable stage and/or the unreliability of the prey population, the vast majority of seeds, seedlings, larvae, or other vulnerable stages "escape" predation.

Escape in space

In this case, the prey population is relatively rare and disperses readily, so that it is not abundant in any local habitat. Since the prey population is so rare, the predator may never develop a search image for it. Furthermore, predators that specialize on this species cannot become very common themselves, given the limited number of opportunities to feed. Such predators, once they find a prey population, however, may wipe it out completely. Although the prey population goes locally extinct, if it disperses rapidly enough it will not go extinct over a larger area. Such an interaction is consistent with the definition of a metapopulation that survives local extinctions as long as there are other local populations in the region. Examples of such interactions are the herbivorous and predatory mites studied in the laboratory by Huffaker (1958) and the *Opuntia stricta–Cactoblastis cactorum* relationship in Australia (Dodd 1940).

Behavioral means

Prey species have a variety of means by which they confuse predators. Some of these are combined with coloration schemes (see below). It is well known that among birds, adults will attempt to draw the attention of predators away from nests or vulnerable offspring by attacking and mobbing them, creating noisy disturbances, or pretending to be injured. Noteworthy also are herding (among mammals) and schooling (among fish) behaviors. In order to hunt effectively, a predator usually needs to focus on a given individual. Animals moving in herds or schools create confusion or disrupt the path of a predator that has focused its attention on a particular prey individual. In other cases, potential prey attempt to intimidate their predators through aggressive displays. The frilled lizard (*Chlamydosaurus kingii*) exposes a skin flap (a "frill") around its neck, opens its mouth widely, and makes aggressive lunges at a potential predator in an attempt to make it look large and fierce. Several species of fish and toads are known to fill themselves with air in order to look larger to potential predators.

Physical and chemical mechanisms

Physical toughness
In animals, spines, scales, quills, and bony plates are used for protection. Among plants there exist spines (modified leaves), thorns (modified stems), epidermal hairs, thick cuticles, heavy cork layers, lignified or silicified cells in the epidermis or cortex, barbed and sticky trichomes (an outgrowth of the epidermis), and the general use of fibers, lignin, crystals, silicon, and cellulose itself to make plant tissues difficult to eat and to digest. When combined with noxious chemical defenses ("secondary compounds") produced by plants, the result is that most plant material, including leaves, bark, stems, and roots, constitutes an unrewarding diet for most animals.

Resistance to chemical degradation
Cellulose is the most abundant compound in plants. Approximately 10 billion tons of carbon is transformed into cellulose per year (Albersheim 1965). Cellulose is a linear polysaccharide, and the molecules are juxtaposed so as to form linear crystals or microfibrils. As such, cellulose is highly resistant to chemical and enzymatic degradation. Numerous bacteria, one termite species, and a few protistans have been shown to manufacture an enzyme, **cellulase**, capable of partially degrading cellulose. But cellulase has rarely been evolved in the animal kingdom. Actually two enzymes are necessary to degrade cellulose, cellulase, which yields the disaccharide cellobiose, and **cellobiase** that hydrolyzes the dimer to yield free hexose.

Wood contains large amounts of **lignin**. Conducting vessels and the fibers that strengthen plants are mostly lignified cell walls. Lignin is a close second to cellulose as the most abundant compound in nature (perhaps 60% of the amount of cellulose). Lignin is a polymer that contains a large number of aromatic side chains, and it is quite resistant to chemical degradation and almost impervious to enzymatic digestion. In fact, lignin is probably the most resistant compound to degradation by enzymes found in nature (Albersheim 1965).

Coloration

1 **Cryptic or camouflage coloration.** Cryptic patterns, i.e., blending in with the environment, are commonly found among arthropods, especially insects, and all groups of vertebrates. Even cephalopods such as octopuses have evolved

cryptic color patterns. Among insects, larval and adult stages have evolved patterns that appear to be sticks, living leaves, dead leaves, flowers (Greene 1989), bark, lichens, and even bird droppings.

2 **Confusing patterns.** Several species of insects have evolved false "heads" or other strange appendages on their rear ends. A possible explanation for such traits is that they distract a predator, confuse it, and the result is an attack that is much less likely to be fatal than one directly on the head.

3 **Startle patterns.** Several tropical butterfly species have color patterns on the underwings that are said to resemble owl's eyes, and many other lepidopteran species have evolved concentric rings resembling vertebrate eyes (Blest 1957). When disturbed by a bird, the butterfly flashes these false eyes, with the result that the bird may temporarily back away. Other insect larvae have evolved morphological patterns and/or colors that make them resemble snakes (Pough 1988).

4 **Flash patterns.** In this situation, a cryptic animal flashes a bright color pattern, which is normally hidden on the underside of the animal or by some other means. For example, white-tailed deer flash their white tails when alarmed. The white patch is highly visible as it runs away from its predator. Subsequently, the deer usually takes a sharp-angled turn; at this point the white tail is no longer visible and the deer, which is highly cryptic otherwise, disappears from view. Having focused on the white tail, the predator has a hard time visually locating where the deer has turned, and has trouble tracking it. Other examples include frogs that are a cryptic green color except for a yellow underbelly. If discovered, the frog jumps and the potential predator is "flashed" the bright yellow. This yellow is concealed as the frog settles again into the green vegetation.

5 **Aposematic patterns.** Here we are interested in the interaction of color patterns with chemical defenses (discussed below). When an animal is dangerous, toxic, poisonous, or odoriferous, it is adaptive if the potential predator has been warned of the consequences of dealing with such an animal. Thus, such chemically protected animals have usually evolved highly colorful patterns (red, orange, or other bright colors) that warn predators with color vision (birds, primates, etc.). Alternatively, they have evolved alternating stripes or spots of light and dark colors (e.g., striped and spotted skunks, *Mephitis mephitis* and *Spilogale putorius*, and Hymenoptera such as bees and wasps) to warn animals that lack color vision. These are known as aposematic or warning color patterns.

Chemical defenses

Plants produce a multitude of compounds that are not part of the primary metabolism and are therefore called "secondary compounds." We will discuss these in the next chapter. At this point we will simply note that the great diversity of physical and chemical defenses among plants is one line of evidence suggesting that herbivores have greatly influenced the evolution of plants. The chemical diversity of plants has evidently resulted in a similar diversity in chemical defenses among animals. While some animals are able to synthesize their own chemical defenses, many acquire them from the plants they consume. Such animals have not only evolved mechanisms to detoxify or tolerate the chemicals found in their diet, but they are also able to sequester or modify these chemicals so as to provide them with protection from their own predators. Some animals have active chemical defenses, for example skunks, bombardier beetles (Coleoptera) (Eisner and Aneshansley 1982) and stinkbugs (Hemiptera). These animals are able to spray their potential predators with noxious compounds. Other animals have a more passive defense in that they taste bitter or are toxic if eaten.

Mimicry

The combination of poisonous host plants and chemically defended herbivores with aposematic coloration set the stage for the evolution of mimicry. In **Batesian mimicry**, the color pattern of an unpalatable model species is copied or mimicked by another species that is not itself chemically protected. The classic case of the monarch (*Danaus plexippus*) and the viceroy butterflies (*Limenitis archippus*) was described by Brower and his colleagues (Brower and Brower 1964, Brower *et al.* 1967). The larvae of monarch butterflies feed on milkweed (*Asclepias* spp.) plants that contain cardiac glycosides in a milky sap. The larvae are able to feed upon the glycosides without ill effect and store them in their tissues. The adult monarchs have an aposematic color pattern that is mimicked by the viceroy. In the classic interpretation, viceroys are palatable to birds (but see below). When a bird ingests a monarch, however, it often reacts violently to the cardiac glycosides, which cause vomiting and seizures. Once a bird has this experience it never eats either a monarch or a viceroy again.

In another type of mimicry (known as **Muellerian mimicry**) a group of species (again, often butterflies) all are chemically protected and converge on a common color pattern. The theory is that once a predator learns that a butterfly with this pattern is poisonous, it generalizes the lesson to all of the species that have this color pattern. Recent experiments have suggested that under some circumstances the viceroy is chemically protected, and that the viceroy–monarch story may actually be one of Muellerian rather than Batesian mimicry (Ritland and Brower 1991).

Although mimicry is most famous among members of the Lepidoptera, it should be noted that many other groups of insects can be mimics. Many flies (Diptera), for example, mimic bees and wasps (Hymenoptera) (Pough 1988).

Although discussions of mimicry usually focus on invertebrates, Pough (1988) described many cases of mimicry among vertebrates. Among amphibians, the unpalatable salamander, *Plethodon jordani*, is mimicked by palatable species of *Desmognathus*, and the noxious red salamanders *Pseudotriton ruber* and *P. montanus* are probably Muellerian mimics. Among frogs, the terrestrial diurnal frogs of the genus *Phyllobates* (Dendrobatidae) are protected by curare secretions in the skin. There is evidence that several species of frogs in the family Leptodactylidae are Batesian mimics of *Phyllobates*.

Among reptiles, the most famous examples involve several species of snakes in the family Colubridae that are mimics of the coral snakes (genus *Micrurus*) (Greene and McDiarmid 1981). Pough (1988) suggests that snakes, especially venomous snakes, are such extremely noxious models that predators have generalized their characters very broadly. Additionally, if the model is sufficiently noxious, selection may act to produce innate avoidance of the model. Because of the extreme toxicity of poisonous snakebites, there is no opportunity for the trial-and-error learning by the predator that is the basis for mimicry in insects or other less toxic vertebrates. Selection must therefore favor predators that display an innate avoidance of the characteristics of model species, such as coral snakes.

Several studies have shown that, indeed, several vertebrate predators do have an innate fear of the coral snake pattern. Zoo-reared coatimundis (*Nasua narica*) and javelinas (*Tayassu pecari*) were presented with rubber snakes. The coral snake pattern startled both species and individuals fled. Rubber snakes painted green were attacked (Gehlbach 1972). Smith (1975, 1977, 1980) obtained similar results using hand-reared birds. When turquoise-browed motmots (*Eumomota superciliosa*) and great kiskadees (*Pitangus sulphuratus*) were presented with wooden dowels painted in various neutral colors, both species readily attacked. Both species reacted with alarm calls and fled from dowels painted yellow with red rings. They also fled from a coral snake pattern of yellow, black, and red rings. On the other hand, birds from the temperate zone, such as sparrows, blackbirds, and jays, attacked all patterns presented, including the coral snake pattern. Thus, aversion to aposematic colors does not seem to be an innate behavior of all birds. Innate aversion to the coral snake pattern is, instead, a property of those species of birds potentially exposed to coral snakes.

10.13 Conclusions

What have we learned about predator–prey interactions? Combining what we know from theory, laboratory and field studies, we can conclude that predator–prey interactions are stabilized by: (i) heterogeneity of the environment; (ii) nonrandom hunting by the predator; (iii) density dependence in the prey population unrelated to the predator population; (iv) a prey refuge; (v) a less efficient predator; (vi) providing the prey with less than optimal resources to slow its growth rate and thereby avoiding the "paradox of enrichment;" (vii) a complex of predators including enough generalists to dampen the destabilizing effects of the specialist predators; and (viii) the evolution of effective physical, chemical, or behavioral means by the prey species to limit predation. On the other hand, simple predator–prey communities, time lags, and specialist predators that overwhelm prey defenses tend to promote cycles or extinctions. Furthermore, although generalist predators may stabilize some ecosystems, the introduction of either a specialist or a generalist predator into a "naïve" prey community may have devastating consequences. A functional response by the predator which includes a significant handling/digestion time (energy) component can stabilize a predator–prey interaction if the prey population is kept at low densities. Once the prey population "escapes" and becomes so large that the functional response of the predator has reached the "plateau," the effect is destabilizing. Unless there is also a numerical response by the predator population, the overall predation rate would become inversely density-dependent: the more prey, the lower the proportion of prey being consumed. Finally, some prey species are controlled by predators, producing a top-down effect on the community. As we will explore in the next chapter, other herbivorous prey species are controlled by plant productivity, resulting in a bottom-up control of the community.

11

Plant–herbivore interactions

- Classes of chemical defenses
- Constitutive versus induced defense
- Plant communication and plant–parasitoid communication
- Novel defenses/herbivore responses
- Detoxification of plant compounds by herbivores
- Plant apparency, soil fertility, and chemical defense
- The optimal defense theory
- Modeling plant–herbivore population dynamics
- The complexities of plant–herbivore interactions

11.1 Introduction

One of the most important fields of recent ecological inquiry has been that of plant–herbivore interactions. There are several reasons for the rapid developments in this field. (i) Herbivores are ecologically important. Globally it is estimated that herbivores consume from 7% (Pimentel 1988) to 18% (Cyr and Pace 1993) of leaf area in terrestrial ecosystems and from 30% to 79% of plant net production in aquatic ecosystems (Cyr and Pace 1993). (ii) Caught between "plants and predators" (Olff *et al.* 1999), herbivores must deal with the chemical and morphological defenses of plants while simultaneously defending themselves from their own predators. The study of plants and their herbivores leads to new understandings of interactions involving more than two trophic levels. (iii) Herbivores have been, and still are, important to the evolution of plants and other animals. Through their activities, herbivores have evolutionarily shaped the plant community, partially determining the diversity, abundance, and life form of plants. This, in turn, affects what other types of animals are present and influences ecosystem processes such as energy flow and nutrient cycling. (iv) Herbivores are economically important. In agricultural systems, herbivores may often consume 50% of net productivity and therefore depress yield. Under the worst of circumstances agricultural crops may be wiped out. Furthermore, the animals we raise for food and other products are almost all herbivores themselves. Thus a better understanding of plant–herbivore relationships is crucial to the success of our agricultural endeavors. (v) Plant–herbivore interactions provide unique opportunities to investigate basic ecological and evolutionary processes such as coevolution, food webs, plant chemistry, animal foraging, competition, and ecosystem processes such as nutrient cycling.

The relationship between plant "secondary compounds" and herbivore–plant relationships was first outlined by Fraenkel (1959). Since the compounds under discussion are not part of the primary metabolism of plants, they were and still are known as "secondary compounds." Fraenkel suggested that plant secondary compounds have evolved as defenses against herbivores and herbivores have strongly affected plant evolution. Since the publication of his paper research has exploded in the fields of plant–herbivore relationships and chemical ecology. Ecologists such as Harborne (1993, 1997) and Hartley and Jones (1997) accept that the major functions of plant secondary compounds are as defenses against herbivores. Skeptics, such as Smith and Smith (2001) have asserted that there is little evidence that plants evolved secondary compounds for a defensive purpose. However, as Hartley and Jones (1997) put it, although "we are still not sure why plants have such a huge array of secondary compounds or how this came to be, we do know that these chemicals are important in keeping the world green."

The basic theory, which has guided ecological thinking concerning plant–herbivore interactions, was set forth by Ehrlich and Raven (1964). At that time the great diversity of chemical compounds produced by plants was already widely recognized. Many chemists and botanists referred to these compounds as waste products of plant metabolism, lacking any adaptive value. Ehrlich and Raven, by contrast, asserted that secondary compounds were the product of coevolution with herbivores. Plants that produced secondary chemicals made their tissues unpalatable or toxic, lowered herbivore damage, and had a selective advantage since they could devote more energy to competition and reproduction. If less energy were lost to herbivores, then plants would grow more rapidly and enjoy competitive success. Any genes that allowed the plant to produce these chemicals would spread throughout the population. By the same argument, herbivores could be expected to adapt to plant defenses. Again, an herbivore that evolved a detoxification enzyme, or other adaptation to allow it to feed on protected plants, would enjoy competitive success compared to individuals unable to feed on these plants. As discussed previously, some herbivores have turned defensive chemicals to their own advantage. That is, by modifying and storing plant toxins, the herbivore itself became unpalatable to predators. Thus an herbivore, which evolved an adaptation for feeding on a chemically defended plant, could potentially enjoy both competitive success and protection from its own predators.

Faced with an increasing number of herbivores adapted to a particular chemical defense, plants have, according to Ehrlich and Raven's theory, counter-adapted by producing additional defensive chemicals. This process of evolution and counter-evolution of chemical defenses is often referred to as the "evolutionary arms race" and is thought to have helped produce the great diversity of both angiosperms and the insects that feed upon them. Taken together, land plants (particularly angiosperms) and insects make up more than half of all known terrestrial species on earth (excluding microorganisms).

The Ehrlich–Raven theory is based on the following assumptions:

1 Herbivore activity is harmful to plants. This would seem to be obvious, but this assumption cannot be accepted unequivocally.
2 Plants are able to evolve defenses that are effective in deterring feeding by herbivores. Note that, although the Ehrlich–Raven theory stresses herbivores,

these plant chemical defenses could just as easily have been evolved to defend plants against attacks by fungi, bacteria, and other microorganisms. Those chemicals that evolved as antifungal defenses, for example, might also be effective against other microorganisms or herbivores.

3 Herbivore feeding activities, growth, reproduction, and evolution have been guided by the ability of plants to defend themselves, both physically and chemically.

4 Although there exist herbivores that feed on plants of many species, genera, or families with seemingly little regard for the identity of the plants, these "generalists" are actually much more selective than they appear. Generalists engage in a broad "sampling program" in which they eat small amounts of material from many plant species. The majority of what they consume, however, comes from a much smaller species list (Rockwood 1976, 1977, Rockwood and Glander 1979). The careful manner in which generalists eat is thought to be consistent with the central importance of secondary compounds in plant–herbivore interactions.

5 The majority of herbivore species are not generalists, but are specialists, feeding on just one plant species, one plant genus, or perhaps one plant family (for example the Cruciferae or mustard family). Such specialization is consistent with the idea that an herbivore that evolves a way of feeding on a particular plant type eventually loses the ability to feed on other plants.

For example, one simple hypothesis, based on these assumptions, is that plants with few specialized chemical defenses will be fed upon primarily by generalist herbivores, while plants with specialized, complex chemical defenses will by fed upon mostly by specialist herbivores. When Berenbaum (1981) examined the chemical defenses in the Apiaceae (the carrot family) she found some species defended by a relatively simple phenolic (coumarin), some by linear furanocoumarins, and some by sophisticated angular furanocoumarins. As the defenses became more complex the herbivores feeding on the plants changed from mostly generalists (defined as feeding on more than three plant families), to mostly specialists (those feeding on only 1–3 genera).

11.2 Classes of chemical defenses

Plants have evolved a wide variety of chemical defenses. Over 40,000 chemical compounds known as allelochemicals have been described from plants (Berenbaum 2002). Allelochemicals are characterized as having a negative impact on herbivores, disease organisms, or other plants. The three main classes of secondary compounds are (i) terpenoids, (ii) phenolics, and (iii) nitrogen-based compounds such as alkaloids. The terpenoids or isoprenoids are formed from acetyl coenzyme A, and are built on five-carbon units into larger molecules. Phenolics include plant pigments such as anthocyanins but also include the bitter-tasting tannins. These compounds are derived from the amino acid phenylalanine or from malonyl coenzyme A, a precursor to fatty acid and lipid biosynthesis (Harborne 1997). The major groups of nitrogen-based compounds are alkaloids, the glucosinolates of the mustard family (Cruciferae), and non-protein amino acids. Glycosides are dealt with below under nitrogen-based defenses, although cardiac glycosides (those found in milkweeds) do not contain nitrogen.

The production of these secondary compounds is metabolically expensive. To synthesize a gram of alkaloid requires 5 g of photosynthetic carbon dioxide, while a comparable figure would be 2.6 g for a phenolic (Gershenzon 1994). On the other hand, compounds like alkaloids are effective at low concentrations such as 2–4% dry weight, while others, such as tannins, require a higher concentration, such as 5% dry weight in leaves, to be effective. Alkaloid concentration is sometimes only 10–20% of the concentration of a comparable phenolic (Harborne 1997).

Nitrogen-based defenses

Alkaloids

Alkaloids are a heterogeneous group of substances that have two common characteristics: they occur in plants and they have an organic base containing a nitrogen atom. They often have a carbon ring structure. They impart a bitter taste to plants, some are very toxic, and some are hallucinogenic. Alkaloids have evolved many times in a variety of plant taxonomic groups. Accordingly, alkaloid biosynthesis has no underlying unity. When Levin (1976) reviewed the literature he found that 16% of temperate-zone plant species and over 35% of tropical plant species contain alkaloids. Approximately 10,000 types of alkaloids are known (Harborne 1988, 1997).

Alkaloids are less important than the phenolics since they are found in only about 20% of angiosperm species worldwide and they are generally absent from mosses, ferns, and gymnosperms (Harborne 1997). One reason why alkaloids, and nitrogen-based defenses in general, are less ubiquitous than carbon-based defenses such as terpenoids and phenolics is that these alkaloids require protein amino acids. Amino acids are, in turn, dependent on a supply of nitrogen, either from the soil or fixed by bacterial mutualists, and nitrogen is usually limiting to plant growth. In legumes, although *Rhizobium*-type bacteria fix nitrogen from the air, the metabolic costs of fixation are high. The fact that nitrogen-based defenses potentially drain needed nitrogen from protein synthesis may have prevented such defenses from evolving more consistently. Nevertheless, there is a great diversity of alkaloids, as described below.

1 Pyridine and piperidine alkaloids. The major example is nicotine.
2 Polyacetyl alkaloids. These alkaloids are found in the plant division Lycopodophyta. Note that although alkaloids have evolved repeatedly in flowering plant families, they can be present in lower plant groups such as the lycopods.
3 Pyrrolidine and tropane alkaloids. The major examples are cocaine (from coca, *Erythroxylon coca*) and atropine (from deadly nightshade, *Atropa belladonna*, a plant in the tomato and potato family).
4 Isoquinoline alkaloids. Examples include pain relievers such as morphine and codeine.
5 Indole alkaloids. Examples include both the anti-malarial drug quinine and the poison strychnine.
6 Pyrrolizidine alkaloids. Senecionine is found in the genus *Senecio* (ragwort) and causes convulsions and liver damage if ingested. Since ragwort often grows in open pasturelands, seneciosis is a common disease of livestock.
7 Pseudoalkaloids. When methylated nucleic acids are degraded, they produce compounds now known as pseudoalkaloids. Examples include caffeine and

theobromine. Caffeine is, of course, the main pseudoalkaloid found in coffee, tea, and cola nuts, while theobromine is found in cocoa. These compounds are all stimulants and diuretics. While they are not very toxic, they give a very bitter taste to seeds and other plant parts.

Glycosides

Glycosides are biologically active forms of steroids, derived from oligosaccharides. **Cardiac glycosides,** known as cardenolides, are C_{23} steroids and, though belonging to the terpenoid group (see below), are convenient to discuss here. They are known from at least 11 plant families, but are especially well known from the plant families Apocynanceae, Asclepiadaceae, and Scrophulariaceae. Often the youngest leaves have the greatest amount of these compounds. Cardiac glycosides are heart poisons, which usually provoke vomiting among animals which ingest them, and can cause death in livestock. The heart drug, digitalis, is derived from a cardiac glycoside. The cardiac glycosides in the milky sap of milkweeds (*Asclepias* spp.) are well documented.

A recent review of the role of **cyanogenic glycosides** in plant–herbivore interactions is instructive (Gleadow and Woodrow 2002). Over 2500 plant species are able to release hydrogen cyanide (HCN) upon ingestion by an herbivore. When an herbivore chews a leaf it disrupts tissues and cells that had previously separated the β-glucosidase enzymes from the cyanogenic glycoside compounds. The resultant degradation of the cyanogenic compound eventually releases HCN.

Examples of poisonings of humans and animals, some leading to death, abound. In addition, there are many cases in which animals have been shown to avoid foods containing cyanogenic glycosides. On the other hand, numerous studies have found that for some herbivores glycosides have little effect or even act as a feeding stimulant (Gleadow and Woodrow 2002).

As explained by Gleadow and Woodrow (2002), in attempting to understand a specific herbivore–plant interaction, four factors must be accounted for. These are: (i) concentration of the presumed toxin; (ii) the status of the herbivore as a "specialist" versus a "generalist"; (iii) the ability of an herbivore to monitor glycosides in its diet; and (iv) the feeding mode of the animal.

In the first case, the cyanogenic compounds must be present at a threshold level in order to be effective. For example, Gleadow and Woodrow (2000) found a significant inverse relationship between concentrations of cyanogenic compounds in young leaves of *Eucalyptus cladocalyx* and herbivore damage. The age of the leaves and the health of the plant often determines the level of defensive compounds in plants, and hence their level of toxicity.

There exist a number of studies that illustrate the principle that, while specialists have evolved the means to tolerate cyanogenic glycosides, plants containing these compounds are avoided by generalist herbivores. For example, Schappert and Shore (1999a) found that the number of herbivore species feeding on the plant *Turnera ulmifolia* in Jamaica was inversely related to the cyanogenic glycoside content. Yet the total amount of herbivory was similar regardless of the glycosides. While the cyanogenic compounds deterred generalist herbivores, the specialists were not deterred, and total leaf area lost was equivalent.

In many cases an animal is able to sequester the cyanogens while preventing the release of HCN. These sequestered compounds are then used as a defense from predators. For example, larvae of *Euptoieta hegesia* accumulate cyanogenic compounds from *Turnera ulmifolia* and become bad-tasting to *Anolis* lizards (Schappert and Shore 1999b).

The third point is simply based on the fact that animals, including humans, are able to monitor the amount of glycosides in their diet due to their bitter taste. Thus it is unlikely that animals will actually poison themselves while feeding on plant material containing cyanogenic glycosides. Hungry animals are more likely to overexpose themselves to plant material containing glycosides. On the other hand, exposure to cyanogenic glycosides leads to an increased ability to tolerate them through greater detoxification ability. For example, humans and other mammals can be induced to produce increased levels of the enzyme rhodanase, which detoxifies HCN as it is released (Gleadow and Woodrow 2002).

Finally, Gleadow and Woodrow (2002) point out that any mode of feeding which minimizes tissue disruption will allow insects and other animals to successfully feed on plants containing cyanogenic glycosides. The main successful tactic involves using sucking mouthparts (Hemiptera, Homoptera) as opposed to chewing mouthparts (Coleoptera).

Gleadow and Woodrow (2002) conclude that, although cyanogenesis is not a totally effective defense, it is toxic and distasteful enough to be effective against a wide variety of herbivores if available in sufficient concentration.

Carbon-based defenses

Phenolic compounds

Phenolic compounds are essentially ubiquitous in the plant kingdom (Swain 1965, Harborne 1988); they are found in ferns and gymnosperms, as well as in angiosperms. Phenolics are defined as compounds with aromatic structures with one or more hydroxyl groups attached (Harborne 1997). They have a common biosynthetic origin from the amino acid phenylalanine. The total number of known phenolic compounds is estimated at 8000 (Harborne 1997). Phenolic compounds range in size from simple phenol, the essential oil in *Pinus sylvestris*, to condensed tannins with molecular weights of 20,000. Phenolics include flavanoids (derivatives of which give color to flowers and fruits), coumarins, and tannins. The tannins are substances present in vegetable extracts that are responsible for converting animal skins into leather by the tanning process. When leaves of many plants are boiled or decompose, tannins give the water its brown color. Tannins are defined as naturally occurring compounds of high enough molecular weight (500–3000), and containing a sufficient number of phenolic hydroxyl groups, to enable them to form cross-links between proteins and other macromolecules. Phenolic compounds of lower molecular weight are apparently too small in size to form effective cross-links, while compounds of molecular weight beyond 3000 may be ineffective tanning agents because they are too large to penetrate between plant collagen fibers. Although tannins are among the most effective types of defensive compounds (Feeny 1976), simple phenols are also effective feeding deterrents.

Phenolic compounds are broadly toxic to most forms of life. Because of their ability to form cross-links with proteins and other polymers, phenols and tannins are capable of inhibiting enzymes, and are also markedly astringent. That is, they cause a dry or puckery sensation in the mouth by reducing the lubricant action of the glycoproteins in the saliva.

There are two distinct types of vegetable tannins. The first class has a core polyhydric alcohol such as glucose. Acids, bases, or enzymes can readily hydrolyze these tannins to yield a carbohydrate and a phenolic acid. Known as **hydrolyzable tannins,** the main effect of these compounds is to inactivate the digestive enzymes of herbivores, especially insects.

Tannins of the second class contain only phenolic nuclei. These are called non-hydrolyzable or **condensed tannins.** Condensed tannins are bound to cellulose and proteins of the cell walls. They make it very difficult for herbivores to extract amino acids from plant material and also defend plants from microbial or fungal attack.

Furanocoumarins

Furanocoumarins are another phenolic group (Harborne 1988) whose biosynthesis can be traced back to the amino acid phenylalanine (Berenbaum and Zangerl 1999). These allelochemicals are found primarily in two plant families, the Apiaceae (the carrot family) and the Rutaceae (the rue family), and thanks to the work of Berenbaum and her colleagues (Berenbaum 1991, 1995, Berenbaum and Zangerl 1996), we know a great deal about the biosynthesis and effects of the furanocoumarins in wild parsnip (*Pastinaca sativa*). Furanocoumarins have biological effects on a wide variety of organisms, from bacteria and fungi to insects and vertebrates. These chemicals "act principally by binding covalently to pyrimidine bases in DNA and interfering with transcription, but these compounds are also capable of causing toxicity by binding to unsaturated lipids in membranes, inactivating enzymes and generating oxyradicals" (Berenbaum and Zangerl 1999, p. 62). Furanocoumarin content of leaves, roots, and stems can be rapidly "induced" to increase by either biotic or abiotic damage. For example, furanocouramin content in leaves can increase over 200% due to mechanical damage. Furthermore, this induced response can be rapid. Mechanical damage can cause xanthotoxin concentration to reach a maximum level within six hours (Berenbaum and Zangerl 1999).

Terpenoids

Isoprenoids and their derivatives constitute a group of naturally occurring plant materials that have in common the fact that are derivatives of the five-carbon compound isoprene, C_5H_8. This group includes essential oils, resins, steroids, carotenoids, and rubber. Terpenoids are classified according to the number of C_5 units they contain. For example, monoterpenoids are C_{10} compounds; sesquiterpenoids are C_{15} compounds, and so on through tetraterpenoids, which are C_{40} compounds.

Produced by over 2000 species of plants in 60 plant families, monoterpenoids are usually pleasant-smelling, as in the characteristic odor of the pine family (Pinaceae). Cells in the resin ducts of pine produce oleoresin, from which turpentine is made via steam distillation. Oils from citrus and *Eucalyptus* are also monoterpenes. When resin ducts are disrupted by herbivores the oleoresin, usually under pressure, spills out and is effective in "gumming up" the mouthparts of insects and even vertebrate herbivores.

The triterpenoids, C_{30} compounds, include the toxic and bitter cucurbitacins produced by the cucumber family (Cucurbitaceae), the cardiac glycosides described above, and the saponins found in over 70 plant families (Harborne 1993).

In summary, the diversity of both physical defenses and of secondary compounds is one line of evidence suggesting that herbivores have greatly influenced the evolution of plants. The chemical diversity of plants, in turn, has aided in the production of chemical defenses by animals. While some animals may have evolved their own chemicals, many are dependent upon their host plant for their defenses.

11.3 Constitutive versus induced defense

Plants usually produce a certain quantity of a chemical defense, a sort of background amount. This is called a **constitutive defense**. After a plant is attacked, however, the amount of these chemicals usually increases. In other cases, entirely new compounds are produced after an attack. Such a reaction, known as an **induced defense** to herbivore attack, can be thought of as a parallel to the immune system in animals.

According to Karban and Myers (1989), any change in a plant following herbivory can be thought of as an induced response. Such changes include not only allelochemical induction (Baldwin 1994), but also increases in physical defenses such as thorn length (Young 1987), spine density, production of trichomes, emission of volatiles that attract predators and parasites, reduction in plant nutritional quality for herbivores, and even increases in extrafloral nectar in plants protected by ants (Agrawal 2000). Induced defenses are not limited to plants. Phytoplankton, rotifers, ciliated protozoans, cladocerans, and even carp (*Cyprinus carpio*) are known to respond morphologically or behaviorally to the presence of herbivores or predators (Tollrian and Harvell 1999). Even marine bryozoans have been shown to have inducible physical defenses. Bryozoans produced spines, which were effective in reducing mortality, after being attacked by nudibranchs (Harvell 1984).

To qualify as an induced defense (or induced resistance), the response must result in a decrease in herbivore or predatory damage, and an increase in fitness must be demonstrated as compared to non-induced controls (Karban and Baldwin 1997). Harvell and Tollrian (1999) identified the following as conditions necessary for the evolution of an inducible defense: (i) the selective pressures should be variable and unpredictable, but sometimes strong; if the inducing species is constant, then permanent, constitutive defenses should be present; (ii) a reliable cue is necessary to activate the defense; (iii) the defense must be effective; (iv) the inducible defense must save energy as compared to a constitutive defense or no defense at all.

The basic hypothesis is that while defenses increase plant fitness when herbivores are present, the energy invested in such defenses results in lowered plant fitness when herbivores are absent (Agrawal *et al.* 1999). Inducible defenses allow an organism to invest in defense when necessary, but avoid costly allocations to defense when the herbivore or predator is absent.

In the 1980s Schultz and Baldwin (1982) and Rossiter *et al.* (1986) showed that defoliation by gypsy moth (*Lymantria dispar*) larvae stimulated oaks to increase the phenolic content of leaves. Rossiter *et al.* found a significant negative correlation between phenolic content of leaves and the size of female gypsy moth pupae. Assuming reproductive output of a female gypsy moth is positively correlated with size, the induction of a high phenolic content in oak leaves would be expected to eventually cause a decline in gypsy moth populations. However, a complication to this seemingly straightforward picture is the fact that another control agent, the gypsy moth nuclear polyhedrosis virus, is inhibited by high levels of phenolic compounds in oak leaves. Even though leaf phenolics depress gypsy moth reproduction, survivorship is enhanced due to suppression of the virus. The two potential control agents work in opposition to each other, making it increasingly difficult to predict the dynamics of gypsy moth populations (Foster *et al.* 1992). Furthermore, throughout the 1990s gypsy moth populations declined to very low levels, apparently due to a fungus. The moth populations in the eastern United States increased during the period 2000–02, but did not reached the population levels common in the early 1990s.

Agrawal (2000) exposed the leaves of the herbaceous plant *Lepidium virginicum* (Brassicaceae or mustard family) to herbivory by the larvae of *Pieris rapae* (Pieridae). Induced plants responded by producing more trichomes per leaf and increasing the glucosinolates (mustard oil) content of the leaves. While induction did not affect the feeding behavior of the specialist *P. rapae*, feeding by generalist caterpillars in the family Noctuidae was reduced.

In a related field study, Agrawal (2000) again induced a response by allowing leaves to be consumed by larvae of *P. rapae*. However, in addition to undamaged controls, he damaged leaves by clipping with scissors. In the field, aphids were important herbivores. Controlled herbivory by *P. rapae* induced resistance to attack by the aphids, but clipping with scissors did not. The number of aphids feeding on control and clipped plants averaged five per plant, while the average number on induced plants was just three per plant. Furthermore, plant survivorship was lowest in the clipped plants. Previous studies have suggested that the saliva of herbivores is necessary for successful defensive induction (Bodnaryk 1992, Mattiacci *et al.* 1995). **Plants therefore may respond differently to leaf losses from herbivores as opposed to leaf damage from storms or other physical causes.**

Other research has focused on the ability of plants to produce **proteinase inhibitors** that inhibit the major digestive enzymes of insects. For example, when attacked by herbivores, sagebrush (*Artemisia tridentate*) produces a compound known as jasmonic acid. Under the influence of jasmonic acid, tobacco (*Nicotiana sylvestris*), tomato (*Lycopersicon esculentum*) (Farmer and Ryan 1990, 1992), and alfalfa (*Medicago sativa*) plants all were induced to produce proteinase inhibitors. More recently it was found that injury to a plant tissue causes the production of a peptide hormone. The hormone stimulates the release of linolenic acid, a fatty acid common to plant cell membranes. Linolenic acid is then converted to jasmonic acid, which in turn stimulates proteinase inhibitors (Chen 1990). In another study, jasmonic acid stimulated the production of nicotine in tobacco plants (Ohnmeiss and Baldwin 2000).

Some induced responses to wounds are considered "systemic." This means that damaged plant tissue may transmit a signal to other areas of the plant, resulting in the induced reaction. Karban and Baldwin (1997) have outlined the requirements for a hypothetical signal to be taken seriously.

1 The signal must be rapidly generated at the wound site;
2 the inducer must be known;
3 the signal must travel through the plant in a time course consistent with the induced response;
4 the signal must stimulate the induced response at concentrations consistent with those known from damaged plants.

Signals that have been proposed include: oligosaccharide fragments from cell walls, systemin (a polypeptide), salicylic acid, ethylene, abscisic acid, jasmonic acid/methyl jasmonate, and electrical signals (Karban and Baldwin 1997). Of these, jasmonic acid and methyl jasmonate are the most likely signal compounds in that they meet the requirements listed above (Karban and Baldwin 1997). Jasmonic acid is derived from common fatty acids and it, along with its methyl ester relative (methyl jasmonate), are commonly found in plants. Both compounds elicit a multitude of responses in plants. Mechanical wounding increases the levels of jasmonic acid, which then move rapidly through the phloem.

Moreover, methyl jasmonate is very volatile. Minute concentrations of gaseous methyl jasmonate can induce the synthesis of proteinase inhibitors, as described above. Other chemicals involved in plant defense such as ethylene, systemin, and several alkaloids increase after plants are exposed to methyl jasmonate. The possibility that the gas methyl jasmonate is a signaling compound provides a potential mechanism for communication among plants (Karban and Baldwin 1997).

11.4 Plant communication

Damage to a plant may result in an induction of chemical defenses in neighboring plants. Several experiments have shown that volatile chemicals are released when plants are damaged. Experiments by Rhoades (1983) and Baldwin and Schultz (1983) suggested that plants which share the same air space with damaged plants increase the production of defensive chemicals even though they are not damaged themselves. In these early experiments, however, the effect was temporary and alternative explanations have been offered. Many scientists have questioned the results and, in other experiments, the evidence for communication among plants was negative (Karban and Baldwin 1999). As outlined above, methyl jasmonate was proposed to be the gaseous carrier, and a number of plant species have responded to this signal by producing proteinase inhibitors. Evidence is increasing, however, that volatile cues from damaged plant tissues may be used by herbivores, and by the predators and parasites of these herbivores, to locate these plants. However, it remains to be seen whether signals released by plants that are damaged are used by un-attacked neighboring plants, to induce defensive responses (Karban and Baldwin 1999).

11.5 Plant–parasitoid communication

Research shows that when a herbivore, such as a caterpillar begins to consume a leaf, the plant releases a volatile, "green leaf," odor that attracts parasitoids. These parasitoids are female wasps which sting the caterpillar, paralyze it, and lay eggs on it. The wasp eggs become larvae, which then consume the caterpillar. Technically, "green leaf" odors are simply mixtures of alcohols, aldehydes, and esters produced by oxidation of fatty acids from plant membranes. Such volatile compounds are released by mechanical damage to plant tissues and are relatively short-lived. But when a plant is infested with herbivores it produces a suite of other compounds such as monoterpenes, homoterpenes, and phenylpropanoids. These odors are long-lived and consistently attract predators. Salivary extracts from potential herbivores, or even from humans, can induce the production of volatile compounds different from those produced by simple mechanical damage (Karban and Baldwin 1999).

DeMoraes et al. (1998) found that both tobacco and maize (Zea mays) plants produce distinct volatile blends in response to damage by two closely related herbivore species (Heliothis virescens, the tobacco budworm, and H. zea, the maize earworm). Oral secretions by the herbivores trigger the production and release of several volatile terpenes. These terpenes are induced, that is, produced de novo in response to insect feeding. The chemical composition of these volatile products varies among plant tissues, species, varieties, and even cultivars! The parasitic wasp Cardiochiles nigriceps detects these odors and distinguishes among them to locate its host, H. virescens, as opposed to H. zea. This study showed not only that these chemicals attract parasitic wasps, but also that the wasps can distinguish among plants infested by different herbivore hosts. The interaction between plants and the natural enemies of their herbivores is surprisingly sophisticated.

11.6 A classic set of data reconsidered

As described in the Chapter 10, the British ecologist Charles Elton proposed that the data on pelt returns of snowshoe hare (*Lepus americanus*) and lynx (*Lynx canadensis*) by the Hudson's Bay Company provided support for the theory that periodical oscillations in populations are inherent in predator–prey interactions as predicted by the Lotka–Volterra equations. Many factors have been proposed to explain the hare–lynx cycles, including forest fires (Fox 1978). However, Bryant and Kuropat (1980) and Bryant *et al.* (1983) proposed that the hare cycle was the result of induction of chemical defenses in the Lare's primary winter foods, willow (*Salix* spp.), alder (*Alnus* spp.), and birch (*Betula* spp.). The chemical pinosylvin methyl ether, a toxic phenolic, deters feeding by snowshoe hares on alder. The chemical is present in the foliar buds and catkins, but not in the internodes. The hares will eat the internodes, but they are higher in fiber and lower in nutrients and carbohydrates as compared to buds and catkins. Willow, alder, and birch all have inducible chemical defenses. When the snowshoe hare damages twigs or leaves, the regrowth contains high quantities of phenolic resins and terpenes. Bryant *et al.* proposed that the decreased amount of palatable browse provides the hares with little high-quality food and results in population declines. Once the hare populations are low, the amount of damage to twigs and leaves is reduced enough that the induced defenses decline. The browse becomes more palatable and the hare population increases again. According to this reinterpretation, the lynx feed well at the peak hare population levels, especially since the hare are in poor condition due to lack of quality food. Lynx populations go up, but the major effect of the lynx is to accelerate the decline in a hare population already on the downswing.

Recently, however, Karban and Baldwin (1997) reviewed the literature and concluded that induced resistance cannot provide the single factor that is necessary and sufficient to produce cyclic dynamics. As was outlined in the previous chapter, Krebs *et al.* (1995, 2001) found that both food availability and predation work synergistically to produce the hare cycles. Karban and Baldwin (1997) feel that it is unclear at this point what role induced resistance plays in driving population cycles of herbivorous species.

11.7 Novel defenses/herbivore responses

Many plants store resins, latexes, gums, and mucilages under pressure in networks of canals around the plant. When these plants are damaged, the fluids are released from the injured tissues. In many plants the canals contain defensive compounds. The secretions also mechanically deter herbivores by interfering with their feeding since the secretions solidify when exposed to air. Sometimes insects are trapped in the secretions. Several species of Lepidoptera, Coleoptera, and Orthoptera circumvent the secretions by cutting trenches in the leaves, biting the veins, or even cutting the petioles. These species often have morphological adaptations for trenching or vein-cutting. For example, Dussourd and Denno (1994) showed that the cabbage looper (*Trichoplusia ni*) is able to cut a trench in leaves of *Lactuca serriola*, thereby disrupting the flow of latex to areas of the leaf distal to the trench. Not only is the cabbage looper able to feed on leaves distal to the trench, but so also is the yellow-striped armyworm (*Spodoptera ornithogalli*), a generalist herbivore. Lacking the ability to trench on its own, however, the armyworm is unable to feed successfully on plants with intact canals.

Becerra (1994) has reported on a squirt-gun defense of the succulent deciduous shrub *Bursera schlechtendalii*, found in Mexican deserts. This plant has simple leaves that produce terpenes distributed in a network of resin canals located in the stem cortex and throughout the leaves. When a leaf is cut, a syringe-like squirt of terpenes is released. The squirt travels 5–150 cm and persists for several seconds. Some leaves release large amounts of terpenes without the squirt response. This "rapid bath response" covers the entire surface of the leaf. Other leaves do not release resins when damaged. A specialized beetle (genus *Blepharida*, Chrysomelidae) feeds only on the leaves of this species. The larvae are able to sever the resin canals. They bite the mid-vein to eliminate the squirt or bath response. If larvae are bathed or squirted they withdraw from the leaf and try to clean themselves, remaining inactive for many hours.

Becerra (1994) showed that there is a great deal of variation in terpene response within the *Bursera* population. In most plants, 80% of the leaves released resins when damaged. Some had extremely high responses, with resin squirting from mid-veins and from lateral veins. The number of plants with high response increased throughout the season. Larvae growing on plants with high rates of response had higher mortality rates. Early-instar larvae are unable to sever veins, and feed by leaf mining. Sometimes they inadvertently rupture canals. When this happens they die, covered with resins. Larvae also grow at a slower rate on plants with a higher frequency of leaf response. High frequency of leaf response is correlated with leaf scars, probably indicating an induced response from a prior attack. The frequency of vein-cutting behavior is also associated with frequency of leaf response.

In summary, canals can be effective barriers against non-specialized insects. Even specialized trenching insects have higher mortality on highly responsive plants. It can take 1.5 hours for a larva to deactivate a resin canal of a leaf. The consumption of a leaf after disarming takes only 10–20 minutes. Thus there is high handling cost. Meanwhile, the plant's response is not fixed but depends on water potential and plant health.

11.8 Detoxification of plant compounds by herbivores

Animals are by no means helpless in the face of plant secondary compounds. Both invertebrate and vertebrate herbivores have evolved detoxification mechanisms. There are basically three parts to detoxification. Since these toxins are usually attracted to lipids (lipophilic), the first step is usually for the herbivore to convert the compound to a water-soluble form, usually an alcohol, so that it can be excreted. So-called phase I enzymes are found in the livers of vertebrates and involve cytochrome P450. These enzymes are capable of acting on a wide variety of chemicals and are known as PSMOs (polysubstrate mixed-function oxidases), or simply mixed-function oxidases (MFOs). PSMOs are known from many groups of animals. In insects they are found in the gut, fat bodies, and Malpighian tubules. In addition to oxidation, other reactions such as reduction and/or hydrolysis may occur (Harborne 1993).

In the second stage of detoxification, the metabolite from stage I is conjugated with a sugar or sulfate anion. This means that the compound from stage I is united with another molecule to form an inactive, non-toxic, product. These reactions require energy. Therefore, although an herbivore may not suffer damage from ingesting plant material containing plant toxins, the detoxification process is an energy drain. In the last stage, the conjugated chemical is excreted through bile or urine. There is a great deal of variation in the detoxification process, both between and within species. For example, Millburn (1978) found that benzoic acid is excreted as a conjugate of glycine in mammals, amphibians, fish, and insects, as an ornithine conjugate in birds and reptiles, and as an arginine conjugate by arachnids. As pointed out by Harborne (1993), 14% of the British people cannot metabolize the alkaloid betanin, the red pigment in beetroot (*Beta vulgaris*). The result is red-colored urine after ingesting beets (beeturia). The major point to bear in mind is that herbivores are capable of detoxifying defensive chemicals, but there are high metabolic costs in doing so.

11.9 Plant apparency and chemical defense

A general theory developed by Feeny (1976) stated that the amount and the type of defense a plant has evolved is dependent on the likelihood of its being found and consumed by herbivores. His hypotheses are as follows:

1 Plants which are large, woody, persistent, have clumped distributions, and which are abundant are "bound to be found" by herbivore populations. Feeny called such species "**apparent**," and asserted that they would have a relatively large investment in chemical defenses. Such plants produce chemicals such as phenolics that reduce the ability of all herbivores to digest the plant parts that contain them. Such "digestibility-reducing compounds" often bind to proteins in such a way that herbivores have great difficulty eating, swallowing, and digesting this vegetation. These compounds are more effective in larger doses and take more time and energy to produce than some other defensive compounds. Thus Feeny termed them "**quantitative defenses**."

2 "**Unapparent**" plants are small, herbaceous, and ephemeral, have highly dispersed distributions, and/or are rare. If toxins are present they are effective at low concentrations, are rapidly produced, and require less total energy in their formation. Alkaloids and glycosides are examples of such compounds, termed "**qualitative defenses**" by Feeny. These compounds are effective in deterring generalist or non-adapted herbivores. They are subject to counter-adaptation and often are consumed by host-specific herbivorous insects. *r*-selected plants, which exist in temporary environments, where growth rates can be very high due to lack of competition, primarily use the **qualitative** defenses, while *K*-selected plants should favor the **quantitative** defenses. *K*-selected plants normally use carbon-based defenses (especially phenolics). This may be because "apparent" plants are often members of a community in which nutrients eventually become limiting to growth. On the other hand, defenses found among *r*-selected species include nitrogen-based alkaloids, glycosides, and carbon-based coumarins.

3 Defenses vary with time in plant parts. New leaves are ephemeral and somewhat unapparent, in that their characteristics change. Young leaves are
usually less tough, higher in water and protein content, and lower in defensive compounds, particularly of the quantitative type (Feeny 1970, Rockwood
1976). As a leaf ages it increases its quantitative defenses, becomes tougher
through changes in fiber or lignin content, and its protein and moisture contents drop. At the other extreme, as leaves age they often become senescent
and more susceptible to herbivore or fungal attack (lilac, *Syringa vulgaris*, leaves,
for example). As explained in the section on optimal defense theory, plants
invest chemical defenses in the most valuable tissues. Nitrogen-based
defenses are withdrawn from less valuable tissues. Thus new leaves have higher
amounts of qualitative defenses than older leaves. For example, in *Coffea
arabica* leaves, purine alkaloids are present at those periods when the plant
is most susceptible to herbivory. During leaf development the concentration
of the alkaloid increases to 4% of dry weight in new leaves. The rate of synthesis decreases as the leaf matures. At senescence the leaf contains only
traces of the alkaloids (Harborne 1993).

11.10 Soil fertility and chemical defense

Janzen (1974) observed that tropical plant communities growing on nutrient-poor white-
sand soils always had fewer herbivores than nearby plant communities based on fertile
soils. He hypothesized that plants growing on infertile soils would have a difficult
time replacing tissues lost to herbivores, due to their slow growth rates. Therefore, the
permanent or "apparent" plants found on such soils should be heavily protected by
quantitative defenses. Indirect evidence in favor of this hypothesis is that rivers draining
such white-sand environments contain tea-colored waters. These brown- or black-water
rivers contain large amounts of phenolic compounds that have leached from the surrounding plant communities. The famous Rio Negro of the Amazon basin drains such a
white-sand area.

McKey and his colleagues (McKey *et al.* 1978) tested this idea in Africa. Plants from a
white-sand region in Uganda had twice the phenolic compounds as compared with plants
from a nutrient-rich soil in Cameroon.

Finally, Coley and others (Coley 1980, Coley *et al.* 1985) have shown that plants not
limited by a physiological stress of some kind are fast-growing, have higher leaf protein
contents, shorter leaf lifetimes, higher herbivory rates, lower amounts of defensive
metabolites, and higher turnover rates of defenses as compared with plants that grow slowly.
Furthermore, herbivores, especially generalist herbivores, have a definite preference for the
fast-growing plant species.

11.11 The optimal defense theory

The distribution of defensive compounds within a plant is perhaps best explained by the
"optimal defense theory." This theory predicts that tissues of high fitness value, specifically those whose contributions to fitness lie in the future, will be better defended than
tissues of less value. This theory assumes that defense is costly in terms of energy and

nutrients consumed, and that trade-offs exist between defense and other functions such as growth and reproduction. Furthermore, the theory predicts that within-plant allocation of defenses will be a function of the relative fitness values of the plant parts, as well as their probability of being attacked. In most cases the youngest leaves and the tissues associated with reproduction have the highest fitness values and are well protected by secondary compounds such as alkaloids and glycosides.

Ohnmeiss and Baldwin (2000) tested the optimal defense theory using tobacco plants. A great deal of research has been done on tobacco, and the plant has several characteristics that are ideal for this type of research. First, it is known that this defense is costly. Nicotine is one of the most energetically costly secondary metabolites known, and after induction a plant produces enough nicotine that it consumes 5–8% of the plant's nitrogen budget (Baldwin *et al.* 1994). Second, the mechanisms for inducible nicotine defense are known. Herbivory and leaf wounding induce jasmonic acid (JA) synthesis in the leaves at the wound site. JA is then transported from the wound site via the phloem to the roots, where nicotine synthesis occurs. Nicotine is subsequently transported to the shoot via xylem. Third, since research has already established the physiological responses of the plant to nitrogen limitation, it is possible to test whether changes in leaf value are mirrored by changes in the distribution of the defense. Finally, nicotine is used as a proxy for energy devoted to defense, although it is likely that the plant produces proteinase inhibitors and other defensive metabolites.

Ohnmeiss and Baldwin wounded or removed leaves of different ages (young, mature, old) and determined the effects of nicotine allocation and fitness (viable seed production). Results were: (i) Leaf removal at the elongation stage of growth results in a significant decrease in seed production, but leaf removal at the rosette or flowering stages does not. Apparently the plants can compensate, and damage to rosette leaves can be replaced. By the time of flowering, plants have already allocated resources for seed production. Within the elongation stage, the least amount of seed production was lost when old leaves were removed, but significantly fewer seeds were produced if young or mature leaves were lost. (ii) The relative value of leaves decreases from young and mature to old leaves. (iii) Leaf damage significantly increases the whole-plant nicotine contents of rosette-stage plants, but not of later stage plants. (iv) After damage, younger leaves are more heavily defended than older leaves at the elongation and flowering stages. (v) Under nitrogen limitation, tissues of the highest value are defended at the expense of other leaves. Overall, Ohnmeiss and Baldwin (2000) concluded that plants distribute nicotine among leaves in accordance with their fitness value, supporting the optimal defense theory.

11.12 Modeling plant–herbivore population dynamics

Most plant–herbivore models are for **grazers**. As pointed out by Turchin (2003, p. 112), a grazer is a "consumer that scores low on both intimacy and lethality." That is, models are usually based on generalist herbivores, which do not usually kill their resources, and which consume small amounts from many different individuals and species of plants. Specialist herbivores, especially sucking insects, are functionally parasites and score high on the "intimacy" scale. As we will see, grazer–vegetation models tend to be similar to predator–prey models when they focus on the quantity of vegetation harvested by grazers. In these models we assume that plant quality does not change in spite of attack by herbivores. In another approach, the assumption is made that plant quality is the variable, while quantity is held constant.

Crawley has summarized several possible models for vegetation–herbivore population interactions of the first type (the quantity of the vegetation is the variable). To begin, we can use the basic ideas gleaned from predator–prey models described in the previous chapter. Density-independent plant growth can be described as $r_v V$, where r_v is maximum growth (equivalent to the intrinsic rate of increase) and V represents plant abundance or biomass.

For density-dependent plant growth we can simply use the logistic equation:

$$dV/dt = r_v V \frac{K_v - V}{K_v}$$

where K_v is a carrying-capacity term for the plant population.

We then subtract plant losses due to herbivore feeding. This is equivalent to the herbivore functional response and can be of several forms. Since herbivores are not usually killing the vegetation, we measure the grazer's functional response in terms of amounts of biomass removed. We also have to be careful how we define V for the herbivore since most herbivores cannot eat entire plants, but specialize on leaves, stems, buds, etc. Recall from the last chapter that we used E as a coefficient measuring the efficiency of predation. Here F will be used to measure the efficiency of herbivore removal of vegetation biomass. If the herbivore functional response is simply modeled as FNV, where N is the number of herbivores, we have the equivalent of the Lotka–Volterra equation for prey. This is essentially a type I functional response, linear and unlimited, meaning that there is no handling component, and that each herbivore has an unlimited appetite. However, we can add either a type II or type III functional response as we did in Chapter 10. The type II functional response would be:

$$\frac{FNV}{1 + Fh_2 V}$$

where h is the handling-time component. The type III functional response is:

$$\frac{FNV^2}{1 + Fh_3 V^2}$$

and finally, the functional response with a half-saturation constant for the herbivore–plant interaction is:

$$\frac{fNV}{b + V}$$

The term f equals the maximum consumption or grazing rate when the search and capture components have been minimized, and b equals half of the maximum consumption rate.

Using the functional response with the half-saturation constant, the plant growth equation becomes:

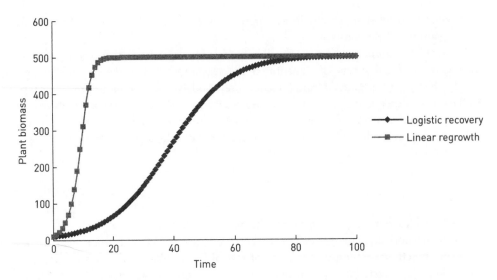

Figure 11.1 Vegetation recovery from grazing: logistic recovery versus linear regrowth.

$$dV/dt = r_v V \left(\frac{K_v - V}{K_v} \right) - \frac{fNV}{b + V} \tag{11.1}$$

Before we continue by modeling the herbivore or grazer population, we should stop and consider a point raised by Turchin (2003) and others. Using the logistic for growth of the vegetation assumes that when V is close to zero, growth rate is rather slow at first, then accelerates until $V = K_v/2$, before slowing down to zero at K_v. The problem with this is that many plants, especially grasses, store 80–90% of their biomass underground (Wielgolaski 1975). When we are modeling V, we are usually only measuring above-ground vegetation (e.g., leaves and stems). As Turchin points out, when re-growth occurs after an herbivore attack, the initial re-growth pattern is more likely to follow a linear path to K, as opposed to a logistic pattern (Figure 11.1).

Using this logic, the equation describing re-growth in the absence of herbivore losses simplifies to Equation 11.2:

$$dV/dt = u_0 \left(1 - \frac{V}{K_v} \right) \tag{11.2}$$

where u_0 represents plant growth rate when V is close to zero, and V specifically represents above-ground biomass.

Originally proposed by Turchin and Batzli (2001), this is known as the **linear initial re-growth** or simply the **re-growth** model. In contrast to the logistic model, V increases linearly to K with no initial slow-growth phase. Turchin (2003) points out that both models are simplistic, but different vegetation systems are reasonably characterized by one of these two models. For example, when lemmings feed on mosses, the logistic best describes the interaction since most of the moss biomass is vulnerable to direct consumption. On the other hand, grasses and sedges are best modeled with the linear re-growth equation, in that most of the biomass is underground and escapes herbivory. The exception would be if herbivores such as root voles (*Microtus oeconomus*) consume rhizomes and other underground parts (Pitelka 1957, Tast 1974).

The herbivore population is modeled with growth based on a positive numerical response to increases in plant net production, coupled with a density-independent or density-dependent mortality rate. The numerical response can be of several forms. Following the Lotka–Volterra format for the predator population, with no limits based on a functional response, the numerical response is simply $\chi_h fNV$, where f is maximum grazing rate and χ_h is the herbivore's assimilation rate. The product $\chi_h f$ describes the maximum rate at which plant material is turned into herbivores, and is the equivalent of $\chi_p c$ in Equation 10.16. We can again add either a type II or type III functional response to the numerical response. Alternatively, we add the numerical response based on the half-saturation constant (Equation 10.16):

$$\frac{\chi_h fNV}{b + V}$$

The herbivore death rate can simply be a density-independent constant, m_h. Or we can add a coefficient, θ, which is density-independent when equal to one, but which increases the herbivore death rate at high densities if $\theta > 1$. The resultant equation (11.3) describes the dynamics of the herbivore population with a functional response and a density-dependent death rate:

$$dN/dt = \frac{\chi_h fNV}{b + V} - m_h N^\theta \tag{11.3}$$

Crawley (1997) makes the interesting point that functional responses often work more in terms of the amount of food per herbivore as opposed to the amount of food per area. This point has also been made for predator–prey relationships. Therefore the functional response might be more appropriately modeled as **ratio-dependent**. Using Equation 11.3, if we assume $\theta = 1$ and the functional response is dependent on the ratio of V/N, we have:

$$dN/dt = \chi_h fN\left(\frac{V}{N}\right) - m_h N \tag{11.4}$$

If we assume both populations of herbivores and plants have stopped growing, they reach an equilibrium when $dN/dt = 0$ in Equation 11.4. By canceling the Ns, the result $\chi_h fV = m_h N$ can be solved for V^* (Eqn. 11.5), which is the equilibrium plant abundance. V^* turns out to be directly proportional to the herbivore equilibrium population size, N^*, and the herbivore death rate, but inversely proportional to herbivore foraging and assimilation efficiency. Obviously high plant abundance is favored when herbivore foraging efficiency is low and/or the herbivore population has a high death rate. Less intuitively obvious is the positive relationship between herbivore equilibrium population size and plant abundance. In fact, if we solve Equation 11.5 for N^*, we have Equation 11.6, which implies that an increase in vegetation results in an increase in herbivores. But Equation 11.5 seems to say that an increase in herbivores leads to an increase in vegetation biomass. As pointed out by Turchin (2003), these equations, which are the equivalent of the Rosenzweig and MacArthur model in Chapter 10, lead to the so-called "paradox of enrichment" in which the system is highly unstable when there is an abundance of vegetation.

$$V^* = \frac{m_h N^*}{\chi_h f} \tag{11.5}$$

$$N^* = \frac{\chi_h f V^*}{m_h} \tag{11.6}$$

If we replace the logistic equation for vegetation growth with the linear re-growth equation, we have the herbivore/re-growth model of Turchin and Batzli (2001) (Eqns. 11.7 and 11.8). The herbivore equation is based on Equation 10.16b, in which χ_h is again the assimilation efficiency of the herbivore, f represents the maximum grazing rate and b is the half-maximal grazing rate by the herbivores. The parameter μ_h is the minimum food intake necessary for the herbivore to survive, and u_0 represents plant growth rate when V is close to zero:

$$dV/dt = u_0\left(1 - \frac{V}{K_v}\right) - \frac{fNV}{b+V} \tag{11.7}$$

$$dN/dt = \chi_h N\left(\frac{fV}{b+V} - \mu_h\right) \tag{11.8}$$

According to Turchin (2003) this model produces stability, mainly because the logistic model has an inherent time lag built into it, while the linear model does not. Also, the fact that most of the biomass of the plants is protected below ground means that the vegetation has a refuge, invulnerable (normally) to grazing. As we saw in Chapter 10, the presence of a refuge has a stabilizing effect on a consumer–resource system.

To this point we have exclusively focused on the **quantity** of the vegetation available to grazers, but based on the theory of inducible defenses discussed above, the dynamics of herbivore–plant interactions include changes in the **quality** of the resource for the grazer. By preferentially grazing on better-quality plants and plant parts, and by inducing defenses, heavy grazing by an herbivore population results in lower–quality resources for itself.

Edelstein-Keshet (Edelstein-Keshet 1984, Edelstein-Keshet and Rausher 1989) developed a model that includes plant **quality** as a variable. She assumed that, without herbivory, plants devote less energy to defenses and overall plant quality increases. By contrast, grazing tends to decrease plant quality for herbivores. A simplified form of her models is presented below (from Turchin 2003):

$$dQ/dt = q - fQN(N-n) \tag{11.9}$$

$$dN/dt = r_h N\left(\frac{Q/q - N}{Q/q}\right) = r_h N\left(1 - \frac{qN}{Q}\right) \tag{11.10}$$

In this model, Q is average plant quality, N is herbivore density, n is a specific density of herbivores, and r_h is the maximum growth rate of the herbivore population. Equation 11.9

assumes that in the absence of grazers, Q increases at the constant rate, q. The second term represents the loss of quality due to the presence of herbivores. The phrase $(N - n)$ describes a tipping point. When $N < n$, the plant quality is increased faster than in the absence of herbivores (see below). When $N > n$, plant quality is decreased by grazing. Equation 11.10 is simply the logistic equation for the herbivore, but the carrying capacity of the herbivore depends on average plant quality (Q/q).

This model predicts that an increase in grazing leads to dampened oscillations leading to a stable point. Furthermore, the interactions produce lower average plant quality when grazing intensity is high.

Turchin (2003) developed a discrete-time model for herbivore–plant interactions that encompasses plant quality. He then applied this model to the interaction between larch trees (*Larix* sp.) and the larch budworm (*Zeiraphera diniana*) in the Swiss Alps. The moth population displays regular oscillations, which cover five orders of magnitude. Time-series data are available from 1959 through 1977 for this moth population. As Turchin noted, plant quality is currently the "reigning" explanation for the larch budworm oscillations. When he analyzed the time-series data, however, he found that plant quality explained only 31% of the variance in the budworm populations. Secondly, he found that parasitoids (ichneumonid wasps) likely played a role in determining larch budworm rates of change. Finally, Turchin (2003) has found that a tritrophic model combining both plant quality and parasitism is superior to any single explanation. His conclusion mirrors what we have found previously: a model involving three trophic levels, combining plant quality and parasitism, did the best job in matching the observed dynamics of the larch budworm populations.

Tritrophic (three trophic level) models can be traced to Oksanen *et al.* (1981), who combined the plant–herbivore and predator–prey models of Rosenzweig and MacArthur into a three-equation set. Turchin proposed that the vegetation model be based on his herbivore–plant re-growth equations described above (Eqns. 11.7 and 11.8). The herbivore (prey) growth rate is diminished by the term $cNP/(d + N)$ as described in Chapter 10. r_h is the maximum growth rate of the herbivore (prey) population, while χ_p is the predator's assimilation rate. The predator equation is identical to Equation 10.16b, but Turchin has added a predator self-limitation term, $\left(\dfrac{s_0 P^2}{k}\right)$. s_0 and κ represent the rate of increase and a carrying capacity due to territoriality.

$$dV/dt = u_0\left(1 - \frac{V}{K_v}\right) - \frac{fNV}{b + V} \tag{11.11}$$

$$dN/dt = r_h N\left(\frac{\chi_h fV}{b + V} - \mu_h\right) - \left(\frac{cNP}{d + N}\right)$$

$$dP/dt = \chi_h P\left(\frac{cN}{d + N} - \mu_p\right) - \left(\frac{s_0 P^2}{k}\right)$$

In summary, (i) the addition of self-limitation terms adds stability to both plant–herbivore and predator–prey systems; (ii) modeling producers with the re-growth, rather than the logistic model, also produces a more stable outcome (Turchin 2003).

11.13 Conclusions: the complexities of herbivore–plant interactions

Herbivores can affect individual plants, plant populations, and plant communities in subtle, complex, and unexpected ways.

Distribution. Herbivores have the ability to limit the distribution of plants geographically or by habitat. Reef fishes limit the distribution of turtle grass (*Thalassia testudinum*) to areas away from the protection provided to the fish by the reefs. Randall (1965) observed a conspicuous band of bare sand averaging 10 m in width that separated coral reefs from beds of sea grasses (*Thalassia testudinum* and *Cymodocea manatorum*). The klamath weed (*Hypericum perforatum*) introduced to California from Europe, became so widespread and common that it was considered a pest. The beetle *Chrysolina quadrigemina*, which feeds upon it in Europe, was introduced. This introduction, a successful case of biological control, limited the klamath weed so well that it is now found only in shady moist areas where the beetles do not reproduce (Holloway 1964).

Attack by other organisms. Although attack by herbivores may result in an increase in induced defenses, heavy attacks can leave the plant in such a weakened condition that it becomes subject to further attack. Rockwood (1974) described one variation on this theme. The calabash tree (*Crescentia alata*) was defoliated by hand in order to examine the effects of defoliation on reproduction (see below). At the time of defoliation it was noted that mature leaves had very little herbivore damage, but a beetle was consuming newly produced leaves at the top of the tree. After the hand defoliation the entire tree produced a crop of new leaves. Beetles flew in from a wide geographic area and began feasting on the new leaves. The tree was completely defoliated again, while neighboring control trees, full of mature leaves, were untouched by the beetles. The trees actually flushed a crop of leaves again and were once again defoliated by the beetles. This cycle of destruction ended when the beetles went dormant in the dry season.

Effects of herbivores on productivity and reproduction. One of the tenets of plant–herbivore theory is that herbivore activity has a negative effect on plant growth and reproduction. Such an obvious relationship is, however, more complex than it appears. Several authors have proposed that herbivory, up to a certain level, may be beneficial to plant growth and reproduction. Let us begin with the common recommendation that roses and fruit trees are kept productive by a constant pruning. Next we should note that grasslands are kept at high productivity by either allowing grazing by herbivores, or by a consistent mowing. Finally, Paige and Whitham (1987) and Whitham *et al.* (1991) found that when the scarlet gilia (*Ipomopsis aggregata*), an herbaceous plant from the southwestern United States, was browsed such that the apical meristem was destroyed, it responded by producing multiple inflorescences and up to three times as many flowers, fruits, and seeds. In all of these studies the plants appear to compensate for herbivore damage through re-growth. In some cases the re-growth appears to be great enough to surpass the growth in the un-browsed plants. This is known as **overcompensation**.

Is overcompensation possible, and is it real? In theory, low levels of herbivory can be beneficial to plants such as monocots, in which the meristems are near ground level, for the following reasons: (i) herbivores remove old, non-productive tissues while allowing light to penetrate to the ground where new leaves are found; (ii) herbivores speed up the cycling of mineral nutrients (old leaves are consumed and deposited on the ground as manure, which is quickly broken down by fungi and bacteria, releasing mineral nutrients); (iii) the saliva of herbivores contains growth-promoting substances or encourages an induced defense that protects the plant from further herbivore attacks. McNaughton (1986) endorsed overcompensation, asserting that plants have the capacity to compensate for herbivory and overcompensate for damage so as to increase fitness. Crawley (1997), however, stresses that there is actually no evidence that herbivory can increase Darwinian fitness. He believes that overcompensation has yet to be proven from a well-designed, controlled experiment. In grasses, for example, although above-ground production is maximized by frequent mowing or grazing, it is at the expense of energy storage in roots, or results in suppression of flowering and fruiting. In his critique of Paige and Whitham (1987), Crawley (1997) finds it likely that their result was due to large plants being allocated to the grazed group and small plants to the ungrazed controls.

If overcompensation does occur, the following factors must be explored: (i) timing of herbivory: if a plant is browsed early in the growing season, there is ample evidence that many plants are able to recover; (ii) nutrient, water, and light availability: overcompensation can only occur if the plants are not stressed by a lack of resources needed for photosynthesis; (iii) competition: if there are few competitors in the area, then a plant can recover, again assuming adequate resources are available; under heavy competitive stress, however, a plant that is damaged by an herbivore will lose its position in the community to other plants that are not damaged; (iv) type of tissue lost: as seen in the paper by Ohnmeiss and Baldwin (2000), damage to certain tissues, such as new leaves, is of much greater importance to plant productivity than is damage to other tissues.

Certain plant species have little potential for compensatory responses. For example, species with a physiology that limits new growth, or that live in resource-limited environments, will be severely affected by herbivore attacks. On the other hand, species such as *r*-selected annuals and perennials, with rapid growth rates and a physiology that allows a rapid response to herbivore damage, are likely candidates for overcompensation. Long-lived, woody perennials, however, are unlikely to easily recover from severe herbivore damage.

High levels of herbivore damage severely limit reproduction in many species. Rockwood (1973) showed that heavy defoliation virtually eliminated reproduction in six species of tropical shrubs and small trees. Subsequently, Marquis (1984, 1992) and Rockwood and Lobstein (1994) have demonstrated a graduated response to differing levels of herbivory. At low levels of herbivory, reproduction is little reduced. However, the timing of the defoliation, its local intensity, and the amount of competition from other plants all modify the reproduction responses to defoliation. Less than 50% defoliation had little effect on herbaceous plants in northern Virginia (Rockwood and Lobstein 1994), and the effects of defoliation were often expressed a year later. Marquis (1992) discovered that a given branch of the tropical shrub *Piper arieianum* suffered an 80% reduction in seed production from a mere 10% leaf removal when it was concentrated on that one branch. Finally, when the plant *Abutilon theophrasti* was grown at low densities, up to 75% defoliation had no effect on reproductive fitness. At high densities, however, the same amount of defoliation reduced reproduction by 50% (Lee and Bazzaz 1980).

Community-level effects. Certain herbivores and predators have been described as **keystone species**. The mere presence or absence of such species is critical to community organization and ecosystem functioning. A simple example is the effect of elephants (*Loxodonta africana*) in East Africa. Because of their browsing activities and their ability to destroy even the largest trees, a large population of elephants can convert shrub land into a habitat dominated by grasses. Conversely, if elephants are removed from an area it may change back to heavy brush. Darwin discovered that the grazing of cattle on the English heath prevented forests from being established. Upon close examination, under the heath stems he found small fir trees, one of which was 26 years old according to its growth rings. When this common land was enclosed, ending a tradition dating to the Middle Ages, the heath quickly converted to forest (Kingsland 2004). The introduction of Nile perch (*Lates niloticus*) into Lake Victoria and a species of bass into Lake Gatun in Panama virtually eliminated many species of smaller plankton-feeding fish. This resulted in increases in zooplankton populations, higher consumption rates of phytoplankton, and a decline in overall productivity in the lakes. In some cases a group of ecologically related species, known as a guild, can be considered as keystones. Brown and Heske (1990) reported that eliminating three species of seed-eating kangaroo-rats in the Southwestern United States led to an increase in large-seeded winter annuals. Rescued from rodent predation, the large-seeded species eventually out-competed the small-seeded annuals.

Other rodents such as prairie dogs (*Cynomys ludovicianus*) can have complex effects on the plant community. In Texas, Weltzin (1991) found that the elimination of prairie dogs was usually followed by an increase in the shrub mesquite. He excluded cattle from an area containing a prairie dog colony. The rodents removed pods and seeds from mesquite and stripped bark from young plants. Such activities help reduce mesquite establishment from around their colonies. He concluded that elimination of prairie dogs in the past had allowed mesquite to spread throughout the cattle ranges.

On the other hand, at Wind Cave National Park in South Dakota, it was shown that prairie dogs favored the establishment of herbaceous dicots over grasses. With no prairie dogs present the herbaceous community consisted of 87% grasses and 13% herbaceous dicots. With prairie dogs present it shifted to 47% grasses and 53% herbs.

Paine (1966) carried out the classic study on keystone predators. In the rocky intertidal zone off the coast of Washington state there were 15 species of coexisting invertebrates. The dominant predator was the starfish *Pisaster*. The community consisted of species of chitons, limpets, bivalves, barnacles, and a marine snail. Paine experimentally removed starfish from half of the experimental areas. In those areas where the starfish was removed, the bivalve *Mytilus* and the barnacles became the dominant competitors. They crowded out several other species and the community declined to eight coexisting species. The starfish, when present, preyed consistently on the dominant competitors, preventing them from crowding out the other species.

Paine's work lead to the general hypothesis that predators restrict populations of competitively dominant species and allow coexistence of a greater number of species than would occur in the absence of the predators. This is known as the "top-down" control of communities. As has been recently shown by Robles and Desharnais (2002), this view was too simplistic. A variety of factors, including the interplay of the physical environment with prey refuges, prey dispersal, and prey production determine prey populations and community structure. Nevertheless, Paine's classic work was an important milestone in the history of ecology.

The relationship between herbivores and plant diversity is muddled. Though some studies have shown an increase in plant diversity with herbivory (Belsky 1992), some have shown a decrease (Milton 1940), and still others no effect whatsoever (Crawley 1989). Again, we must stress that herbivore populations are themselves affected by their own predators and parasites (Hartley and Jones 1997). Thus, the control herbivores might have on plant communities can be neutralized by top-down forces. Herbivores themselves, of course, are most likely controlled by a combination of bottom-up and top-down forces.

Multiple-trophic-level effects. As mentioned in Chapter 10, tritrophic interactions involving plants, herbivores, and predators can produce "trophic cascades" by which predators affect prey populations to the extent that the herbivore–plant interactions are fundamentally altered. Predators can thereby influence plant productivity and community composition (Marquis and Whelan 1994). Evidence for trophic cascades includes the reintroduction of wolves (*Canis lupus*) to Yellowstone National Park described in Chapter 10. In other cases, the effect of predators on herbivore–plant interactions has been demonstrated by experimentally removing predators. In one such experiment, Marquis and Whelan (1994) caged white oak (*Quercus alba*) saplings to eliminate bird predation on insect herbivores over a two-year period. Birds were allowed free access to control plants and to a third treatment in which they used insecticides to estimate plant growth with minimal insect damage. In the first year, caged plants suffered 25% leaf area loss, as compared to 13% in plants where birds preyed upon insect herbivores. Sprayed plants suffered only 6% leaf damage. Figures in the second year were 34%, 24%, and 9%, respectively. Prior to this study, most examples of trophic cascades (the effect of predators and parasites on plant productivity and composition) had come from aquatic ecosystems. For terrestrial forests, these results mirror those of Turchin (2003) (larch budworms and their parasitoids) described above. In this case, however, we are dealing with the entire insect herbivore community and its interaction with the insect-consuming bird community.

In conclusion, herbivores have had significant effects on the evolution of plants and continue to exert considerable selective pressure on plants today. Yet herbivores themselves have been under "the gun" from their own predators, so to speak, in both contemporary and evolutionary time. Several lines of evidence tell us that the most effective way to investigate herbivores is to evaluate both their food resources and their predators. Of course the roles of herbivores in an ecosystem are also undoubtedly affected by the physical environment, including the soil and the local climate.

In this book we began with simple growth models for single populations. These models became increasingly complex as we added time lags and stochastic effects. In the second half of the book we began with relatively simple competitive interactions involving only two species. We then moved on to consider interactions between different trophic levels, again starting with simple two-species models (one predator, one prey). But you should now realize that all interspecific interactions must be analyzed in the context of the entire community of organisms, not simply in terms of one competing or one predatory species. Plants are attacked by a multitude of herbivores; herbivores exist in a community of both competitors and predators; and predators themselves are bedeviled with parasites while they compete with other predators of their own species, as well as with individuals of other species. All of this takes place on a complex physical landscape and in an ever-changing climate. Attempting to understand, much less model, these complexities should keep population ecology (indeed all types of ecology) fresh and challenging for a long time to come.

Appendix 1

Problem sets

Here are six exercises to help you see how some of the population models in the book work in practice. The first five relate to Chapter 1, the sixth to Chapter 4. Answers have been provided for problem 6 only.

Problem 1

You survey an annual insect and find 5000 females per acre. One year later, you census the population and find 6000 females per acre. What is the net replacement rate, R, for this population? What size will the population be three years from the original census if the population continues to grow at the same rate? Five years later? Ten years later? In what year would the population reach 100,000?

Problem 2

The birth rate for Asia in 1982 was 30 per thousand, while the death rate was 11 per thousand. What was the intrinsic rate of increase, assuming a stable age distribution? If the population size was 2.67 billion, what was the projected population in 2004? Between 1982 and 2001 the population increased from 2.67 billion to 3.72 billion. What was the actual r during that time period? Given this r-value, what is the doubling time? What is λ?

Problem 3

In 1995 the human population of the world was 5.7 billion. The population of China was 1.2 billion. If its birth rate is 17 per thousand and its death rate is 10 per thousand, what was China's intrinsic rate of increase? In what year would the population of China exceed the 2001 population of the world (6.137 billion) if it continued to grow at the 1995 rate?

Problem 4

In 1981, the world human population was 4.5 billion. The birth rate was 28 per thousand (= 0.028 per capita) and the death rate, 11 per thousand (= 0.011). Thus the r-value was 0.017. Using these figures, project the population for 20 years to 2001. What was the expected human population based on those figures? Compare this to the actual human population of 6.1 billion in 2001. Now project the population 119 years to the year 2100. What is the projected human population for the year 2100 based on the 1981 figures? How does this compare to the 11 billion projected by the UN and Lomborg (2001)?

Problem 5

A yeast population grows as shown in the table below. Time is measured in hours. Find the r_{max} (per hour) and r_a (actual r over the entire time period) for this population.

Time in hours	N (cells per mL)	ln N	ln[(K − N)/N] Use K = 320
0	10		
1	20		
2	40		
3	75		
4	140		
5	210		
6	273		
7	300		
8	310		
9	315		
10	318		
11	319		
12	317		
13	318		
14	319		
15	315		
16	319		

Hint: Find the natural log of N and complete column 3. Then find the natural log of $(K − N)/N$ and then complete column 4. Prepare graphs of columns 2, 3, and 4 versus time. For r_{max} find the slope during the exponential phase of growth (hours 0–4, for example). For r_a, find the slope of the values computed in column 4 versus time.

Problem 6

A population has the life table shown below. Find the gross reproductive rate (GRR), the net reproductive rate (R_0), p_x, and q_x. The p_x and q_x columns have been filled in for you. Would r be predicted to be positive, negative, or zero? Find r, λ, the stable age distribution, the life expectancy by age class, and reproductive value by age class. Given the population by age class at time zero, project this population one year into the future. Repeat this process, using the Leslie matrix.

Age, x	l_x	m_x	p_x	q_x	n_x at $t = 0$	n_x at $t = 1$
0	1.00	0	0.40	0.60	1000	900
1	0.40	1.0	0.50	0.50	300	400
2	0.20	3.0	0.25	0.75	100	150
3	0.05	2.0	0	1.00	60	25
4	0	0	–	–	0	0
Σ		6.0			1460	1475

$GRR = \sum m_x = 6$
$R_0 = \sum l_x m_x = (0.40)(1) + (0.20)(3) + (0.05)(2) = 1.10$
Since $R_0 > 1$, r should be positive
To estimate r, we begin by estimating G (generation time)
$G = \sum x l_x m_x / R_0 = 1.90/1.10 = 1.73$
Estimate of $r = \ln R_0/G = \ln(1.1)/1.73 = 0.055$
Next, confirm r with the Euler equation. We use $r = 0.055$ in column 4 (below) and $r = 0.056$ in column 5. Since 0.056 produces a sum closer to 1.000, we will use 0.056.
Since $\lambda = e^r$, $\lambda = 1.057$

Age, x	$l_x m_x$	$x l_x m_x$	$l_x m_x e^{-rx}$ $r = 0.055$	$l_x m_x e^{-rx}$ $r = 0.056$
0	0	0	0	0
1	0.40	0.40	0.379	0.378
2	0.60	1.20	0.538	0.536
3	0.10	0.30	0.085	0.085
4	0	0	0	0
Σ	1.10	1.90	1.002	**0.999**

$$
\begin{array}{c}
\text{Matrix} \qquad\qquad t=0 \qquad t=1 \\[4pt]
\begin{vmatrix} 0.40 & 1.50 & 0.50 & 0 \\ 0.40 & 0 & 0 & 0 \\ 0 & 0.50 & 0 & 0 \\ 0 & 0 & 0.25 & 0 \end{vmatrix}
\begin{vmatrix} 1000 \\ 300 \\ 100 \\ 60 \end{vmatrix}
=
\begin{vmatrix} 900 \\ 400 \\ 150 \\ 25 \end{vmatrix} \\[4pt]
\Sigma = 1460 \quad \Sigma = 1475
\end{array}
$$

Stable age distribution with $r = 0.056$

Age, x	$e^{-rx}l_x$	c_x for stable age distribution
0	1.00	0.625
1	0.378	0.236
2	0.179	0.112
3	0.042	0.026
Σ	1.599	0.999

Expectation of life

Age, x	l_x	L_x	T_x	e_x
0	1.00	0.700	1.150	1.150
1	0.40	0.300	0.450	1.125
2	0.20	0.125	0.150	0.750
3	0.05	0.025	0.025	0.500
4	0	0	0	0
		$\Sigma = 1.150$		

Reproductive value. To calculate reproductive value, we need the following information:

Age, x	l_x	m_x	$e^{-rx}l_x m_x$	$e^{-rx}l_x$
0	1.00	0	0	1.000
1	0.40	1.0	0.378	0.378
2	0.20	3.0	0.537	0.179
3	0.05	2.0	0.085	0.043
4	0	0	0	0
Sum			1.000	

Calculation of reproductive value, V_x

Age, x	Method one	Method two
0	$(0 + 0.378 + 0.537 + 0.085)/1.000$ = **1.000**	$(0.378 + 0.537 + 0.085)/1.000$ = **1.000**
1	$(0.378 + 0.537 + 0.085)/0.378$ = **2.64**	$(0.537 + 0.085)/0.378$ = **1.64**
2	$(0.537 + 0.085)/0.179 = $ **3.47**	$(0.085)/0.179 = $ **0.47**
3	$(0.085)/0.043 = $ **1.98**	$(0)/0.132 = $ **0**

Appendix 2

Matrix algebra: the basics

For more information on matrices the reader should consult Vandermeer and Goldberg (2003), Caswell (1989), Searle (1966) or any basic matrix algebra textbook.

What is a matrix? A matrix is a rectangular array of numbers. Each number in a matrix is known as an **element**. An ordinary single number is known as a **scalar**. The elements of a matrix are usually made up of scalars. A matrix is made up of rows and columns. If there is a single column of numbers, the array is known as a **column vector**. A single row of numbers is known as a **row vector**. In a matrix, if the number of rows equals the number of columns, the result is known as a **square matrix**. Although the rows must be of equal length and the columns must be of equal length in a matrix, the number of rows only equals the number of columns in a square matrix. The size of a matrix is known as its **order** and is known by the number of rows (r) first, followed by the number of columns (c). Thus a matrix is known as "r by c." For example below we have a 4×3 matrix, $|A|$ followed by a column vector (x) and a row vector (y). A scalar, by the way, can be thought of as the element of a 1×1 matrix.

$$
|A| = \begin{vmatrix} 1 & 2 & 3 \\ 4 & 5 & 6 \\ 7 & 8 & 9 \\ 0 & 10 & 11 \end{vmatrix} \qquad x = \begin{vmatrix} 1 \\ 4 \\ 7 \\ 0 \end{vmatrix} \qquad y = \begin{vmatrix} 1 & 2 & 3 \end{vmatrix}
$$

In a square matrix, if all of the non-diagonal elements are zero, the matrix is described as a **diagonal matrix**. For example, a 4×4 diagonal matrix $|D|$ is shown below:

$$
|D| = \begin{vmatrix} 1 & 0 & 0 & 0 \\ 0 & 2 & 0 & 0 \\ 0 & 0 & 3 & 0 \\ 0 & 0 & 0 & -4 \end{vmatrix}
$$

Matrix operations

Addition and subtraction

To add or subtract two matrices, the corresponding elements from the row and column positions are added (or subtracted). For example, matrix $|A|$ and $|B|$ are added as follows:

$$\text{If } |A| = \begin{vmatrix} 1 & 2 & 3 \\ 4 & 0 & 7 \end{vmatrix} \quad \text{and} \quad |B| = \begin{vmatrix} 2 & 5 & 0 \\ 1 & 8 & 2 \end{vmatrix}$$

$$\text{then } |A| + |B| = \begin{vmatrix} 1+2 & 2+5 & 3+0 \\ 4+1 & 0+8 & 7+2 \end{vmatrix} = \begin{vmatrix} 3 & 7 & 3 \\ 5 & 8 & 9 \end{vmatrix}$$

Similarly, if matrix $|B|$ is subtracted from matrix $|A|$, we have the following:

$$|A| - |B| = \begin{vmatrix} 1-2 & 2-5 & 3-0 \\ 4-1 & 0-8 & 7-2 \end{vmatrix} = \begin{vmatrix} -1 & -3 & 3 \\ 3 & -8 & 5 \end{vmatrix}$$

It should be obvious that that matrix addition and subtraction can only occur when the two matrices involved are of the same order. Two matrices that are of the same order are said to be "conformable for addition (or subtraction)."

The null matrix and the identity matrix

The null (or zero) matrix, is a matrix equivalent of the scalar number zero. There are many possible null matrices, however, of different orders. Below are a 3×2 null matrix and a square 3×3 null matrix.

$$\begin{vmatrix} 0 & 0 \\ 0 & 0 \\ 0 & 0 \end{vmatrix} \quad \begin{vmatrix} 0 & 0 & 0 \\ 0 & 0 & 0 \\ 0 & 0 & 0 \end{vmatrix}$$

The equivalent of the number one in algebra is the identity matrix (or the unit matrix). The identity matrix must be a square matrix with all diagonal elements equal to one and all off-diagonal elements equal to zero. The identity matrix is identified by the letter I, sometimes with a subscript for the order. For example, I_3 is shown below:

$$\begin{vmatrix} 1 & 0 & 0 \\ 0 & 1 & 0 \\ 0 & 0 & 1 \end{vmatrix}$$

Mutiplication

1 By a scalar
Multiplication of a matrix by a scalar simply involves multiplying each element, in turn, by the scalar, producing a matrix of the same order as the original matrix. See the example below:

$$|A| = \begin{vmatrix} a_{11} & a_{12} & a_{13} & a_{14} \\ a_{21} & a_{22} & a_{23} & a_{24} \\ a_{31} & a_{32} & a_{33} & a_{34} \\ a_{41} & a_{42} & a_{43} & a_{44} \end{vmatrix}$$

$$3|A| = \begin{vmatrix} 3a_{11} & 3a_{12} & 3a_{13} & 3a_{14} \\ 3a_{21} & 3a_{22} & 3a_{23} & 3a_{24} \\ 3a_{31} & 3a_{32} & 3a_{33} & 3a_{34} \\ 3a_{41} & 3a_{42} & 3a_{43} & 3a_{44} \end{vmatrix}$$

If

$$|A| = \begin{vmatrix} 4 & 5 & 6 & 1 \\ 7 & 8 & 9 & 10 \\ 2 & -1 & -3 & 10 \\ 0 & 1 & -2 & 1 \end{vmatrix}$$

Then

$$3|A| = \begin{vmatrix} 12 & 15 & 18 & 3 \\ 21 & 24 & 27 & 30 \\ 6 & -3 & -9 & 30 \\ 0 & 3 & -6 & 3 \end{vmatrix}$$

2 Multiplication of a row vector by a column vector
To multiply a row by a column, the two vectors must be of the same order. The result is a scalar, which is found by using the following formula:

Given row vector $a = |a_1 \ a_2 \ a_3|$, and column vector $x = \begin{vmatrix} x_1 \\ x_2 \\ x_3 \end{vmatrix}$

Then the product $ax = a_1 x_1 + a_2 x_2 + a_3 x_3$. For example:

if $|a| = |2 \quad 4 \quad 6|$ and $|x| = \begin{vmatrix} -5 \\ 1 \\ 0 \end{vmatrix}$, the result is: $[-10 + 4 + 0] = -6$

3 Multiplication of a matrix by a column vector

In order to multiply a matrix by a column vector we repeat the above for each row. That is, each row is multiplied independently by the column vector. The result is a column vector of the same order as the previous column vector. In order to be conformable for multiplication, the number of elements in the row (equal to the number of columns) must equal the number of elements in the column vector. For example:

$$|A| = \begin{vmatrix} 5 & 6 & 8 \\ 2 & 5 & 12 \\ 1 & 3 & 9 \end{vmatrix} \quad y = \begin{vmatrix} 1 \\ 2 \\ 3 \end{vmatrix}$$

$$|A|y = \begin{vmatrix} (5 \times 1) + (6 \times 2) + (8 \times 3) \\ (2 \times 1) + (5 \times 2) + (12 \times 3) \\ (1 \times 1) + (3 \times 2) + (9 \times 3) \end{vmatrix} = \begin{vmatrix} 41 \\ 48 \\ 34 \end{vmatrix}$$

In general, if $\quad |A| = \begin{vmatrix} a_{11} & a_{12} & a_{13} \\ a_{21} & a_{22} & a_{23} \\ a_{31} & a_{32} & a_{33} \end{vmatrix}$ and $y = \begin{vmatrix} x_1 \\ x_2 \\ x_3 \end{vmatrix}$

then $\quad |A|y = \begin{vmatrix} a_{11} x_1 + a_{12} x_2 + a_{13} x_3 \\ a_{21} x_1 + a_{22} x_2 + a_{23} x_3 \\ a_{31} x_1 + a_{32} x_2 + a_{33} x_3 \end{vmatrix}$

4 The product of two matrices

Multiplying two matrices can be thought of as a repetitive exercise in multiplying a matrix by as many column vectors as are present in the second matrix. That is, think of matrix $|B|$ as a series of column vectors. Again, to be conformable for multiplication the number of columns in matrix $|A|$ must equal the number of rows in matrix $|B|$.

For example, if matrix $|A|$ is as above, and matrix $|B|$ consists of two columns and three rows, the product of $|A||B|$ is shown below:

$$|B| = \begin{vmatrix} x_1 & z_1 \\ x_2 & z_2 \\ x_3 & z_3 \end{vmatrix}$$

$$|A||B| = \begin{vmatrix} a_{11} x_1 + a_{12} x_2 + a_{13} x_3 & a_{11} z_1 + a_{12} z_2 + a_{13} z_3 \\ a_{21} x_1 + a_{22} x_2 + a_{23} x_3 & a_{21} z_1 + a_{22} z_2 + a_{23} z_3 \\ a_{31} x_1 + a_{32} x_2 + a_{33} x_3 & a_{31} z_1 + a_{32} z_2 + a_{33} z_3 \end{vmatrix}$$

For example:

$$|A||B| = \begin{vmatrix} 1 & 0 & 2 \\ 3 & 1 & 7 \\ 4 & 5 & 6 \end{vmatrix} \begin{vmatrix} -1 & 4 & 8 \\ -2 & 0 & 9 \\ -5 & 1 & 5 \end{vmatrix}$$

$$= \begin{vmatrix} (1 \times -1) + (0 \times -2) + (2 \times -5) & (1 \times 4) + (0 \times 0) + (2 \times 1) & (1 \times 8) + (0 \times 9) + (2 \times 5) \\ (3 \times -1) + (1 \times -2) + (7 \times -5) & (3 \times 4) + (1 \times 0) + (7 \times 1) & (3 \times 8) + (1 \times 9) + (7 \times 5) \\ (4 \times -1) + (5 \times -2) + (6 \times -5) & (4 \times 4) + (5 \times 0) + (6 \times 1) & (4 \times 8) + (5 \times 9) + (6 \times 5) \end{vmatrix}$$

$$= \begin{vmatrix} -11 & 6 & 18 \\ -40 & 19 & 68 \\ -44 & 22 & 107 \end{vmatrix}$$

The laws of algebra

Matrices follow the associative laws of addition and multiplication provided the matrices are conformable. That is, $(A + B) + C = A + B + C = A + (B + C)$. Similarly, $A(BC) = (AB)C = ABC$.

The distributive law also holds for matrices provided B and C are conformable for addition (necessarily of the same order) and matrices A and B are conformable for multiplication (hence A and C are also conformable). Thus: $A(B + C) = AB + AC$.

Addition of matrices follows the commutative rules provided they are conformable. If both matrices are of the same order we can write: $A + B = B + A$.

However, multiplication of matrices does not usually follow the commutative rule. In general, $AB \neq BA$. There are particular cases, where $AB = BA$, but this is not true as a general rule.

The inverse of a matrix

Division in the usual sense does not exist in matrix algebra. The concept of dividing by a matrix is replaced by the idea of multiplication by the inverse of a matrix. In algebra, if we want to divide a number by two, we have the option of simply multiplying by the inverse of two, that is, 1/2 (0.5). In matrices, instead of dividing y by x, we multiply y by $1/x$. Another property of the inverse is that if we multiply x by its inverse the result is the number one. In matrix algebra, if we multiply $|A|$ by $|A|^{-1}$ the result should be the identity matrix, I, which will be of the same order as $|A|$ and $|A|^{-1}$. The above is only true if $|A|$ and $|A|^{-1}$ are square matrices. The inverse of $|A|$ only exists if $|A|$ has a determinant (see next section). It is often difficult to solve for the inverse of a matrix, but computer programs are now available that do the hard work for you.

The determinant

The concept of a determinant also does not exist in traditional algebra. It is also difficult to explain the meaning of a determinant in any intuitive fashion. A determinant is only defined for a square matrix and is the result of a series of calculations resulting in a **scalar**. It is the sum of a series of products of the elements of the matrix and each product is multiplied by +1 or −1 according to certain rules. For example, for a 2×2 matrix, the determinant is found by taking the difference between the products of the diagonals versus the off-diagonals. That is, the determinant of matrix $|A| = (a_{11}{}^{*}a_{22}) - (a_{12}{}^{*}a_{21})$.

If $|A| = \begin{vmatrix} 7 & 3 \\ 4 & 6 \end{vmatrix}$ the determinant is calculated as $(7 \times 6) - (4 \times 3) = 30$.

A determinant for a 3×3 matrix follows similar, but more complicated, rules. As the order of the matrices gets larger the calculations become increasingly complex. Again, however, there are computer programs to evaluate the determinant of a matrix.

Mathematical symbols used
in this book

In all chapters the following are conventionally used:

e	the base of natural logarithms
K	a carrying-capacity term
N	population size
N_t	population size at time t
r	the intrinsic rate of increase $= b - d =$ growth rate per individual per unit time
r_a	the actual intrinsic rate of increase in a given environment
r_m	the Malthusian parameter, or r_{max}; the density-independent, maximum growth rate of a genotype in a given environment
R^2	the amount of variance explained by a statistical regression model
t	time

Symbols used in specific chapters are as follows:

Chapter 1

λ (*lambda*)	the finite rate of increase or the growth rate per unit time $= N_{t+1}/N_t = e^r$ when there is a stable age distribution
b	birth rate per individual per time period (per capita birth rate)
B	the number of births per unit time; the litter size or the clutch size
d	death rate per individual per time period (per capita death rate)
D	the number of deaths per unit time
E	the number of emigrants per unit time
I	the number of immigrants per unit time
p_i	probability of a given λ when computing the geometric mean
$P_{0,t}$	the probability of extinction at time t
R	the net replacement rate or net growth rate per generation in a population with discrete generations

Chapter 2

θ (*theta*)	parameter describing non-linear responses of the logistic equation when $\theta \neq 1$
T (*tau*)	measures a lag-time effect in the logistic equation
a'	$(R-1)/K$; also a carrying-capacity parameter
a	a constant of integration
$b*$	the exponent that relaxes the assumption that population growth follows exact (linear) density dependence (exact compensation). When $b* = 1$ there is "exact compensation," but when $b* > 1$ there is "overcompensation," meaning that growth decreases more rapidly than expected with an increase in population density. When $b* < 1$ there is "undercompensation," meaning that growth decreases less rapidly than expected with an increase in population density.
C	the final constant yield
K	the carrying capacity of the environment for a given population
MVP	minimum viable population size
R_A	the density-dependent or "actual" net growth parameter in a population with discrete generations; the net replacement rate or net growth rate per generation
R or R_I	the density-independent growth parameter in a population with discrete generations
$\bar{w}$	mean mass per plant
w_m	the maximum potential mass per plant

Chapter 3

D	the number of deaths due to density-independent factors per unit time

Chapter 4

$|A|$	a matrix
A	mature adult stages
c_x	proportion of the population belonging to an age category x
e_x	age-specific life expectancy; estimated as $e_x = \dfrac{T_x}{l_x}$
G	mean generation time
GRR	gross reproductive rate; the average number of female offspring produced by a female living through all of the reproductive age classes.
I	identity matrix
l_x	the age-specific survivorship; the proportion of the population, measured from age zero, to live to a given age class x
L_x	mean survivorship for any particular age interval; it assumes that, on average, an organism dies half way between two age classes. $L_x = \dfrac{l_x + l_{x+1}}{2}$

m_x	age-specific fertility; the mean number of female offspring produced by a female of age x
n_x	the number of individuals in an age class x
p_x	age-specific probability of living to the next age class. $\quad p_x = \dfrac{l_{x+1}}{l_x}$
q_x	age-specific probability of death prior to the next age class. $q_x = 1 - p_x$
R_0	net reproductive rate in a population with age classes and overlapping generations; the mean number of females produced per female in the population per generation.
S_x	the number of survivors to a given age, x, based on a cohort of 1000
T_x	the area under the survivorship curve for an individual of a given age, x, to the age, w, at which the oldest individual dies. T_x is estimated as

$$T_x = \sum_x^w L_x = \sum_x^w \frac{l_x + l_{x+1}}{2}$$

V_x	reproductive value of an individual of the age x
x	age class
Y	young adult stages

Chapter 5

α (*alpha*)	the dispersal ability of each individual organism
β (*beta*)	the number of individuals dispersing
ε (*epsilon*)	extinction rate in a metapopulation
A	the area of an island or habitat
A_i	the area of the habitat patch i
c	the colonization rate in a metapopulation
C	the y-intercept in the MacArthur and Wilson species–area relationship
C_i	the colonization rate or probability per unit time in a metapopulation
d_i	the distance from a source population i
d_{ij}	the distance between patches i and j
D	a coefficient of diffusion in a random walk movement model
e	a parameter related to the probability of extinction per unit time in a patch of a given size
E_i	the extinction rate per unit time in a metapopulation
H	the total number of available habitat patches
J_i	the long-term probability of a patch being occupied (the incidence)
M_i	the sum of individuals arriving at patch i, from all of the surrounding habitat patches
P	the proportion of available habitat patches occupied by a population
P_j	= 0 for an unoccupied patch and 1 for an occupied patch in the incidence function model
P'	the number of habitat patches occupied by a population
$\hat{P}$	the equilibrium value of P
R_e	the total number of possible source species found on the mainland
S	the number of species on an island

S_i	a measure of connectivity between patches
T	the total number of habitat patches available
T_L	expected time to local extinction
T_M	long-term persistence of a metapopulation
X	the rate of change of extinction per unit time with increasing patch size (a measure of environmental stochasticity)
y	efficiency of colonization
z	a constant representing the slope of the line in a log–log plot of area and species number

Chapter 6

α (alpha)	the age at first reproduction
δ (delta)	age of maximum reproduction
ω (omega)	the age at last reproduction
b	the allometric constant
B	litter size
C	constant final yield
E	activation energy
G	mean generation time
I	whole-organism metabolic rate as a function of mass (M) and a normalization constant, I_0 or i_0
k	the Boltzmann constant
M	mass
S	total reproductive output
T	absolute temperature in degrees Kelvin
V_x	$l_x m_x$
$\overline{w}$	mean mass per plant
Y	a physiological rate or some other variable dependent on mass; and Y_0 is normalization constant

Chapter 7

α_{ij}	the competition coefficient: the effect of species j on species i
ε_i	the mortality or extinction rate of species i
μ (mu)	the growth rate on a given substrate of resource R, in the Michaelis–Menton equation
μ_{max}	the maximum growth rate of the population
b	the efficiency by which each individual converts the resource into new individuals
b_i	maximum cell division rate ($= r_{max}$) for species i
c_i	the colonisation rate of species i
C_j	the availability of the resource j
K_μ or K_i	the half-saturation constant for the resource i; the concentration of the resource that produces half the maximum growth rate.

K_1	the carrying capacity of species one
K_2	the carrying capacity of species two
k_{Ri}	the supply rate of resource R_i
m	death or mortality rate
N_1	the number of individuals of species one
N_2	the number of individuals of species two
P_i	the proportion of available habitat sites or patches occupied by species i
$\hat{P}$	the proportion of habitat sites occupied at equilibrium
$\hat{P}_i$	the proportion of habitat sites occupied at equilibrium by species i
q	the rate at which each individual must consume the resource in order to maintain itself
r_1	the intrinsic rate of increase of species one
r_2	the intrinsic rate of increase of species two
R^*	the amount of a resource needed to just sustain the population. That is, R^* is the level of the resource needed to balance mortality (growth just offsets mortality)
R_i	the quantity or concentration of a resource i

Chapter 8

c_i and c_j	mutualism coefficients that replace the competition coefficients in the Lotka–Volterra competition equations. The term c_i measures the rate at which an individual of N_j benefits the growth rate of population N_i.
K_i^*	the carrying capacity for species i after a mutualistic interaction with species j.

Chapter 9

α (*alpha*)	disease-induced mortality rate
β (*beta*)	transmission rate of disease from one host to another
γ (*gamma*)	rate at which recovered individuals lose their immunity. That is, the rates at which individuals return to the susceptible class (S) from the recovered (R) class.
δ (*delta*)	probability that an infected disperser infects the resident population in a susceptible patch
v (*nu*)	recovery rate, or the per capita rate of passage from the infected (I) to the recovered (R) classes. This is usually the inverse of the average infectious period.
ψ (*psi*)	the migration rate between susceptible and infected populations
A	the average age at which an individual in a population will become infected with a particular disease
b	host birth rate
D	the length of the infectious period
g	infection rate from an "outside source," that is, from another species in the patch

I	infected host density or the proportion of infected patches (both host population and disease present)
m	natural host mortality rate unrelated to disease mortality
N	total host population density
p	the fraction of the population that has been immunized
R	recovered (immune) host density
R_0	the basic reproductive number (BSR) or parameter for a disease. It represents the mean number of new infections caused by a single infective individual, and equals βSD.
S	susceptible host density or proportion of susceptible host patches in the metapopulation model
S_T	the minimum size for an epidemic. $S_T = 1/\beta D$
x_I	extinction rate of an infected population in a habitat patch
x_S	extinction rate of a susceptible population in a habitat patch

Chapter 10

λ (*lambda*)	the finite rate of increase per generation for the prey in the Nicholson–Bailey model
μ_p	the minimum rate of prey consumption necessary for a predator to survive and replace itself
ρ_0	the zero term of the Poisson distribution
χ_p	the efficiency by which food (prey) is turned into new predator individuals, or the assimilation efficiency of the predator
a	constant for the parasitoid searching efficiency
c	the maximum predation or killing rate when the search and capture components have been minimized; replaces the parameter E
d	half-saturation parameter equal to half of the maximum predation rate
E	the searching and capturing efficiency of the predator; a functional-response term
E_1, E_2, E_3	search and capture components for type I, II and III functional responses
f or f_n	functional-response term; equal to EN in a type I functional response
h_1, h_2, h_3	handling and digestion-rate components of the type I, II, and III functional responses
K_n	carrying-capacity term for the prey population
K_R	the half-saturation constant for a resource
m_p	instantaneous, density-independent mortality rate of the predator population
M	mass
M^*	prey population never exposed to predation (prey refuge)
n	the number of parasitoids emerging from each host individual
N	prey population
N^*	the equilibrium number of prey when both prey and predator populations have stopped changing ($dN/dt = 0$ and $dP/dt = 0$)
N_a	the number of hosts actually parasitized
N_e	the number of encounters between host or prey species and their parasitoids
P	the predator population or parasitoid population

$P*$	the equilibrium number of predators when both prey and predator populations have stopped changing ($dN/dt = 0$ and $dP/dt = 0$)
R	the concentration of a resource
r_n	the maximum growth rate (intrinsic rate of increase) for the prey population

Chapter 11

θ (*theta*)	a density-dependent parameter affecting the herbivore death rate; $= 1$ at low herbivore densities, but > 1 and increases the herbivore death rate at high population densities
κ (*kappa*)	the carrying-capacity term for the predator population due to territoriality
μ_h	the minimum rate of food intake necessary for an herbivore to survive and replace itself
χ_h	the assimilation efficiency of the herbivore population
χ_p	the assimilation efficiency of the predator population
b	the half-saturation parameter for the herbivore population, equal to half of the maximum consumption rate, f
c	the maximum killing rate for the predator population when the search and capture components have been minimized
d	half-saturation parameter equal to half of the maximum predation rate, c
F	the functional-response term for searching efficiency of the herbivore
f	the maximum consumption or grazing rate for the herbivore when the search and capture components have been minimized; replaces the parameter F
h	handling and digestion-rate components of the functional response of the herbivore
K_v	carrying-capacity term for the vegetation (plant population)
m_h	instantaneous, density-independent mortality rate of the herbivore population
n	the number of grazers that determines whether plant quality increases or decreases
N	the number of herbivores or grazers
$N*$	the herbivore equilibrium population size when both plant and herbivore populations have stopped changing ($dV/dt = 0$ and $dN/dt = 0$)
P	the number of predators
Q	average plant quality
q	the increase in plant quality in the absence of grazers
r_h	herbivore (prey) maximum growth rate
r_v	plant maximum growth rate
s_0	predator rate of increase limited by a carrying capacity due to a territory
u_0	plant growth rate when V is close to zero and V represents above-ground biomass
V	plant abundance or biomass
$V*$	the equilibrium plant abundance when both plant and herbivore populations have stopped changing ($dV/dt = 0$ and $dN/dt = 0$)

References

Agrawal, A.A. 2000. Benefits and costs of induced plant defenses for *Lepidium virginicum* (Brassicaceae). *Ecology* 87: 1804–1813.

Agrawal, A.A., S.Y. Strauss, and M.J. Stout. 1999. Costs of induced responses and tolerance to herbivory in male and female fitness components of wild radish. *Evolution* 55: 1093–1104.

Agutter, P.S. and D.N. Wheatley. 2004. Metabolic scaling: consensus or controversy? *Theoretical Biology and Medical Modelling* 1: 1–13. www.tbiomed.com/content/1/1/13.

Albersheim, P. 1965. Biogenesis of the cell wall. Pages 298–321 *in* J. Bonner and J.E. Varner, editors, *Plant Biochemistry*. Academic Press, New York, NY.

Alberts, S.C. and J. Altmann. 2003. Matrix models for primate life history analysis. Pages 66–102 *in* P.M. Kappeler and M.E. Pereira, editors, *Primate Life Histories and Socioecology*. University of Chicago Press, Chicago, IL.

Allee, W.C. 1931. *Animal Aggregations: a Study in General Sociology*. University of Chicago Press, Chicago, IL.

Alliende, M.C. and J.L. Harper. 1989. Demographic studies of a dioecious tree: colonization, sex, and age-structure of a population of *Salix cinerea*. *Journal of Ecology* 77: 1029–1048.

Alstad, D. 2001. *Basic Populus Models of Ecology*. Prentice Hall, Upper Saddle River, NJ.

Anderson, J.D., D.D. Hassinger, and G.H. Dalrymple. 1971. Natural mortality of eggs and larvae of *Ambystoma t. tigrinum*. *Ecology* 52: 1107–1112.

Anderson, R.M. 1982. Transmission dynamics and control of infectious disease agents. *In* R.M. Anderson and R.M. May, editors, *Population Biology of Infectious Diseases* (Dahlem Conference Report). Springer-Verlag, New York, NY.

Anderson, R.M. and R.M. May. 1979. Population biology of infectious diseases. Part I. *Nature* 280: 361–367.

Anderson, R.M. and R.M. May. 1981. The population dynamics of microparasites and their invertebrate hosts. *Philosophical Transactions of the Royal Society of London B* 291: 451–524.

Anderson, R.M. and R.M. May. 1982. Directly transmitted infectious diseases: control by vaccination. *Science* 215: 1053–1060.

Andrewartha, H.G. 1961. *Introduction to the Study of Animal Populations*. University of Chicago Press, Chicago, IL.

Andrewartha, H.G. and L.C. Birch. 1954. *The Distribution and Abundance of Animals*. University of Chicago Press, Chicago, IL.

Anonymous. 1981–2004. *Population Data Sheets*. Population Reference Bureau, Washington, DC.

Anonymous. 1988. *National Center for Health Statistics: Vital Statistics of the United States, 1986, Vol. II, Sec. 6, Life Tables*. DHHS Pub. No. 88–1147. Public Health Service, Washington, DC.

Anonymous. 2003. *Atlantic Flyway Mute Swan Management Plan 2003–2013*. Atlantic Flyway Council. Maryland Department of Natural Resources. Posted July 28, 2003. www.dnr.state.md.us/wildlife/afcmuteplan.html.

Arcese, P. and J.N.M. Smith. 1988. Effects of population density and supplemental food on reproduction in song sparrows. *Journal of Animal Ecology* 57: 119–136.

Ashmole, N.P. 1963. The regulation of numbers of tropical oceanic birds. *Ibis* 103b: 458–473.

Atkins, P. 1999. *Sunday Times* Supplement (October 24). London.

Atwood, J.L., M.J. Elpers, and C.T. Collins. 1990. Survival of breeders in Santa Cruz Island and mainland California scrub jay populations. *Condor* 92: 783–788.

Baker, M.C., L.M. Mewaldt, and R.M. Stewart. 1981. Demography of white-crowned sparrows (*Zonotrichia leucophrys nuttalli*). *Ecology* 62: 636–644.

Baldwin, I.T. 1994. Chemical changes rapidly induced by folivory. Pages 1–23 *in* E. Bernays, editor, *Insect–Plant Interactions. V.* CRC Press, New York, NY.

Baldwin, I.T. and J.C. Schultz. 1983. Rapid changes in tree leaf chemistry induced by damage: evidence for communication among plants. *Science* 221: 277–279.

Baldwin, I.T., M.J. Karb, and T.E. Ohnmeiss. 1994. Allocation of ^{15}N from nitrate to nicotine after leaf damage: production and turnover of a damage-induced mobile defense. *Ecology* 75: 1703–1713.

Baldwin, N.S. 1964. Sea lamprey in the Great Lakes. *Canadian Audubon Magazine*, November–December: 142–147.

Barkalow, F.S., R.B. Hamilton, and R.F. Soots. 1970. The vital statistics of an unexploited gray squirrel population. *Journal of Wildlife Management* 34: 489–500.

Beauchamp, R.S.A. and P. Ullyott. 1932. Competitive relationships between certain species of freshwater triclads. *Journal of Ecology* 20: 200–208.

Becerra, J.X. 1994. Squirt-gun defense in *Bursera* and the Chrysomelid counter ploy. *Ecology* 75: 1991–1996.

Begon, M., J.L. Harper, and C.L. Townsend. 1986. *Ecology: Individuals, Populations and Communities*. Sinauer Associates, Sunderland, MA.

Beissinger, S.R. 1990. Experimental brood manipulation and the monoparental threshold in snail kites. *American Naturalist* 136: 20–38.

Beissinger, S.R. and D.R. McCullough. 2002. *Population Viability Analysis*. University of Chicago Press, Chicago, IL.

Belsky, A.J. 1992. Effects of grazing, competition, disturbance and fire on species composition and diversity in grassland communities. *Journal of Vegetation Science* 3: 187–200.

Berenbaum, M.R. 1981. Patterns of furanocoumarins distribution and insect herbivory in the Umbelliferae: plant chemistry and community structure. *Ecology* 62: 1254–1266.

Berenbaum, M.R. 1991. Coumarins. Pages 221–249 *in* G.A. Rosenthal and M.R. Berenbaum editors, *Herbivores: Their Interactions with Secondary Plant Metabolites*. Vol. 1. 2nd edition. Academic Press, New York, NY.

Berenbaum, M.R. 1995. Metabolic detoxification of plant prooxidants. Pages 181–209 *in* S. Ahmad, editor, *Oxidative Stress and Antioxidant Defense in Biology*. Chapman and Hall, New York, NY.

Berenbaum, M.R. 2002. Postgenomic chemical ecology: from genetic code to ecological interactions. *Journal of Chemical Ecology* 28: 873–896.

Berenbaum, M.R. and A.R. Zangerl. 1996. Phytochemical diversity: adaptive or random variation? *Recent Advances in Phytochemistry* 30: 1–24.

Berenbaum, M.R. and A.R. Zangerl. 1999. Coping with life as a menu option: inducible defenses of the wild parsnip. Pages 10–32 *in* R. Tollrian and C.D. Harvell, editors, *The Ecology and Evolution of Inducible Defenses*. Princeton University Press, Princeton, NJ.

Berg, Å., T. Lindbergh, and K. Gunner. 1992. Hatching success of lapwings on farmland: differences between habitats and colonies of different sizes. *Journal of Animal Ecology* 61: 469–476.

Beverton, R.J.H. and S.J. Holt. 1957. On the dynamics of exploited fish populations. *Fisheries Investigations* 19: 1–533.

Birkhead, T.R. 1977. The effect of habitat and density on breeding success in the common guillemot (*Uria aalge*). *Journal of Animal Ecology* 46: 751–764.

Birkhead, T.R., B.J. Hatchwell, R. Lindner, D. Blomquist, E.J. Pellatt, R. Giffiths, and J.T. Lifjeld. 2001. Extra-pair paternity in the Common Murre. *Condor* 103: 158–162.

Blest, A.D. 1957. The function of eyespot patterns in the Lepidoptera. *Behaviour* 11: 209–256.

Blomquist, D., B. Kempenaers, R.B. Lanctot, and B.K. Sandercock. 2002. Genetic parentage and mating guarding in the Arctic-breeding Western Sandpiper. *The Auk* 119: 228–233.

Blueweiss, L., H. Fox, V. Kudzma, D. Nakashima, R. Peters, and S. Sams. 1978. Relationships between body size and some life history parameters. *Oecologia* 37: 257–272.

Bodnaryk, R.P. 1992. Effects of wounding on glucosinolates in the cotyledons of oilseed rape and mustard. *Phytochemistry (Oxford)* 31: 2671–2677.

Borror, D.J., C.A. Triplehorn, and N.F. Johnson. 1989. *Introduction to the Study of Insects*. 6th edition. Saunders College Publishing, Philadelphia, PA.

Botkin, D.B. 1990. *Discordant Harmonies*. Oxford University Press, Oxford.

Botkin, D.B. and R.S. Miller. 1974. Mortality rates and survival of birds. *American Naturalist* 108: 181–192.

Boyce, M.S. 2002. Reconciling the small-population and declining-population paradigms. Pages 41–49 *in* S.R. Beissinger and D.R. McCullough, editors, *Population Viability Analysis*. University of Chicago Press, Chicago, IL.

Boycott, A.E. 1930. A re-survey of the fresh-water mollusca of the parish of Aldenham after ten years with special reference to the effect of drought. *Transactions of the Hertfordshire Natural History Society* 19: 1–25.

Branch, G.M. 1975. Intraspecific competition in *Patella coclear* Born. *Journal of Animal Ecology* 44: 263–281.

Bronstein, J.L. 2000. Fig pollination and seed-dispersal mutualisms. *In* N.M. Nadkarni and N.T. Wheelwright, editors, *Monteverde: Ecology and Conservation of a Tropical Cloud Forest*. Oxford University Press, New York, NY.

Bronstein, J.L. 2001. The costs of mutualism. *American Zoologist* 41: 825–839.

Bronstein, J.L. 2002. Profiles. Page 139 *in* Peter Stiling. *Ecology: Theories and Applications*. Prentice Hall, Upper Saddle River, NJ.

Brower, L.P. and J.V.Z. Brower. 1964. Birds, butterflies, and plant poisons: a study in ecological chemistry. *Zoologica* 49: 137–159.

Brower, L.P., J.V.Z. Brower, and J.M. Corvino. 1967. Plant poisons in a terrestrial food chain. *Proceedings of the National Academy of Sciences (USA)* 57: 1059–1066.

Brown, J.H. 1971a. Mechanisms of competitive exclusion between two species of chipmunks. *Ecology* 52: 305–311.

Brown, J.H. 1971b. Mammals on mountaintops: non-equilibrium insular biogeography. *American Naturalist* 105: 467–478.

Brown, J.H. 1978. The theory of insular biogeography and the distribution of boreal birds and mammals. *Great Basin Naturalist Memoirs* 2: 209–227.

Brown, J.H. and E.J. Heske. 1990. Control of a desert-grassland transition by a keystone rodent guild. *Science* 250: 1705–1707.

Brown, J.H. and A. Kodric-Brown. 1977. Turnover rates in insular biogeography: effect of immigration on extinction. *Ecology* 58: 445–449.

Brown, J.H., J.F. Gillooly, A.P. Allen, V.M. Savage, and G.F. West. 2004. Toward a metabolic theory of ecology. *Ecology* 85: 1771–1789.

Brown, J.S., J.W. Laundre, and M. Gurung. 1999. The ecology of fear: optimal foraging, game theory, and trophic interactions. *Journal of Mammalogy* 80: 385–399.

Brown, W.L. and E.O. Wilson. 1956. Character displacement. *Systematic Zoology* 5: 49–94.

Bryant, J.P. and P.J. Kuropat. 1980. Selection of winter forage by subarctic browsing vertebrates: the role of plant chemistry. *Annual Review of Ecology and Systematics* 11: 261–286.

Bryant, J.P., G.D. Wieland, P.B. Reichardt, V.E. Lewis, and M.C. McCarthy. 1983. Pinosylvin methyl ether deters snowshoe hare feeding on green alder. *Science* 222: 1023–1025.

Calder, W.A., III. 1984. *Size, Function and Life History*. Harvard University Press, Cambridge, MA.

Caraco, T. and L.L. Wolf. 1975. Ecological determinants of group sizes for foraging lions. *American Naturalist* 109: 343–352.

Case, T.J. 2000. *An Illustrated Guide to Theoretical Ecology*. Oxford University Press, Oxford.

Caswell, H. 1989. *Matrix Population Models: Construction, Analysis and Interpretation*. Sinauer Associates, Sunderland, MA.

Cattadori, I.M., D.T. Haydon, and P.J. Hudson. 2005. Parasites and climate synchronize red grouse populations. *Nature* 433: 737–741.

Caughley, G. 1966. Mortality patterns in mammals. *Ecology* 47: 9067–9018.

Caughley, G. 1970. Eruption of ungulate populations, with emphasis on Himalayan thar in New Zealand. *Ecology* 51: 51–72.

Caughley, G. 1977. *Analysis of Vertebrate Populations*. Wiley, New York, NY.

Charnov, E.L. and W.M. Schaffer. 1973. Life history consequences of natural selection: Cole's result revisited. *American Naturalist* 107: 791–793.

Chen, I. 1990. Plants bite back. *Science News* 138: 408–410.

Chitty, D. 1996. *Do Lemmings Commit Suicide? Beautiful Hypotheses and Ugly Facts*. Oxford University Press, Oxford.

Cliff, A.D., P. Haggett, and J.K. Ord. 1986. *Spatial Aspects of Influenza Epidemics*. Pion, London.

Clutton-Brock, T.H. 1984. Reproductive effort and terminal investment in iteroparous animals. *American Naturalist* 123: 212–229.

Clutton-Brock, T.H., F.E. Guinness, and S.D. Albon. 1982. *Red Deer: Behavior and Ecology of Two Sexes*. University of Chicago Press, Chicago, IL.

Clutton-Brock, T.H., S.D. Albon, and F.E. Guinness. 1989. Fitness costs of gestation and lactation in wild mammals. *Nature* 337: 260–262.

Cody, M.L. 1966. A general theory of clutch size. *Evolution* 20: 174–184.

Cole, L.C. 1954. The population consequences of life history phenomena. *Quarterly Review of Biology* 29: 103–137.

Coley, P.D. 1980. Effects of leaf age and plant life history patterns on herbivory. *Nature* 284: 545–546.

Coley, P.D., J.P. Bryant, and F.S. Chapin, III. 1985. Resource availability and plant antiherbivore defense. *Science* 230: 895–899.

Connell, J.H. 1961a. Effects of competition, predation by *Thais lapillus* and other factors on natural populations of the barnacle, *Balanus balanoides*. *Ecological Monographs* 31: 61–104.

Connell, J.H. 1961b. The influence of interspecific competition and other factors on the distribution of the barnacle, *Chthamalus stellatus*. *Ecology* 42: 710–723.

Connell, J.H. 1983. On the prevalence and relative importance of interspecific competition: evidence from field experiments. *American Naturalist* 122: 661–696.

Cook, R.R. and I. Hanski. 1995. On expected lifetimes of small and large species of birds on islands. *American Naturalist* 145: 307–315.

Craig, T. 2003. Wildlife agency offers to halt swan killings. *Washington Post* September 18, 2003: B01.

Crawley, M.J. 1989. The relative importance of vertebrate and invertebrate herbivores in plant population dynamics. Pages 45–71 *in* E.A. Bernays, editor, *Insect–Plant Interactions*, Vol. 1. CRC Press, Boca Raton, FL.

Crawley, M.J. 1992. Population dynamics of natural enemies and their prey. Pages 40–89 *in* M.J. Crawley, editor, *Natural Enemies: the Population Biology of Predators, Parasites and Diseases*. Blackwell Scientific, Oxford.

Crawley, M.J. 1997. Plant herbivore dynamics. Pages 401–474 *in* M.J. Crawley, editor, *Plant Ecology*. Blackwell Scientific, Oxford.

Crombie, A.C. 1942. The effect of crowding upon the oviposition of grain-infesting insects. *Journal of Experimental Biology* 19: 311–340.

Crombie, A.C. 1944. On intraspecific and interspecific competition in larvae of graminivorous insects. *Journal of Experimental Biology* 20: 135–151.

Crombie, A.C. 1945. On competition between different species of graminivorous insects. *Proceedings of the Royal Society of London B*, 132: 362–395.

Crombie, A.C. 1946. Further experiments on insect populations. *Proceedings of the Royal Society B* 133: 76–109.

Crombie, A.C. 1947. Interspecific competition. *Journal of Animal Ecology* 16: 44–73.

Crone, E.E., D. Doak, and J. Pokki. 2001. Ecological influences on the dynamics of a field vole meta-population. *Ecology* 82: 831–843.

Crowl, T.A. and A.P. Covich. 1990. Predator-induced life history shifts in a freshwater snail. *Science* 247: 949–951.

Culver, D.C., J.R. Holsinger, and R. Barogdy. 1973. Toward a predictive cave biogeography: the Greenbrier Valley as a case study. *Evolution* 27: 689–695.

Cyr, H. and M.L. Pace. 1993. Magnitude and patterns of herbivory in aquatic and terrestrial eco-systems. *Nature* 361: 148–150.

Darveau, C.A., R.K. Suarez, R.D. Andrews, and P.W. Hochachka. 2002. Allometric cascade as a unifying principle of body mass effects on metabolism. *Nature* 417: 166–170.

Dash, M.C. and A.K. Hota. 1980. Density effects on the survival, growth rate, and metamorphosis of *Rana tigrina* tadpoles. *Ecology* 61: 1025–1028.

Davidson, J. 1938. On the growth of the sheep population in Tasmania. *Transactions of the Royal Academy of South Australia* 62: 342–346.

Davidson, J. and H.G. Andrewartha. 1948. The influence of rainfall, evaporation and atmospheric temperature on fluctuations in the size of a natural population of *Thrips imaginis* (Thysanoptera). *Journal of Animal Ecology* 17: 200–222.

Davies, N.B. 1992. *Dunnock Behavior and Social Evolution*. Oxford University Press, Oxford.

Davies, N.B. 1995. Backyard battle of the sexes. *Natural History* 104: 68–73.

Davies, N.B. 2000. *Cuckoos, Cowbirds and Other Cheats*. Academic Press, New York, NY.

Dayan, T., D. Simberloff, E. Tchernov, and Y. Yom-Tov. 1990. Feline canines: community-wide character displacement among the small cats of Israel. *American Naturalist* 136: 39–60.

Deevy, E.S. 1947. Life tables for natural populations of animals. *Quarterly Review of Biology* 22: 283–314.

DeMoraes, C.M., W.J. Lewis, P.W. Pare, H.T. Alborn, and J.H. Tumlinson. 1998. Herbivore-infested plants selectively attract parasitoids. *Nature* 393: 570–573.

Deshmukh, I. 1986. *Ecology and Tropical Biology*. Blackwell Science, Oxford.

Diamond, J.M. 1975. The island dilemma: lessons of modern biogeographic studies for the design of natural reserves. *Biological Conservation* 7: 129–146.

Diamond, J.M. 1978. Niche shifts and the rediscovery of interspecific competition. *American Scientist* 66: 322–331.

Diamond, J.M. 1983. Laboratory, field and natural experiments. *Nature* 304: 586–587.

Dingle, H. 1990. The evolution of life histories. Pages 267–289 *in* K. Wohrmann and S.K. Jain, editors, *Population Biology: Ecological and Evolutionary Viewpoints*. Springer-Verlag, Berlin.

Dittus, W.P.J. 1977. The social regulation of population density and age–sex distribution in the toque monkey. *Behavior* 63: 281–322.

Doak, D.F. and L.S. Mills. 1994. A useful role for theory in conservation. *Ecology* 75: 615–626.

Dobson, A.P. 1982. Comparisons of some characteristics of the life histories of microparasites, macro-parasites, parasitoids, and predators. Page 5 *in* R.M. Anderson and R.M. May, editors, *Population Biology of Infectious Diseases*. Springer-Verlag, New York, NY.

Dobson, A.P. and P.J. Hudson. 1992. Regulation and stability of a free-living host–parasite system: *Trichostrongylus tenuis* in red grouse. II. Population models. *Journal of Animal Ecology* 61: 487–498.

Dobson, A.P., P.J. Hudson, and A.M. Lyles. 1992. Macroparasites: worms and others. Pages 329–348 *in* M.J. Crawley, editor, *Natural Enemies: the Population Biology of Predators, Parasites and Diseases*. Blackwell Scientific, Oxford.

Dodd, A.P. 1940. *The Biological Campaign Against Prickly Pear*. Commonwealth Prickly Pear Board. Government Printer, Brisbane, Australia.

Dussourd, D.E. and R.F. Denno. 1994. Host range of generalist caterpillars: trenching permits feeding on plants with secretory canals. *Ecology* 75: 69–78.

Edelstein-Keshet, L. 1984. *Mathematical Theory for Plant–Herbivore Systems*. Lefschetz Center for Dynamical Systems, Providence, RI.

Edelstein-Keshet, L. and M.D. Rausher. 1989. The effects of inducible plant defenses on herbivore populations. I. Mobile herbivores in continuous time. *American Naturalist* 133: 787–810.

Ehrlich, P.R. and I. Hanski. 2004. *On the Wings of Checkerspots*. Oxford University Press, New York, NY.

Ehrlich, P.R. and P.H. Raven. 1964. Butterflies and plants: a study in co-evolution. *Evolution* 18: 586–608.

Eisner, T. and D.J. Aneshansley. 1982. Spray aiming in bombardier beetles: jet deflection by the coanda effect. *Science* 215: 83–85.

Elmhagen, B. and A. Angerbjörn. 2001. The applicability of metapopulation theory to large mammals. *Oikos* 94: 94–100.

Elton, C.S. 1924. Periodic fluctuations in the numbers of animals: their causes and effects. *Journal of Experimental Biology* 2: 119–163.

Elton, C.S. 1927. *Animal Ecology*. Macmillan, New York, NY.

Elton, C.S. and M. Nicholson. 1942. The ten-year cycle in numbers of lynx in Canada. *Journal of Animal Ecology* 11: 215–244.

Emlen, D.J. 2000. Integrating development with evolution: a case study with beetle horns. *Bioscience* 50: 403–418.

Enquist, B.J., J.H. Brown, and G.B. West. 1998. Allometric scaling of plant energetics and population density. *Nature* 395: 163–165.

Enquist, B.J., G.B. West, E.L. Charnov, and J.H. Brown. 1999. Allometric scaling of production and life-history variation in vascular plants. *Nature* 401: 907–911.

Enquist, B.J., G.B. West, and J.H. Brown. 2000. Quarter-power allometric scaling in vascular plants: functional basis and ecological consequences. Pages 167–198 *in* J.H. Brown and G.B. West, editors, *Scaling in Biology*. Oxford University Press, Oxford.

Errington, P.L. 1946. Predation and vertebrate populations. *Quarterly Review of Biology* 21: 144–177, 221–245.

Estes, J.A., K. Crooks, and R. Holt. 2001. Predators, ecological role of. Pages 857–878 *in* S. Levin, editor, *Encyclopedia of Biodiversity*, Vol. 4. Academic Press, San Diego, CA.

Faber, W. 1967. Beiträge zur Kenntnis sozialparasitischer Ameisen, 1: *Lasius* (*Austrolasius* n.sg.) *reginae* n.sp., eine neue temporär sozialparasitische Erdameise aus Österreich (Hym. Formicidae). *Pflanzenschutz-Berichte* 36 (5–7): 73–107.

Farmer, E.E. and C.A. Ryan. 1990. Interplant communication: airborne methyl jasmonate induces synthesis of proteinase inhibitors in plant leaves. *Proceedings of the National Academy of Sciences (USA)* 87: 7713–7716.

Farmer, E.E. and C.A. Ryan. 1992. Octadecanoid precursors of jasmonic acid activate the synthesis of wound-inducible proteinase inhibitors. *Plant Cell* 4: 129–134.

Feeny, P.P. 1970. Seasonal changes in oak leaf tannins and nutrients as a cause of spring feeding by winter moth caterpillars. *Ecology* 51: 565–581.

Feeny, P.P. 1976. Plant apparency and chemical defense. Pages 1–40 *in* J. Wallace and R.L. Mansell, editors, *Biochemical Interactions Between Plants and Insects*. Plenum Press, New York, NY.

Fenchel, T. 1975. Character displacement and coexistence in mud snail (Hydrobiidae). *Oecologia* 20: 1–17.

Fisher, R.A. 1930. *The Genetical Theory of Natural Selection*. Clarendon Press, Oxford.

Fons, R., F. Poitevin, J. Catalan, and H. Croset. 1997. Decrease in litter size in the shrews *Crocidura suaveolens* (Mammalia, Insectivora) from Corsica (France): evolutionary response to insularity? *Canadian Journal of Zoology* 75: 954–958.

Foppen, R.P.B., J.P. Chardon, and W. Liefveld. 2000. Understanding the role of sinkpatches in source–sink metapopulations: reed warblers in an agricultural landscape. *Conservation Biology* 14: 1881–1892.

Forbes, S., S. Thornton, B. Glassey, M. Forbes, and N.J. Buckley. Why parent birds play favorites. *Nature* 390: 351–352.

Foster, M.A., J.C. Schultz, and M.D. Hunter. 1992. Modelling gypsy moth–virus–leaf chemistry interactions: implications of plant quality for pest and pathogen dynamics. *Journal of Animal Ecology* 61: 509–520.

Fowler, C.S. 1981. Density dependence as related to life history strategy. *Ecology* 62: 602–610.

Fox, J.F. 1978. Forest fires and the snowshoe hare–Canada lynx cycle. *Oecologia* 31: 349–374.

Fraenkel, G.S. 1959. The raison d'être of secondary plant substances. *Science* 129: 1466–1470.

Fretwell, S.D. 1986. Distribution and abundance of the dickcissel. *Current Ornithology* 4: 211–242.

Fritts, T.H. 1988. The brown tree snake, *Boiga irregularis*, a threat to Pacific islands. *US Fish and Wildlife Service Biological Report* 88 (31). 36 pp.

Gause, G.F. 1932. Experimental studies on the struggle for existence. I. Mixed populations of two species of yeast. *Journal of Experimental Biology* 9: 399–402.

Gause, G.F. 1934. *The Struggle for Existence*. Williams and Wilkins, Baltimore, MD. Reprinted in 1964 by Macmillan (Hafner Press), New York, NY.

Gehlbach, F.R. 1972. Coral snake mimicry reconsidered: the strategy of self-mimicry. *Forma Functio* 5: 311–320.

Gershenzon, J. 1994. The cost of plant chemical defense against herbivory: a biochemical perspective. Pages 105–173 *in* E.A. Bernays, editor, *Insect–Plant Interactions*, Vol. 5. CRC Press, Boca Raton, LA.

Gibbons, J.W. 1987. Why do turtles live so long? *Bioscience* 73: 262–269.

Gilg, O., I. Hanski, and B. Sittler. 2003. Cyclic dynamics in a simple vertebrate predator–prey community. *Science* 302: 866–868.

Gilpin, M.E. and F.J. Ayala. 1973. Global models of growth and competition. *Proceedings of the National Academy of Sciences (USA)* 70: 3590–3593.

Gleadow, R.M. and I.E. Woodrow. 2000. Temporal and spatial variation in cyanogenic glycosides in *Eucalyptus cladocalyx*. *Tree Physiology* 20: 591–598.

Gleadow, R.M. and I.E. Woodrow. 2002. Constraints on effectiveness of cyanogenic glycosides in herbivore defense. *Journal of Chemical Ecology* 28: 1301–1313.

Gog, J., R. Woodroffe, and J. Swinton. 2002. Disease in endangered metapopulations: the importance of alternative hosts. *Proceedings of the Royal Society of London B* 269: 671–676.

Gould, S.J. 1979. An allometric interpretation of species–area curves: the mean of the coefficient. *American Naturalist* 114: 335–343.

Grant, P.R. 1999. *Ecology and Evolution of Darwin's Finches*. Princeton University Press, Princeton, NJ.

Grant, P.R. and B.R. Grant. 1992. Demography and genetically effective sizes of two populations of Darwin's finches. *Ecology* 73: 766–784.

Greene, E. 1989. A diet-induced developmental polymorphism in a caterpillar. *Science* 243: 643–645.

Greene, H.W. and R.W. McDiarmid. 1981. Coral snake mimicry: does it occur? *Science* 213: 1207–1212.

Grime, J.P. 1977. Evidence for the existence of three primary strategies in plants and its relevance to ecological and evolutionary theory. *American Naturalist* 111: 1169–1194.

Grinnell, J. 1917. The niche-relationships of the California thrasher. *The Auk* 34: 427–433.

Grover, J.P. 1997. *Resource Competition*. Chapman and Hall, London.

Gurevitch, J., L.L. Morow, A. Wallace, and J.S. Walsh. 1992. A meta-analysis of competition in field experiments. *American Naturalist* 140: 539–572.

Gurney, W.S.C. and R.M. Nisbet. 1978. Single species population fluctuations in patchy environments. *American Naturalist* 112: 1075–1090.

Gurney, W.S.C. and R.M. Nisbet. 1998. *Ecological Dynamics*. Oxford University Press, New York, NY.

Gyllenberg, M., G. Soderbacka, and S. Ericsson. 1993. Does migration stabilize local population dynamics? Analysis of a discrete metapopulation model. *Mathematical Biosciences* 118: 25–49.

Hairston, N.G., F.E. Smith and L.B. Slobodkin. 1960. Community structure, population control and competition. *American Naturalist* 94: 421–5.

Hamer, W.H. 1906. Epidemic disease in England. *Lancet* 1: 733–739.

Hansen, S.R. and S.P. Hubbell. 1980. Single-nutrient microbial competition: qualitative agreement between experimental and theoretically forecast outcomes. *Science* 207: 1491–1493.

Hanski, I. 1993. Dynamics of small mammals on islands. *Ecography* 16: 372–375.

Hanski, I. 1994a. A practical model of metapopulation dynamics. *Journal of Animal Ecology* 63: 151–162.

Hanski, I. 1994b. Patch-occupancy dynamics in fragmented landscapes. *Trends in Ecology and Evolution* 9: 131–135.

Hanski, I. 1997. Predictive and practical metapopulation models: the incidence function approach. Pages 21–45 *in* D. Tilman and P Kareiva, editors, *Spatial Ecology*. Princeton University Press, Princeton, NJ.

Hanski, I. 1998. Metapopulation dynamics. *Nature* 396: 41–49.

Hanski, I. 1999. *Metapopulation Ecology*. Oxford University Press, Oxford.

Hanski, I. 2001. Spatially realistic theory of metapopulation ecology. *Naturwissenschaften* 88: 372–381.

Hanski, I. 2002. Metapopulations of animals in highly fragmented landscapes and population viability analysis. Pages 86–108 *in* S.R. Beissinger and D.R. McCullough, editors, *Population Viability Analysis*. University of Chicago Press, Chicago, IL.

Hanski, I. and M.E. Gilpin. 1997. *Metapopulation Biology: Ecology, Genetics and Evolution*. Academic Press, London.

Hanski, I. and M. Kuussaari. 1995. Butterfly metapopulation dynamics. Pages 149–171 *in* N. Cappuccino and P.W. Price, editors, *Population Dynamics: New Approaches and Synthesis*. Academic Press, San Diego, CA.

Hanski, I., T. Pakkala, M. Kuussaari, and G. Lei. 1995. Metapopulation persistence of an endangered butterfly in a fragmented landscape. *Oikos* 72: 21–28.

Hanski, I., P.R. Ehrlich, M. Nieminen, D.D. Murphy, J.J. Hellmann, C.L. Boggs, and J.F. McLaughlin. 2004. Checkerspots and conservation biology. Pages 264–287 *in* P.R. Ehrlich and I. Hanski, editors, *On the Wings of Checkerspots*. Oxford University Press, New York, NY.

Harborne, J.B. 1988. *Introduction to Ecological Biochemistry*. 3rd edition. Academic Press, London.

Harborne, J.B. 1993. *Introduction to Ecological Biochemistry*. 4th edition. Academic Press, London.

Harborne, J.B. 1997. Plant secondary metabolism. Pages 132–155 *in* M.J. Crawley, editor, *Plant Ecology*. Blackwell Scientific, Oxford.

Harder, L.D. and S.C.H. Barrett. 1992. The energy cost of bee pollination for *Pontederia cordata* (Pontederiaceae). *Functional Ecology* 6: 226–233.

Hardin, G. 1960. The competitive exclusion principle. *Science* 131: 1292–1297.

Harrison, S. 1991. Local extinction in a metapopulation context: an empirical evaluation. *Biological Journal of the Linnean Society* 42: 73–88.

Harrison, S. 1994. Metapopulations and conservation. Pages 111–128 *in* J.P. Edwards, R.M. May, and N.R. Webb, editors, *Large-Scale Ecology and Conservation Biology*. Blackwell, London.

Hartley, S.E. and C.G. Jones. 1997. Plant chemistry and herbivory, or why the world is green. Pp. 284–324 *in* M.J. Crawley, editor, *Plant Ecology*. Blackwell Scientific, Oxford.

Harvell, C.D. 1984. Predator-induced defense in a marine bryozoan. *Science* 224: 53–54.

Harvell, C.D. and R. Tollrian. 1999. Why inducible defenses? Pp. 3–9 *in* R. Tollrian and C.D. Harvell, editors, *The Ecology and Evolution of Inducible Defenses*. Princeton University Press, Princeton, NJ.

Hassell, M.P. 1975. Density-dependence in single species populations. *Journal of Animal Ecology* 44: 382–95.

Hassell, M.P. 1976. Arthropod predator–prey systems. Pages 71–93 *in* R.M. May, editor, *Theoretical Ecology: Principles and Applications*. W.B. Saunders, Philadelphia, PA.

Hassell, M.P. 1978. *The Dynamics of Arthropod Predator–Prey Systems*. Princeton University Press, Princeton, NJ.

Hastings, A. 1980. Disturbance, coexistence, history, and competition for space. *Theoretical Population Biology* 18: 363–373.

Heath, S.A. 2002. Wintering waterfowl on the Northern Virginia Piedmont: an unstudied population. MS thesis, George Mason University, Fairfax, VA.

Heinsohn, R. and C. Packer. 1995. Complex cooperative strategies in group-territorial African lions. *Science* 269: 1260–1262.

Heller, H.C. 1971. Altitudinal zonation of chipmunks: interspecific aggression. *Ecology* 52: 312–319.

Hemmingsen, A.M. 1960. Energy metabolism as related to body size and respiratory surfaces, and its evolution. *Reports to the Steno Memorial Hospital and Nordisk Insulin Laboratorium* 9: 6–110.

Hess, G. 1996. Disease in metapopulations: implications for conservation. *Ecology* 77: 1617–1632.

Hethcote, H.W. 2000. The mathematics of infectious diseases. *SIAM Review* 42: 599–653.

Holldobler, B. 1986. Food robbing in ants, a form of interference competition. *Oecologia* 69: 12–15.

Holldobler, B. and E.O. Wilson. 1990. *The Ants*. Belknap Press, Cambridge, MA.

Holling, C.S. 1959. The components of predation as revealed by a study of small mammal predation of the European pine sawfly. *Canadian Entomologist* 91: 293–320.

Holling, C.S. 1961. Principles of insect predation. *Annual Review of Entomology* 6: 163–182.

Holling, C.S. 1966. The functional response of invertebrate predators to prey density. *Memoirs of the Entomological Society of Canada* 48: 1–86.

Holling, C.S. 1973. Resilience and stability of ecological systems. *Annual Review of Ecology and Systematics* 4: 1–24.

Holloway, J.K. 1964. Host specificity of a phytophagous insect. *Weeds* 12: 25–27.

Horn, H.S. 2004. Commentary on Brown *et al*. "Toward a metabolic theory of ecology." *Ecology* 85: 1816–1818.

Houston, D.B. 1982. *The Northern Yellowstone Elk: Ecology and Management*. MacMillan, New York, NY.

Hubbell, S.P. 2001. *The Unified Theory of Biodiversity and Biogeography*. Princeton University Press, Princeton, NJ.

Hubbell, S.P. and R. Foster. 1986a. Biology, chance and history and the structure of tropical rainforest tree communities. Pages 314–329 *in* J. Diamond and T.J. Cawe, editors, *Community Ecology*. Harper and Row, New York, NY.

Hubbell, S.P. and R. Foster. 1986b. Commonness and rarity in a Neotropical forest: implications for tropical tree conservation. Pages 205–231 *in* M.E. Soule, editor, *Conservation Biology: The Science of Scarcity and Diversity*. Sinauer Associates, Sunderland, MA.

Hubbell, S.P. and P.A. Werner. 1979. On measuring the intrinsic rate of increase of populations with heterogeneous life histories. *American Naturalist* 113: 277–293.

Hubbell, S.P., R.B. Foster, S.T. O'Brien, K.E. Harms, R. Condit, B. Wechsler, S.J. Wright and S. Loo de Lao. 1999. Light-gap disturbances, recruitment limitation, and tree diversity in a Neotropical forest. *Science* 283: 554–557.

Hudson, P.J. and O.N. Bjørnstad. 2003. Vole stranglers and lemming cycles. *Science* 302: 797–798.

Hudson, P.J., A.P. Dobson, and D. Newborn. 1998. Prevention of population cycles by parasite removal. *Science* 282: 2256–2258.

Huff, D.E. and J.D. Varley. 1999. Natural regulation in Yellowstone National Park's northern range. *Ecological Applications* 9: 17–29.

Huffaker, C.B. 1958. Experimental studies on predation: dispersion factors and predator–prey oscillations. *Hilgardia* 27: 343–383.

Humphrey, S.R., C.H. Courtney, and D.J. Forrester. 1978. Community ecology of helminth parasites of the brown pelican. *The Wilson Bulletin* 90: 587–598.

Hutchinson, G.E. 1957. Concluding remarks. *Cold Spring Harbor Symposium on Quantitative Biology* 22: 415–427.

Hutchinson, G.E. 1959. Homage to Santa Rosalia, or why are there so many kinds of animals? *American Naturalist* 93: 145–159.

Hutchinson, G.E. 1961. The paradox of the plankton. *American Naturalist* 95: 137–147.

Hutchinson, G.E. 1978. *An Introduction to Population Ecology.* Yale University Press, New Haven, CT.

Janzen, D.H. 1966. Coevolution of mutualism between ants and acacias in Central America. *Evolution* 20: 249–275.

Janzen, D.H. 1974. Blackwater rivers, animals, and mast fruiting by the Dipterocarpaceae. *Biotropica* 6: 69–103.

Janzen, D.H. 1976. Why bamboos wait so long to flower. *Annual Review of Ecology and Systematics* 7: 347–91.

Janzen, D.H. 1983. *Costa Rican Natural History.* University of Chicago Press, Chicago, IL.

Janzen, D.H. 1985. The natural history of mutualisms. Pages 40–99 *in* D.H. Boucher, editor, *The Biology of Mutualisms.* Oxford University Press, Oxford.

Jarman, P.J. and M.V. Jarman. 1973. Social behavior, population structure and reproduction in impala. *East African Wildlife Journal* 11: 329–38.

Jarman, P.J. and S.M. Wright. 1993. Macropod studies at Wallaby Creek. IX. Exposure and responses of eastern grey kangaroos to dingos. *Wildlife Research* 20: 833–43.

Jenni, D.A. 1983. *Jacana spinosa.* Pages 584–586 *in* D.H. Janzen, editor, *Costa Rican Natural History.* University of Chicago Press, Chicago, IL.

Johnson, M.L. and M.S. Gaines. 1990. Evolution of dispersal: theoretical models and empirical tests using birds and mammals. *Annual Review of Ecology and Systematics* 21: 449–480.

Johnson, N.C., J.H. Graham, and F.A. Smith. 1997. Functioning of mycorrhizal associations along with the mutualism–parasitism continuum. *New Phytologist* 135: 575–585.

Johnston, J.P., W.J. Peach, R.D. Gregory, and S.A. White. 1997. Survival rates of tropical and temperate passerine birds: A Trinidadian perspective. *American Naturalist* 150: 771–789.

Karban, R. 1997. Evolution of prolonged development: a life table analysis for periodical cicadas. *American Naturalist* 150: 446–461.

Karban, R. and I.T. Baldwin. 1997. *Induced Responses to Herbivory.* University of Chicago Press, Chicago, IL.

Karban, R. and J.H. Myers. 1989. Induced plant responses to herbivory. *Annual Review of Ecology and Systematics* 20: 331–348.

Kendall, B.E., J. Prendergast, and O.N. Bjørnstad. 1998. The macroecology of population dynamics: taxonomic and biogeographic patterns in population cycles. *Ecology Letters* 1: 160–164.

Kenward, R.E. 1978. Hawks and doves: factors affecting success and selection in goshawk attacks on wood pigeons. *Journal of Animal Ecology* 47: 449–460.

Kermack, W.O. and A.G. McKendrick. 1927. Contributions to the mathematical theory of epidemics, part I. *Proceedings of the Royal Society of London A* 115: 700–721.

Kindvall, O. 1995. The impact of extreme weather on habitat preference and survival in a metapopulation of the bush cricket *Metrioptera bicolor* in Sweden. *Biological Conservation* 73: 51–58.

Kindvall, O. 2000. Comparative precision of three spatially realistic simulation models of metapopulation dynamics. *Ecological Bulletins* 48: 101–110.

Kingsland, S. 1995. *Modeling Nature: Episodes in the History of Population Ecology.* 2nd edition. University of Chicago Press, Chicago, IL.

Kingsland, S. 2004. Conveying the intellectual challenge of ecology: an historical perspective. *Frontiers in Ecology and the Environment* 2: 267–374.

Kleiber, M. 1932. Body size and metabolism. *Hilgardia* 6: 315–322.

Korpimaki, E. and K. Norrdahl. 1998. Experimental reduction of predators reverses the crash phase of small-rodent cycles. *Ecology* 79: 2448–2455.

Krebs, C.J. 1994. *Ecology.* 4th edition. Harper Collins, New York, NY.

Krebs. C.J., S. Boutin, R. Boonstra, A.R.E. Sinclair, J.N.M. Smith, M.R.T. Dale, K. Martin, and R. Turkington. 1995. Impact of food and predation on the snowshoe hare cycle. *Science* 269: 1112–1115.

Krebs, C.J., R. Boonstra, S. Boutin, and A.R.E. Sinclair. 2001. What drives the 10-year cycle of snowshoe hares? *Bioscience* 51: 25–35.

Lack, D. 1947. *Darwin's Finches.* Cambridge University Press, Cambridge.

Lack, D. 1954. *The Natural Regulation of Animal Numbers.* Oxford University Press, London.

Lack, D. 1966. *Population Studies of Birds.* Clarendon Press, Oxford.

Lack, D. 1968. *Ecological Adaptations for Breeding in Britain.* Methuen, London.

Lanciani, C.A. 1998. A simple equation for presenting reproductive value to introductory biology and ecology classes. *ESA Bulletin* 79: 192–3.

Laughlin, R. 1965. Capacity for increase: a useful population statistic. *Journal of Animal Ecology* 34: 77–91.

Laurance, S.G.W. and W.F. Laurance. 2003. Bandages for wounded landscapes: faunal corridors and their role in wildlife conservation in the Americas. Pages 313–325 in G.A. Bradshaw and P.A. Marquet, editors, *How Landscapes Change.* Ecological Studies 162. Springer-Verlag, Berlin.

Law, R. 1988. Some ecological properties of intimate mutualisms involving plants. Pages 315–342 *in* A.J. Davy, M.J. Hutchings and A.R. Watkinson, editors. *Plant Population Ecology.* Blackwell Scientific, Oxford.

Lawrey, J.D. 1991. The species–area curve as an index of disturbance in saxicolous lichen communities. *The Bryologist* 94: 377–382.

Lawrey, J.D. 1992. Natural and randomly-assembled lichen communities compared using the species–area curve. *The Bryologist* 95: 137–141.

Lee, T.D. and F.A. Bazzaz. 1980. Effects of defoliation and competition on growth and reproduction in the annual plant *Abutilon theophrasti. Journal of Ecology* 68: 813–821.

Lefkovitch, L.P. 1965. The study of population growth in organisms grouped by stages. *Biometrics* 21: 1–18.

Lehman, C.L. and D. Tilman. 1997. Competition in spatial habitats. Pages 185–203 *in* D. Tilman and P. Kareiva, editors, *Spatial Ecology.* Princeton University Press, Princeton, NJ.

Leirs, H., N.C. Stenseth, J.D. Nichols, J.E. Hines, R. Verhagen, and W. Verheyen. 1997. Stochastic seasonality and nonlinear density-dependent factors regulate population size in an African rodent. *Nature* 389: 176–180.

Leopold, A. 1943. Deer irruptions. *Wisconsin Conservation Department Publication* 321: 3–11.

Leopold, A. 1949. *A Sand County Almanac, and Sketches Here and There.* Reprinted 1977. Oxford University Press, Oxford.

Leslie, P.H. 1945. On the use of matrices in certain population mathematics. *Biometrika* 35: 183–212.

Leslie, P.H., J.S. Tener, M. Vizoso, and H. Chitty. 1955. The longevity and fertility of the Orkney vole, *Microtus orcadensis*, as observed in the laboratory. *Proceedings of the Zoological Society of London* 125: 115–125.

Lessells, C.M. 1991. The evolution of life histories. Pages 32–68 *in* J.R. Krebs and N.B. Davies, editors, *Behavioral Ecology.* 3rd edition. Blackwell, Oxford.

Lett, P.F., R.K. Mohn, and D.F. Gray. 1981. Density-dependent processes and management strategy for the northwest Atlantic harp seal populations. Pages 135–158 *in* C.W. Fowler and T.D. Smith, editors, *Dynamics of Large Mammal Populations.* Wiley, New York, NY.

Leverlich, W.J. and D.A. Levin. 1979. Age specific survivorship and reproduction in *Phlox drummondii. American Naturalist* 113: 881–903.

Levin, D.A. 1976. The chemical defenses of plants to pathogens and herbivores. *Annual Review of Ecology and Systematics* 7: 121–159.

Levins, R. 1968. *Evolution in Changing Environments.* Princeton University Press, Princeton, NJ.

Levins, R. 1969. Some demographic and genetic consequences of environmental heterogeneity for biological control. *Bulletin of the Entomological Society of America* 15: 237–240.

Levins, R. 1970. Extinction. Pages 77–104 in M. Gerstenhaver, editor, *Some Mathematical Problems in Biology*. American Mathematical Society, Providence, RI.

Levins, R. and D. Culver. 1971. Regional coexistence of species and competition between rare species. *Proceedings of the National Academy of Sciences (USA)* 68: 1246–1248.

Levins, R., M.L. Pressick, and H. Heatwole. 1973. Coexistence patterns in insular ants. *American Scientist* 61: 463–472.

Lewontin, R.C. 1965. Selection for colonizing ability. Pages 77–94 in H.G. Baker and G.L. Stebbins, editors, *The Genetics of Colonizing Species*. Academic Press, New York, NY.

Lieberburg, I, P.M. Kranz and A. Seip. 1975. Bermudian ants revisited: the status and interaction of *Pheidole megacephala* and *Iridomyrmex humilis*. *Ecology* 56: 473–478.

Liebig, J.F. 1840. *Chemistry in its Application to Agriculture and Physiology*. Taylor and Walton, London.

Lindstrom, E.R., H. Andrén, P. Angelstam *et al*. 1994. Disease reveals the predator: sarcoptic mange, red fox predation, and prey populations. *Ecology* 75: 1042–1049.

Lomborg, B. 2001. *The Skeptical Environmentalist*. Cambridge University Press, Cambridge.

Lonsdale, W.M. 1990. The self-thinning rule: dead or alive? *Ecology* 71: 1373–1388.

Lotka, A.J. 1925. *Elements of Physical Biology*. Reprinted 1956. Dover, New York, NY.

Luckinbill, L.L. 1973. Coexistence in laboratory populations of *Paramecium aurelia* and its predator, *Didinium nasutum*. *Ecology* 54: 1320–1327.

MacArthur, R.H. 1958. Population ecology of some warblers of northeastern coniferous forests. *Ecology* 39: 599–619.

MacArthur, R.H. 1972. *Geographical Ecology: Patterns in the Distribution of Species*. Harper and Row, New York, NY.

MacArthur, R.H. and E.O. Wilson. 1963. An equilibrium theory of insular zoogeography. *Evolution* 17: 373–387.

MacArthur, R.H. and E.O. Wilson. 1967. *The Theory of Island Biogeography*. Princeton University Press, Princeton, NJ.

Major, P. 1978. Predator–prey interactions in two schooling fishes, *Caranax ignobilis* and *Stalephorus purpureus*. *Animal Behaviour* 26: 760–777.

Marquis, R.J. 1984. Leaf herbivores decrease fitness of a tropical plant. *Science* 226: 537–539.

Marquis, R.J. 1992. A bite is a bite? Constraints on response to folivory in *Piper arieianum* (Piperaceae). *Ecology* 73: 143–152.

Marquis, R.J. and C.J. Whelan. 1994. Insectivorous birds increase growth of white oak through consumption of leaf-chewing insects. *Ecology* 75: 2007–2014.

Martin, T.E., P.R. Martin, C.R. Olson, B.J. Heidinger, and J.J. Fontaine. 2000. Parental care and clutch sizes in North and South American birds. *Science* 287: 1482–1485.

Mattiacci, L., M. Dicke, and M.A. Posthumus. 1995. Betaglucosidase: an elicitor of herbivore-induced plant odor that attracts host-searching parasitic wasps. *Proceedings of the National Academy of Sciences (USA)* 92: 2036–2040.

May, R.M. 1973. Stability in randomly fluctuating versus deterministic environments. *American Naturalist* 107: 621–650.

May, R.M. 1974. Biological populations with non-overlapping generations: stable points, stable cycles and chaos. *Science* 186: 645–647.

May, R.M. 1975a. *Stability and Complexity in Model Ecosystems*. 2nd edition. Princeton University Press, Princeton, NJ.

May, R.M. 1975b. Biological populations obeying difference equations. Stable points, stable cycles and chaos. *Journal of Theoretical Biology* 49: 511–524.

May, R.M. 1976a. Estimating *r*: a pedagogical note. *American Naturalist* 110: 496–499.

May, R.M. 1976b. Models for single populations. Pages 4–25 in R.M. May, editor, *Theoretical Ecology: Principles and Applications*. W.B. Saunders, Philadelphia, PA.

May, R.M. 1976c. Models for two interacting populations. Pages 49–70 *in* R.M. May, editor, *Theoretical Ecology: Principles and Applications.* W.B. Saunders, Philadelphia, PA.

May, R.M. 1981a. Models for single populations. Pages 5–29 *in* R.M. May, editor, *Theoretical Ecology: Principles and Applications.* 2nd edition. Blackwell Scientific, Oxford.

May, R.M. 1981b. Models for two interacting populations. Pages 78–104, *in* R.M. May, editor, *Theoretical Ecology: Principles and Applications.* 2nd edition. Blackwell Scientific, Oxford.

May, R.M. 1983. Parasitic infections as regulators of animal populations. *American Scientist* 71: 36–45.

May, R.M. 1999. Crash tests for real. *Nature* 398: 371–372.

May, R.M. and R.M. Anderson. 1979. Population biology of infectious diseases: Part II. *Nature* 280: 455–461.

May, R.M. and G.F. Oster. 1976. Bifurcations and dynamic complexity in simple ecological models. *American Naturalist* 110: 573–589.

McCullough, D.R. 1981. Population dynamics of the Yellowstone grizzly. Pages 173–196 *in* W. Fowler and T.D. Smith, editors, *Dynamics of Large Mammal Populations.* Wiley, New York, NY.

McCullough, D.R. 1997. Irruptive behavior in ungulates. Pages 69–98 *in* W.J. McShea, H.B. Underwood, and J.H. Rappole, editors, *The Science of Overabundance.* Smithsonian Institution Press, Washington, DC.

McKey, D.B., P.G. Waterman, C.N. Mbi, J.S. Gartlin, and T.T. Strusaker. 1978. Phenolic content of vegetation in two African rainforests: ecological implications. *Science* 202: 61–64.

McNaughton, S.J. 1986. On plants and herbivores. *American Naturalist* 128: 765–770.

Meagher, M.M. and D.B. Houston. 1998. *Yellowstone and the Biology of Time: Photographs Across a Century.* University of Oklahoma Press, Norman, OK.

Mech, L.D. 1966. *The Wolves of Isle Royale.* Fauna of the National Parks of the United States, Fauna Series, Washington, DC.

Mech, L.D., R.E. McRoberts, R.O. Peterson, and R.E. Page. 1987. Relationship of deer and moose populations to previous winter's snow. *Journal of Animal Ecology* 56: 615–625.

Merendino, M.T., C.O. Ankey, and D.G. Dennis. 1993. Increasing mallards, decreasing black ducks: More evidence for cause and effect. *Journal of Wildlife Management* 57: 199–208.

Mertz, D.B. 1970. Notes on methods used in life history studies. Pages 4–17 *in* J.H. Connell, D.B. Mertz, and W.W. Murdoch, editors, *Readings in Ecology and Ecological Genetics.* Harper and Row, New York, NY.

Messier, F. 1991. The significance of limiting and regulating factors on the demography of moose and white-tailed deer. *Journal of Animal Ecology* 60: 377–393.

Messier, F. 1994. Ungulate population models with predation: a case study with the North American moose. *Ecology* 75: 478–488.

Millburn, P. 1978. Biotransformation of xenobiotics by animals. Pages 35–76 *in* J.B. Harborne, editor, *Biochemical Aspects of Plant and Animal Coevolution.* Academic Press, London.

Miller, P.S., and R.C. Lacy. 2003. Appendix I: an overview of population viability analysis using VORTEX. Pages 101–121 in VORTEX: a Stochastic Simulation of the Extinction Process. Conservation Breeding Specialist Group (SSC/IUCN), Apple Valley, MN.

Milton, W. 1940. The effects of manuring, grazing, and cutting on the yield, botanical and chemical composition of natural hill pastures. *Journal of Ecology* 28: 326–356.

Moczek, A.P. and D.J. Emlen. 1999. Proximate determination of male horn dimorphism in the beetle *Onthophagus taurus* (Coleoptera: Scarabaeidae). *Journal of Evolutionary Biology* 12: 27–37.

Moilanen, A., A.T. Smith, and I. Hanski. 1998. Long-term dynamics in a metapopulation of the American pika. *American Naturalist* 152: 530–42.

Monod, J. 1950. La technique de culture continue, théorie et applications. *Annales d'Institut Pasteur* 79: 390–410.

Morris, E.O. 1856. *A History of British Birds.* Broombridge, London.

Morris, W.F. and D.F. Doak. 2002. *Quantitative Conservation Biology: Theory and Practice of Population Viability Analysis.* Sinauer Associates, Sunderland, MA.

Murdoch, W.W. 1994. Population regulation in theory and practice. *Ecology* 75: 271–287.

Murdoch, W.W. and S.J. Walde. 1989. Analysis of insect population dynamics. Pages 113–140 *in* P.J. Grubb and J.B. Whittaker, editors, *Towards a More Exact Ecology*. Blackwell Scientific, Oxford.

Murie, A. 1934. The moose of Isle Royale. *University of Michigan Museum of Zoology Miscellaneous Publications* 25. Ann Arbor, MI.

Murray, K.G., S. Kinsman, and J.L. Bronstein. 2000. Plant–animal interactions. Pages 245–267 *in* N.M. Nadkarni and N.T. Wheelwright, editors, *Monteverde: Ecology and Conservation of a Tropical Cloud Forest*. Oxford University Press, New York, NY.

National Research Council (NRC), Board on Biology Commission on Life Sciences, Committee on Management of Wolf and Bear Populations in Alaska. 1997. *Wolves, Bears and Their Prey in Alaska: Biological and Social Challenges in Wildlife Management*. National Academy Press, Washington, DC.

Nee, S. and R.M. May. 1992. Dynamics of metapopulations: habitat destruction and competitive coexistence. *Journal of Animal Ecology* 61: 37–40.

Neill, W.E. 1974. The community matrix and interdependence of the competition coefficients. *American Naturalist* 108: 399–408.

Neyman, J., T. Park, and E.J. Scott. 1956. Struggle for existence. The *Tribolium* model: biological and statistical aspects. Pages 41–79 *in* Proceedings of the Third Berkeley Symposium on Mathematical Statistics and Probability. Vol. IV. University of California Press, Berkeley, CA.

Nicholson, A.J. 1954. An outline of the dynamics of animal populations. *Australian Journal of Zoology* 2: 9–65.

Nicholson, A.J. 1957. The self-adjustment of populations to change. *Cold Spring Harbor Symposium on Quantitative Biology* 22: 153–172.

Nicholson, A.J. and V.A. Bailey. 1935. The balance of animal populations, part I. *Proceedings of the Zoological Society of London* 3: 551–598.

Nieminen, M., M. Siljander, and I. Hanski. 2004. Structure and dynamics of *Melitaea cinxia* metapopulations. Pages 63–91 *in* P.R. Ehrlich and I. Hanski, editors, *On the Wings of Checkerspots*. Oxford University Press, New York, NY.

Nisbet, R.M. and W.S.C. Gurney. 1982. *Modeling Fluctuating Populations*. John Wiley, New York, NY.

Nokes, D.J. 1992. Microparasites: viruses and bacteria. Pages 349–376 *in* M.J. Crawley, editor, *Natural Enemies: the Population Biology of Predators, Parasites and Diseases*. Blackwell Scientific, Oxford.

Norrdahl, K. and E. Korpimaki. 1995. Effects of predator removal on vertebrate prey populations: birds of prey and small mammals. *Oecologia* 103: 241–248.

O'Donoghue, M., E. Hofer, and F.E. Doyle. 1995. Predator versus predator. *Natural History* 104: 6–9.

O'Dowd, D.J. 1979. Foliar nectar production and ant activity in a Neotropical tree, *Ochroma pyramidale*. *Oecologia* 43: 233–248.

O'Dowd, D.J. 1980. Pearl bodies of a Neotropical tree, *Ochroma pyramidale*: Ecological implications. *American Journal of Botany* 67: 543–549.

Ohnmeiss, T.E. and I.T. Baldwin. 2000. Optimal defense theory predicts the ontogeny of an induced nicotine defense. *Ecology* 81: 1765–1783.

Oksanen, L., S.D. Fretwell, J. Arruda, and P. Niemela. 1981. Exploitation ecosystems in gradients of primary productivity. *American Naturalist* 118: 240–261.

Okubo, A. 1980. *Diffusion and Ecological Problems: Mathematical Models*. Springer-Verlag. Berlin.

Olff, H., V.K. Brown, and R.H. Drent. 1999. *Herbivores: Between Plants and Predators*. Blackwell Science, Oxford, UK.

Paige, K.N. and T.G. Whitham. 1987. Overcompensation in response to mammalian herbivory. *American Naturalist* 129: 407–416.

Paine, R.T. 1966. Food web complexity and species diversity. *American Naturalist* 100: 65–75.

Park, T. 1948. Experimental studies of interspecific competition. I. Competition between populations of the flour beetles, *Tribolium confusum* Duvall and *Tribolium castaneum* Herbst. *Ecological Monographs* 18: 267–307.

Park, T. 1954. Experimental studies of interspecific competition. II. Temperature, humidity, and competition in two species of *Tribolium*. *Physiological Zoology* 27: 177–238.

Park, T. 1962. Beetles, competition, and populations. *Science* 138: 1369–1375.

Park, T., P.H. Leslie, and D.B. Mertz. 1964. Genetic strains and competition in populations of *Tribolium*. *Physiological Zoology* 37: 97–162.

Passera, L., E. Roncin, B. Kaufman, and L. Keller. 1996. Increased soldier production in ant colonies exposed to intraspecific competition. *Nature* 379: 630–631.

Pastor, J., R.J. Naiman, B. Dewey, and P. McInnes. 1988. Moose, microbes, and the boreal forest. *Bioscience* 38: 770–777.

Pearl, P. 1927. The growth of populations. *Quarterly Review of Biology* 2: 532–548.

Pearl, P. and J.L. Reed. 1920. On the growth of the population of the United States since 1790 and its mathematical representation. *Proceedings of the National Academy of Sciences (USA)* 6: 275–288.

Peltonen, A. and I. Hanski. 1991. Patterns of island occupancy explained by colonization and extinction rates in shrews. *Ecology* 72: 1698–1708.

Peters, G.L. and R.P. Larkin. 1989. *Population Geography*. 3rd edition. Kendall/Hunt, Dubuque, IA.

Peterson, D.L. and V.T. Parker. 1998. *Ecological Scale*. Columbia University Press, New York, NY.

Peterson, R.O. 1999. Wolf–moose interaction on Isle Royale: the end of natural regulation? *Ecological Applications* 9: 10–16.

Peterson, R.O., R.E. Page, and K.M. Dodge. 1984. Wolves, moose, and the allometry of population cycles. *Science* 224: 1350–1352.

Peterson, R.O., N.J. Thomas, J.M. Thurber, J.A. Vucetich, and T.A. Waits. 1998. Population limitation and wolves of Isle Royale. *Journal of Mammalogy* 79: 828–841.

Petranka, J.W. and A. Sih. 1986. Environmental instability, competition, and density-dependent growth and survivorship of a stream-dwelling salamander. *Ecology* 67: 729–736.

Petrie, M. and B. Kempenaers. 1998. Extra-pair paternity in birds: Explaining variation between species and populations. *Trends in Ecology and Evolution* 13: 52–58.

Pianka, E. 1970. On *r* and *K* selection. *American Naturalist* 104: 592–597.

Pielou, E.C. 1977. *Mathematical Ecology*. 2nd edition. Wiley-Interscience, New York, NY.

Pimentel, D. 1988. Herbivore population feeding pressure on plant hosts: feedback evolution and host conservation. *Oikos* 53: 289–302.

Pitelka, F.A. 1957. Some characteristics of micotine cycles in the Arctic. Pages 73–88 *in* H.P. Hansen, editor, *Arctic Biology*. Oregon State University Press, Corvalis, OR.

Pokki, J. 1981. Distribution, demography, and dispersal of the field vole, *Microtus agrestis* (L.), in the Tvärminne archipelago, Finland. *Acta Zoologica Fennica* 164: 1–48.

Porter, K.G., J. Gerritsen, and J.D. Orcutt, Jr. 1982. The effect of food concentration on swimming patterns, feeding behavior, ingestion, assimilation, and respiration by *Daphnia*. *Limnology and Oceanography* 27: 935–949.

Porter, K.G., J.D. Orcutt, Jr., and J. Gerritsen. 1983. Functional response and fitness in a generalist filter feeder, *Daphnia magna* (Crustacea: Cladocera). *Ecology* 64: 735–742.

Pough, H. 1988. Mimicry of vertebrates: are the rules different? *American Naturalist* 131 (suppl.): S67–S102.

Preston, F.W. 1962. The canonical distribution of commonness and rarity. *Ecology* 43: 185–215.

Primack, R.B. and P. Hall. 1990. Cost of reproduction in the pink lady's slipper orchid: a four-year experimental study. *American Naturalist* 136: 638–656.

Pulliam, H.R. 1988. Sources, sinks, and population regulation. *American Naturalist* 132: 652–661.

Pulliam, H.R. 1996. Sources and sinks: empirical evidence and population consequences. Pages 45–70 *in* O.E. Rhodes, Jr., R.K. Chester, and M.H. Smith, editors, *Population Dynamics in Ecological Space and Time*. University of Chicago Press, Chicago, IL.

Pyne, S.J. 1997. *Fire in America: a Cultural History of Wildland and Rural Fire*. University of Washington Press, Seattle, WA.

Quinn, J.F. and S.P. Harrison. 1988. Effects of habitat fragmentation and isolation on species richness. Evidence from biogeographic patterns. *Oecologia* 75: 132–40.

Ralls, K., R.L. Brownell, Jr., and J. Ballou. 1980. Differential mortality by sex and age in mammals, with specific reference to the sperm whale. *Report of the International Whaling Commission* (Special Issue 2): 233–243.

Randall, J.E. 1965. Grazing effect on sea grasses by herbivorous reef fishes in the West Indies. *Ecology* 46: 255–260.

Rasmussen, D.I. 1941. Biotic communities of Kiabab Plateau, Arizona. *Ecological Monographs* 3: 229–275.

Reid, D.G., C.J. Krebs, and A.J. Kenney. 1997. Patterns of predation on noncyclic lemmings. *Ecological Monographs* 67: 89–108.

Reid, W.V. 1987. The cost of reproduction in the glaucous-winged gull. *Oecologia* 74: 458–467.

Reznik, D.N. and J. Endler. 1982. The impact of predation on life history evolution in Trinidadian guppies (*Poecilia reticulata*). *Evolution* 36: 160–177.

Reznik, D.N., F.H. Shaw, H.F. Rodd, and R.G. Shaw. 1997. Evaluation of the rate of evolution in Trinidadian guppies: Genetic basis of observed life history patterns. *Evolution* 36: 1236–1250.

Reznick, D.N., M.J. Bryant, and F. Bashey. 2002. r- and K-selection revisited: the role of population regulation in life-history evolution. *Ecology* 83: 1509–1520.

Rhoades, D.F. 1983. Responses of alder and willow to attack by tent caterpillars and weborms: Evidence for phermonal sensitivity of willows. Pages 55–58 *in* P.A. Hedin, editor, *Plant Resistance to Insects*. Symposium series 208. American Chemical Society, Washington, DC.

Ricker, W.E. 1952. Stock and recruitment. *Journal of the Fisheries Research Board of Canada* 11: 559–623.

Ricklefs, R.E. 1990. *Ecology*. 3rd edition. W.H. Freeman, New York, NY.

Ricklefs, R.E. 1997. *The Economy of Nature*. 4th edition. W.H. Freeman, New York, NY.

Ripple, W.J. and R.L. Beschta. 2003. Wolf reintroduction, predation risk, and cottonwood recovery in Yellowstone National Park. *Forest Ecology and Management* 184: 299–313.

Ripple, W.J. and R.L. Beschta. 2004. Wolves and the ecology of fear: can predation risk structure ecosystems? *Bioscience* 54: 755–766.

Ritland, D.B. and L.P. Brower. 1991. The viceroy is not a Batesian mimic. *Nature* 350: 497–498.

Robles, C.D. and R.A. Desharnais. 2002. History and current development of a paradigm of predation in rocky intertidal communities. *Ecology* 83: 1521–1536.

Rockwood, L.L. 1973. The effect of defoliation on seed production of six Costa Rican tree species. *Ecology* 54: 1363–1369.

Rockwood, L.L. 1974. Seasonal changes in the susceptibility of *Crescentia alata* leaves to the flea beetle, *Oedionychus*, sp. *Ecology* 55: 142–148.

Rockwood, L.L. 1976. Plant selection and foraging in two species of leaf-cutting ants, *Atta*. *Ecology* 57: 48–61.

Rockwood, L.L. 1977. Foraging patterns and plant selection in Costa Rican leaf-cutting ants. *Journal of the New York Entomological Society* 85: 222–233.

Rockwood, L.L. and K.E. Glander. 1979. Howling monkeys and leaf-cutting ants: comparative foraging in a tropical deciduous forest. *Biotropica* 11: 1–10.

Rockwood, L.L. and M.B. Lobstein. 1994. The effects of experimental defoliation on reproduction in four species of herbaceous perennials from Northern Virginia. *Castanea* 59: 41–50.

Rodda, G.H. and T.H. Fritts. 1992. The impact of the introduction of the Colubrid snake *Boiga irregularis* on Guam's lizards. *Journal of Herpetology* 26: 166–174.

Rodda, G.H., T.H. Fritts, and P.J. Conry. 1992. Origin and population growth of the Brown Tree Snake, *Boiga irregularis*, on Guam. *Pacific Science* 46: 46–57.

Roseberry, J.L. and W.D. Klimstra. 1984. *Population Ecology of the Bobwhite*. Southern Illinois University Press, Carbondale, IL.

Rosenzweig, M.L. 1969. Paradox of enrichment: destabilization of exploitation systems in ecological time. *Science* 171: 385–387.

Rosenzweig, M.L. and R.H. MacArthur. 1963. Graphical representation and stability conditions of predator–prey interactions. *American Naturalist* 97: 209–223.

Ross, I. 1994. Lions in winter. *Natural History* 103: 52–59.

Rossiter, M., J.C. Schultz, and I.T. Baldwin. 1986. Relationships among defoliation, red oak phenolics, and gypsy moth growth and reproduction. *Ecology* 69: 267–277.

Roughgarden, J. 1998. *Primer of Ecological Theory*. Prentice Hall, Upper Saddle River, NJ.

Royama, T. 1977. Population persistence and density dependence. *Ecological Monographs* 47: 1–35.

Saether, B., S. Engen, and E. Matthysen. 2002. Demographic characteristics and population dynamical patterns of solitary birds. *Science* 295: 2070–2073.

Sandercock, B.K. and A. Jaramillo. 2002. Annual survival rates of wintering sparrows: assessing demographic consequences of migration. *The Auk* 119: 149–165.

Savidge, J.A. 1988. Food habits of *Boiga irregularis*, an introduced predator on Guam. *Journal of Herpetology* 22: 275–282.

Schappert, P.J. and J.S. Shore. 1999a. Cyanogenesis, herbivory, and plant defense in *Turnera ulmifolia* on Jamaica. *Ecoscience* 6: 511–520.

Schappert, P.J. and J.S. Shore. 1999b. Effects of cyanogensis in polymorphism in *Turnera ulmifolia* on *Euptoieta hegesia* and potential *Anolis* predators. *Journal of Chemical Ecology* 25: 1455–1479.

Schluter, D., T.D. Price, and P.R. Grant. 1985. Ecological character displacement in Darwin's Finches. *Science* 227: 1056–1059.

Schmutz, J.A., R.F. Rockwell, and M.R. Petersen. 1997. Relative effects of survival and reproduction on the population dynamics of emperor geese. *Journal of Wildlife Management* 61: 191–201.

Schoener, T. 1982. The controversy over interspecific competition. *American Scientist* 70: 586–595.

Schoener, T. 1983. Field experiments on interspecific competition. *American Naturalist* 122: 240–285.

Schoener, T.W. and C.A. Toft. 1983. Spider populations: extraordinarily high densities on islands without top predators. *Science* 219: 1353–1355.

Schultz, J.C. and I.T. Baldwin. 1982. Oak leaf quality declines in response to defoliation by gypsy moth larvae. *Science* 217: 149–150.

Seal, U.S. and R.C. Lacy. 1989. *Florida Panther Population Viability Analysis*. Report to the US Fish and Wildlife Service. IUCN SSC Captive Breeding Specialist Group. Apple Valley, MN.

Searle, S. 1966. *Matrix Algebra for the Biological Sciences*. Wiley, New York, NY.

Selous, E. 1933. *Evolution of Habit in Birds*. Constable, London.

Shaffer, M.L. 1981. Minimum population sizes for species conservation. *Bioscience* 1: 131–134.

Silvertown, J.W. and J.L. Doust. 1993. *An Introduction to Plant Population Biology*. Blackwell Scientific, Oxford.

Simberloff, D. and W.J. Boecklen. 1981. Santa Rosalia reconsidered: size ratios and competition. *Evolution* 35: 1206–1228.

Simberloff, D.S. and E.O. Wilson. 1969. Experimental zoogeography of islands: the colonization of empty islands. *Ecology* 50: 278–296.

Simberloff, D.S. and E.O. Wilson. 1970. Experimental zoogeography of islands: a two year record of recolonization. *Ecology* 51: 934–937.

Simon, J. 1996. *The Ultimate Resource 2*. Princeton University Press, Princeton, NJ.

Sinclair, A.R.E. 1977. *The African Buffalo: a Study of Resource Limitation of Populations*. University of Chicago Press, Chicago, IL.

Sinclair, A.R.E. and P. Arcese. 1995. Population consequences of predation-sensitive foraging: the Serengeti wildebeest. *Ecology* 76: 882–891.

Sinclair, A.R.E. and J.M. Gosline. 1997. Solar activity and mammal cycles in the Northern Hemisphere. *American Naturalist* 149: 776–784.

Sinclair, A.R.E., S. Mduma, and J.S. Brashares. 2003. Patterns of predation in a diverse predator–prey system. *Nature* 425: 288–290.

Sjögren-Gulve, P. and C. Ray. 1996. Using logistic regression to model metapopulation dynamics: large-scale forestry extirpates the pool frog. Pages 111–137 *in* D.R. McCullough, editor, *Metapopulations and Wildlife Conservation*. Island Press, Washington, DC.

Skogland, T. 1983. The effects of density dependent resource limitation on size of wild reindeer. *Oecologia* 60: 156–168.

Sladen, W. 2003. The feral mute swan problem in Virginia and Maryland. *Cygnus* 16: 1–3.

Slobodkin, L.B. 1962. *Growth and Regulation of Animal Populations*. Holt, Rinehart, and Winston, New York, NY.

Smith, A.D.M. 1983. Epidemiological patterns in directly transmitted human infections. Pages 333–351 *in* N.A. Croll and J.H. Cross, editors, *Human Ecology and Infectious Diseases*. Academic Press, New York, NY.

Smith, B.R. and J.J. Tibbles. 1980. Sea lamprey (*Petromyzon marinus*) in Lakes Huron, Michigan and Superior: history of an invasion and control, 1936–78. *Canadian Journal of Fisheries and Aquatic Science* 37: 1780–1801.

Smith, D.W. and S.D. Cooper. 1982. Competition among Cladocera. *Ecology* 63: 1004–1015.

Smith, D.W., R.O. Peterson, and D.B. Houston. 2003. Yellowstone after wolves. *Bioscience* 53: 330–340.

Smith, F.E. 1954. Quantitative aspects of population growth. Pages 277–294 *in* E. Boell, editor, *Dynamics of Growth Processes*. Princeton University Press, Princeton, NJ.

Smith, R.L. 1996. *Ecology and Field Biology*. 5th edition. Harper Collins, New York, NY.

Smith, R.L. and T.M. Smith. 2001. *Ecology and Field Biology*. 6th Edition. Benjamin Cummings, San Francisco, CA.

Smith, S.M. 1975. Innate recognition of coral snake pattern by a possible avian predator. *Science* 187: 759–760.

Smith, S.M. 1977. Coral-snake pattern recognition and stimulus generalization by native great kiskadees (Aves: Tyrannidae). *Nature* 265: 535–536.

Smith, S.M. 1980. Responses of naïve temperate birds to warning coloration. *American Midland Naturalist* 203: 346–352.

Soler, J.J., G. Sorci, M. Soler, and A.P. Møller. 1999. Change in host rejection behavior mediated by the predatory behavior of its brood parasite. *Behavioural Ecology* 10: 275–280.

Soler, M., J.G. Martinez, J.G. Soler, J.J. Møller, and A.P. Møller. 1995a. Preferential allocation of food by magpies *Pica pica* to great spotted cuckoo *Clamator glandarius* chicks. *Behavioural Ecology and Sociobiology* 37: 243–248.

Soler, M., J.J. Soler, J.G. Martinez, and A.P. Møller. 1995b. Magpie host manipulation by great spotted cuckoos: evidence for an avian Mafia? *Evolution* 49: 770–775.

Solomon, M.E. 1949. The natural control of animal populations. *Journal of Animal Ecology* 18: 1–35.

Solomon, M.E. 1957. Dynamics of insect populations. *Annual Review of Entomology* 2: 121–142.

Souder, W. 2004. Audubon in Kentucky. *Natural History* 113: 46–52.

Soule, M.E. 1980. Thresholds for survival: maintaining fitness and evolutionary potential. Pages 60–77 *in* M.E. Soule and B.A. Wilcon, editors, *Conservation Biology: an Evolutionary–Ecological Perspective*. Sinauer Associates, Sunderland, MA.

Southern, H.N. 1970. The natural control of a population of tawny owls (*Strix aluco*). *Journal of Zoology* 162: 197–285.

Southwood, T.R.E. 1976. Bionomic strategies and population parameters. Pages 26–48 *in* R.M. May, editor, *Theoretical Ecology: Principles and Applications*. W.B. Saunders, Philadelphia, PA.

Southwood, T.R.E., R.M. May, M.P. Hassell, and G.R. Conway. 1974. Ecological strategies and population parameters. *American Naturalist* 108: 791–804.

Spielman, A. and M. D'Antonio. 2001. *Mosquito: the Story of Man's Deadliest Foe*. Hyperion, New York, NY.

Stanton, M. and T. Young. 1999. Thorny relationships. *Natural History* 108: 28–31.

Stearns, S.C. 1992. *The Evolution Of Life Histories*. Oxford University Press, New York, NY.

Strong, D.R. 1986. Density vagueness: abiding the variance in the demography of real populations. Pages 257–268 *in* J. Diamond and T.J. Case, editors, *Community Ecology.* Harper and Row. New York, NY.

Strong, D.R., Jr., and D.S. Simberloff. 1981. Straining at gnats and swallowing ratios: character displacement. *Evolution* 35: 810–12.

Strong, D.R., Jr., L.A. Szyska, and D. Simberloff. 1979. Test of community-wide character displacement against null hypotheses. *Evolution* 33: 897–913.

Strong, D.R., Jr., D.S. Simberloff, L.G. Abele, and A.B. Thistle (editors). 1983. *Ecological Communities: Conceptual Issues and the Evidence.* Princeton University Press, Princeton, NJ.

Sutherland, W.J. 1996. *From Individual Behaviour to Population Ecology.* Oxford University Press, Oxford.

Swain, T. 1965. The tannins. Pages 552–580 *in* J. Bonner and J.E. Varner, editors, *Plant Biochemistry.* Academic Press, New York, NY.

Tanner, J.T. 1966. Effects of population density on growth rates of animal populations. *Ecology* 47: 733–745.

Tansley, A.G. 1917. On competition between *Galium saxatile* L. (*G. hercynicum* Weig.) and *Galiuim sylvestre* poll. (*G. asperum* Schreb.) on different types of soil. *Journal of Ecology* 5: 173–179.

Taper, M.L. and T.J. Case. 1992. *Coevolution Among Competitors.* Oxford Series in Evolutionary Biology. Oxford University Press, Oxford.

Tast, J. 1974. The food and feeding habits of the root vole, *Microtus oecnomus,* in Finnish Lapland. *Aquilo Serie Zoologica* 15: 25–32.

Tepedino, V.J. and N.L. Stanton. 1976. Cushion plants as islands. *Oecologia* 25: 243–256.

Terborgh, J. 1975. Faunal equilibrium and the design of wildlife preserves. Pages 369–380 *in* F. Golley, and E. Medina, editors, *Tropical Ecological Systems: Trends in Terrestrial and Aquatic Research.* Springer-Verlag, New York, NY.

Thompson, P.M., J.N. Giedd, and R.P. Woods. 2000. Growth patterns in the developing brain detected by using continuum mechanical tensor maps. *Nature* 404: 190–193.

Tilman, D. (published as D. Titman). 1976. Ecological competition between algae: experimental confirmation of resource-based competition theory. *Science* 192: 463–466.

Tilman, D. 1977. Resource competition between planktonic algae: an experimental and theoretical approach. *Ecology* 58: 338–348.

Tilman, D. 1981. Tests of resource competition theory using four species of Lake Michigan algae. *Ecology* 62: 802–815.

Tilman, D. 1982. *Resource Competition and Community Structure.* Princeton University Press, Princeton, NY.

Tilman, D. 1987. The importance of the mechanisms of interspecific competition. *American Naturalist* 129: 769–774.

Tilman, D. 1994. Competition and biodiversity in spatially structured habitats. *Ecology* 75: 2–16.

Tilman, D. 1999. Diversity by default. *Science* 283: 495–496.

Tilman, D. and P. Kareiva. 1997. *Spatial Ecology.* Princeton University Press, Princeton, NJ.

Tilman, D., S.S. Kilham, and P. Kilham. 1982. Phytoplankton community ecology: the role of limiting nutrients. *Annual Review of Ecology and Systematics* 13: 349–372.

Tinkle, D.W. 1969. The concept of reproductive effort and its relation to the evolution of life histories of lizards. *American Naturalist* 103: 501–516.

Toft, C.A. 1986. Communities of species with parasitic life cycles. Pages 445–463 *in* J. Diamond and T.J. Case, editors, *Community Ecology.* Harper and Row, New York, NY.

Tollrian, R. and C.D. Harvell. 1999. *The Ecology and Evolution of Inducible Defenses.* Princeton University Press, Princeton, NJ.

Turchin, P. 1995. Population regulation: old arguments and a new synthesis. Pages 19–40 *in* N. Cappuccino and P.W. Price, editors, *Population Dynamics: New Approaches and Synthesis.* Academic Press, San Diego, CA.

Turchin, P. 1999. Population regulation: a synthetic view. *Oikos* 84: 153–159.

Turchin, P. 2001. Does population ecology have general laws? *Oikos* 94: 17–26.

Turchin, P. 2003. *Complex Population Dynamics: a Theoretical/Empirical Synthesis*. Princeton University Press, Princeton, NJ.

Turchin, P. and G. Batzli. 2001. Availability of food and the population dynamics of arvicoline rodents. *Ecology* 82: 1521–1534.

Turchin, P. and S.P. Ellner. 2000. Living on the edge of chaos: population dynamics of Fennoscandian voles. *Ecology* 81: 3099–3116.

Turchin, P., L. Oksanen, P. Ekerholm, T. Oksanen, and H. Henttonen. 2000. Are lemmings prey or predators? *Nature* 405: 562–565.

Turner, M.G., R.H. Gardner, and R.V. O'Neill. 2001. *Landscape Ecology in Theory and Practice*. Springer. New York.

Turner, M.G., W.H. Romme, and D.B. Tinker. 2003. Surprises and lessons from the 1988 Yellowstone fires. *Frontiers in Ecology and the Environment* 1: 351–358.

Utida, S. 1957. Cyclic fluctuations of population density intrinsic to the host–parasite system. *Ecology* 38: 442–449.

Vandermeer, J.H. 1969. The competitive structure of communities: an experimental approach with protozoa. *Ecology* 50: 362–371.

Vandermeer, J.H. and D.E. Goldberg. 2003. *Population Ecology: First Principles*. Princeton University Press, Princeton, NJ.

Vehrencamp, S.L. 1977. Relative fecundity and parental effort in communally nesting anis, *Crotophaga sulcirostris*. *Science* 197: 403–405.

Vehrencamp, S.L. 1978. The adaptive significance of communal nesting in groove-billed anis (*Crotophaga sulcirostris*). *Behavioral Ecology and Sociobiology* 4: 1–33.

Verhulst, P.F. 1838. Notice sur la loi que la population suit dans son accroissement. *Corresp. Math. Phys.* 10: 113–121.

Volterra, V. 1926. Fluctuations in the abundance of a species considered mathematically. *Nature* 118: 558–560.

Volterra, V. 1931. *Leçons sur la théorie mathématique de la lutte pour la vie*. Gauthiers-Vilars, Paris.

Wahlberg, N., A. Moilanen, and I. Hanski. 1996. Predicting the occurrence of endangered species in fragmented landscapes. *Science* 273: 1536–1538.

Watkinson, A.R. 1997. Plant population dynamics. Pages 395–400 *in* M.J. Crawley, editor, *Plant Ecology*. Blackwell Science, Oxford.

Watkinson, A.R. and W.J. Sutherland. 1995. Sources, sinks, and pseudo-sinks. *Journal of Animal Ecology* 64: 126–130.

Weltzin, J. 1991. Reported in *Bioscience*, December 1991: 753.

Werner, P.A. 1975. Predictions of fate from rosette size in teasel (*Dipsacus fullonum* L.). *Oecologia*. 20: 197–201.

Werner, P.A. and H. Caswell. 1977. Population growth rates and age vs. stage-dependent models for teasel (*Dipsacus sylvestris* Huds.). *Ecology* 58: 1103–1111.

West, G.B., J.H. Brown, and B.J. Enquist. 1997. A general model for the origin of allometric scaling laws in biology. *Science* 276: 122–126.

West, G.B., J.H. Brown, and B.J. Enquist. 1999a. A general model for the structure and allometry of plant vascular systems. *Nature* 400: 664–667.

West, G.B., J.H. Brown, and B.J. Enquist. 1999b. The fourth dimension of life: fractal geometry and allometric scaling of organisms. *Science* 284: 1677–1679.

West, G.B., J.H. Brown, and B.L. Enquist. 2000. The origin of scaling universal: scaling laws in biology. Pages 87–112 *in* J.H. Brown and G.B. West, editors, *Scaling in Biology*. Oxford University Press, Oxford.

Whitham, T.G., J. Maschinski, K.C. Larson, and K.N. Paige. 1991. Plant responses to herbivory: the continuum from negative to positive and underlying physiological mechanisms. Pages 227–256 *in* P.W. Price, T.M. Lewinsohn, G.W. Fernandes, and W.W. Benson, editors, *Plant–Animal Interactions: Evolutionary Ecology in Tropical and Temperate Regions.* Wiley-Interscience, New York, NY.

Wieland, N.K. and F.A. Bazzaz. 1975. Physiological ecology of three co-dominant successional annuals. *Ecology* 56: 681–688.

Wielgolaski, F.E. 1975. Productivity of tundra ecosystems. Pages 1–12 *in* D.E. Reichle, J.F. Frankliin, and D.W. Goodall, editors, *Productivity of World Ecosystems.* National Academy of Sciences, Washington, DC.

Wiens, J.A. 1977. On competition and variable environments. *American Scientist* 65: 5990–5997.

Wilbur, H.M., D.W. Tinkle, and J.P. Collins. 1974. Environmental uncertainly, trophic level, and resource availability in life history evolution. *American Naturalist* 108: 805–817.

Williams, K.S., K.G. Smith, and F.M. Stephen. 1993. Emergence of 13-year periodical cicadas (Cicadidae: *Magicicada*): phenology, mortality, and predator satiation. *Ecology* 74: 1143–1182.

Williamson, P. 1971. Feeding ecology of the red-eyed vireo (*Vireo olivaceus*) and associated foliage-gleaning birds. *Ecological Monographs* 41: 129–152.

Willis, E.O. 1984. Conservation, subdivision of reserves, and the antidismemberment hypothesis. *Oikos* 42: 396–398.

Wilson, E.O. 1961. The nature of the taxon cycle in the Melanesian ant fauna. *American Naturalist* 95: 169–193.

Wilson, E.O. 1992. *The Diversity of Life.* Norton. New York, NY.

Wilson, E.O. and E.O. Willis. 1975. Applied biogeography. Pages 523–534 *in* M.L. Cody and J.M. Diamond, editors, *Ecology and Evolution of Communities.* Harvard University Press, Cambridge, MA.

Wolfe, D.W. 2001a. Out of thin air. *Natural History* 110: 44–53.

Wolfe, D.W. 2001b. *Tales From the Underground: a Natural History of Subterranean Life.* Perseus, New York, NY.

Wolff, J.O. 1988. Maternal investment and sex ratio adjustment in American bison calves. *Behavioral Ecology and Sociobiology* 23: 127–133.

Wright, S. 1940. Breeding structure of populations in relation to speciation. *American Naturalist* 74: 232–248.

Yoda, K., T. Kira, H. Ogawa, and K. Hozumi. 1963. Self-thinning in overcrowded pure stands under cultivated and natural conditions. *Journal of Biology Osaka, City University* 14: 107–129.

Young, T.P. 1987. Increased thorn length in *Acacia depranolobium*: an induced response to browsing. *Oceologia* 71: 436–438.

Yu, D.W. and H.B. Wilson. 2001. The competition–colonization trade-off is dead; long live the competition–colonization trade-off. *American Naturalist* 158: 49–63.

Zahavi, A. 1979. Parasitism and nest predation in parasitic cuckoos. *American Naturalist* 13: 157–159.

Zahavi, A. and A. Zahavi. 1997. *The Handicap Principle.* Oxford University Press, Oxford.

Index

Note: Page numbers in *italics* refer to figures; those in **bold** refer to tables.